Fiber Optics

Fedor Mitschke

Fiber Optics

Physics and Technology

Second Edition

 Springer

Fedor Mitschke
Institut für Physik
Universität Rostock
Rostock, Germany

ISBN 978-3-662-57079-1   ISBN 978-3-662-52764-1   (eBook)
DOI 10.1007/978-3-662-52764-1

This Springer imprint is published by Springer Nature
The registered company is Springer-Verlag GmbH Berlin Heidelberg

Absent a Telephone, a Bicyclist Had to Save the World

On the height of the Cuban missile crisis in 1962, no direct telecommunication line existed between the White House and the Kremlin. All messages going back and forth had to be sent through intermediaries. The world teetered on the brink of nuclear Armageddon when in the evening of October 23 President John F. Kennedy sent his brother, Robert Kennedy, over to the Soviet Embassy for a last-ditch effort to resolve the crisis peacefully. Robert presented a proposal on how both sides could stand down without losing face. Right after the meeting, Ambassador Anatoly Dobrynin hastened to write a report to Nikita Khrushchev in Moscow. A bicycle courier was called in to take this letter to a Western Union telegraph station, and Dobrynin personally instructed him to go straight to the station because the message was important—which was hardly an exaggeration.

That man on the bicycle, in my view, has saved the world, most likely without even knowing it.

A year later, a direct telegraph line was installed which was popularly called the "red telephone." (There never was an actual red telephone sitting in the Oval Office.) A lesson had been learned: communication can be vital when it comes to solving conflicts.

Today the situation is vastly different from what it was less than half a century ago. The world is knit together by a network of connections of economic, political, cultural, and other natures. That is only possible because virtually instantaneous long-distance communication at affordable cost has become ubiquitous. In earlier centuries, important news—say like the outcome of a battle—often was received only several weeks later. Today we are not the least bit astonished when we watch unfolding events in the remotest corner of the planet in real time, living color, and stereophonic sound.

The biggest machine on earth is the international telephone network. It allows you to call this minute, on a lark, your neighbor, your friend in New Zealand, or the Department of Sanitation in Tokyo. And we got used to it! Behind the scenes, of course, there is a substantial investment in technology going into this, and more effort is required to keep up with society's ever-rising demands. Consider international calls: For some time, satellites seemed to be the most efficient and elegant means. Just a decade or two later, they were no more up to the growing task, and a new, earthbound technology took over: optical fiber transmission.

Meanwhile, the amount of data handled by fibers exceeds anything that older technology could have handled ever. Today's Internet traffic would not exist without fiber, and the cost of a long-distance phone call would still be as expensive as it was a quarter century ago.

Optical fibers, mostly made of glass but sometimes also other materials, are the subject of this book. The development toward their maturity we enjoy today was mostly driven by the challenges of telecommunications applications. Research has faced quite a number of questions concerning basic physics of guided-wave optics, and many researchers around the world toiled for answers. As a result, fibers can do more than was anticipated: Besides the obvious application in telecommunications, they have also become useful in data acquisition. This is why engineers and technicians working in either field need to know not only their electrical engineering but increasingly also some optics. At the same time, it emerges that nonlinear physical processes in fibers will lead to exciting new technology.

This book has its origin in lectures for students of physics and engineering which I gave at the universities in Hannover, Münster, and Rostock (all in Germany) and in Luleå (Sweden). The book first appeared in the German language. It was well received, but the German-speaking part of the world is not very big, and I heard opinions that an English version would find a larger audience.

The book presents the physical foundations in some detail, but in the interest of limited mathematical challenges, there is no fully vectorial treatment of the modes. On the other hand, I found it important to devote some space to nonlinear processes on grounds that over the years, they can only become more relevant than they already are. I proceed in outlining the limitation of the data-carrying capacity of fibers as they will be reached in a couple of years, i.e., at a time when the student readers of this book will have entered their professional life as engineers or scientists, dealing with these questions. For the English edition, I have expanded certain sections slightly, to keep up to date with current developments.

It is my hope that both natural scientists and engineers will find the book helpful. Maybe physicists will think that some segments are quite "technical," while engineers may feel that a treatment of nonlinear optics may be not so much for them. My answer to that is that either subject is required to form the full picture. In this context, it is sometimes unfortunate that the structure of our universities emphasizes the distinction between natural scientists and engineers more than is warranted. I envision that, in analogy to electronics engineers, we will see the emergence of photonics engineers. They would have good practical skills on the technical side and at the same time a deep understanding of the underlying physical mechanisms.

About the Second Edition

After the original German-language edition of 2005 (which is now out of print), a first English version of this book was published in 2009 and was quite well received. In the meantime, the field of fiber optics has advanced at a rapid pace. This applies equally to new insights into some phenomena of nonlinear light propagation in fibers as to the state of the art concerning the technology used for high-volume optical data transmission.

The 2009 edition of the book described basics which obviously remain valid. However, it also provided a view of the latest thinking at the time. From the vantage point of today, some important developments have occurred in the meantime but were—naturally—not covered. This led author and publisher to consider a revised version, which is herewith presented. Several segments have been added and some were updated. The opportunity was also taken to correct a few minor errors which, apparently, some mischievous gremlins had arranged to creep in.

May the new edition be useful to its readers!

Rostock, Germany   Fedor Mitschke
March 2016

Contents

Part VI   Appendices

Part I
Introduction

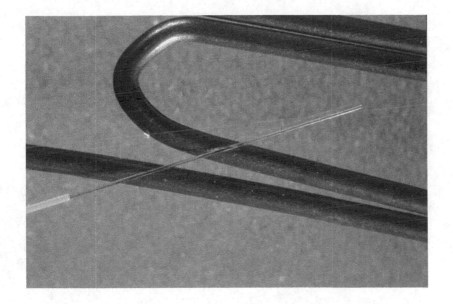

An optical fiber in comparison to a paper clip. On the far left, part of the fiber's plastic coating is visible; mostly the fiber is bare, though. Only a small fraction of its diameter of 125 μm near the fiber axis serves the waveguiding directly.

Part 1
Introduction

Chapter 1
A Quick Survey

Visual, and hence optical, communication is older than language. Hand signals, waving of the arms, and fire and smoke signals are basic means of communication, and except under detrimental environmental conditions like pitch-black darkness or fog, they are useful over longer distances than shouting; besides, they are not thwarted by noises like surf at the seashore.

Normally we communicate verbally. Hence, when optical means are employed, there is a necessity to agree on a *code* that serves to translate the visible signs into a meaningful message.

Certain signs of nontrivial meaning are understood universally and even independent of language: consider the handwaving sign for "come here." On the other hand, the vocabulary of such signs is too limited to convey truly complex messages. Codes that represent smaller units of language—syllables, phonemes, or individual letters—are much more universal. The best-known example may be the Morse alphabet. Of course, it is mandatory that both sender and receiver of the transmitted message have agreed on the code ahead of time. In today's computerized environment, codes of various kinds are of tremendous importance.

The range (maximum distance) of optical transmission of messages can be increased by concatenation of several shorter spans. In the Greek tragedy of *Agamemnon* (part of *The Oresteia*), Aeschylus (ca. 525–456 BCE) mentions how the news about the fall of the city of Troy was transmitted over 500 km to Agamemnon's wife, Clytemnestra [2]. Also, fire and smoke signals were transmitted from post to post along the Great Wall of China as early as several centuries BCE; during the Ming dynasty 1368–1644, this link stretched for over 6000 km from the Jiayuguan Pass outpost to the capital, Beijing (and on to the east). In modern times, the first systematic attempts at optical telecommunication took place in France, where Claude Chappe constructed the first optical telegraph in 1791 [4]. It is little known that Chappe initially worked with electrical devices, but decided that optical ones were advantageous. The French National Convention was initially decidedly disinterested, but in 1794 the first state-operated telegraph line was started between Lille and Paris. Every few kilometers, there were repeater

© Springer-Verlag Berlin Heidelberg 2016
F. Mitschke, *Fiber Optics*, DOI 10.1007/978-3-662-52764-1_1

Fig. 1.1 A semaphore atop the roof of the Louvre. From [1]

stations called semaphores using mechanically movable pointers or hands; they
were observed from neighboring stations, aided by telescopes. This system allowed
to send messages from Paris to Lille in just 6 min—corresponding to twice the
speed of sound. Later, a whole grid of such lines was built across all of France,
eventually reaching a total length of 4800 km (Fig. 1.1). As is often the case with
new technology, the first application was a military use. Napoleon I successfully
used it for his trademark rapid military campaigns and had a portable system built
for his campaign against Russia. Sweden also built a comparable network, and the
UK and other countries followed suit. Around 1840, this technology saw its climax
and was very common. Also the USA had a few lines ("Telegraph Hill" remains a
San Francisco landmark to this day).

However, the age of electric telegraphy dawned by then. Half a century after their
introduction, optical telegraphs were phased out. As it turned out, electric systems
were less prone to service interruption in case of inclement weather. Beginning ca.
1858, progress in the electric technology finally added superior speed as a further
advantage of electric systems.

One should note that the heyday of the electric telegraph coincides with the age
of colonialism. That is relevant insofar as it speaks to the interplay between technical
and political developments. Colonial powers supported the new technology because
it gave them much better control over their dependencies. One hardly overestimates

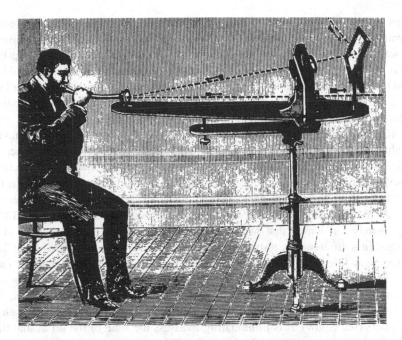

Fig. 1.2 Alexander Graham Bell's photophone: sunlight is directed onto a membrane that vibrates as it is agitated by the sound from the speaker. The modulated light beam is transmitted and eventually demodulated with a Selenium photo cell. Reproduced with permission from Alcatel-Lucent

the importance of fast message transmission for the political situation of the day. We are denizens of the twenty-first century and find it impossible to imagine the absence of electronic means of data transmission.

For a long time, in the development of the technology, optical systems took a back seat. It is therefore amusing to note that the inventor of the telephone, Alexander Graham Bell,[1] was strongly interested in optical means of transmission. In 1880, he introduced what he called the photophone, a contraption in which the sound pressure waves emanating from a speaker's lips moved a mirrored membrane in such a way that a light beam directed onto it got intensity-modulated (Fig. 1.2). On the receiver side, a selenium photocell served as a converter of the received light wave back to an electric current that could be converted to audible sound with an ordinary headphone. Both transmitter and receiver were thus realized with optical means; only at the receiver, electrical gear was also involved.

---

[1]Bell was not the only, indeed not even the first, inventor of the telephone. He filed his patent in 1876, but the Italian technician Antonio Meucci (who lived in New York) had demonstrated a working model as early as 1860 and the German teacher Philip Reis built another version in 1861. The American Elisha Gray had the bad luck of filing his patent all of 2 h after Bell. However, Bell is usually cited as the inventor because he won all legal patent battles, developed the scheme into a marketable product, and had the wherewithal to introduce it to the public.

This system had the unsurmountable disadvantages that a good light source was not available—after all, the sun does not always shine—and that the transmission span was vulnerable to adverse atmospheric conditions: rain, snow, and fog. Bell had no way of knowing, of course, that 100 years later both problems would be solved through the introduction of practical lasers and optical fibers. Only after both these novelties were available, optical data transmission had a new chance. Indeed, the chance turned into a success story probably second to none.

In the 1960s, the laser had just been invented, and the prospect of having workable, practical devices in the near future became realistic. At that time, the propagation of laser beams through the open atmosphere in the presence of various atmospheric conditions was studied systematically and at different wavelengths [3]. As an alternative, there were also attempts to guide light in ducts. This made it necessary to refocus the beam frequently. In one approach, this was attempted with a large number of lenses that were inserted into the beam path in certain short distances. In a different try, researchers experimented with a distributed lens: This involved a gas-filled duct in which a radial temperature gradient was generated and maintained. The temperature gradient, by way of expansion of the warmer gas at the center, gave rise to a refractive index profile that acted as an effective lens. The same basic idea but in a "solid-state" version is used today in the so-called gradient index fibers (see below). It is illuminating to assess the state of the art at that time as described in an account given by Kompfner [6].

There were obvious disadvantages in these approaches: It is not easy to form bends in such light guides—the bend radius had to be hundreds of meters! Also, installation and operation were quite costly. Only a few years later there were optical fibers: thin strands of glass, flexible enough to be coiled around a finger, and as inexpensive as copper wire, with no maintenance cost because the light-guiding index profile is built right into the fiber structure!

At that time, it was well known that microwaves can be sent through waveguides that are easy to produce. It was also known that glass can be spun into thin threads, that such threads are flexible, and even that they can guide light. However, transmission of information through such fibers was impossible due to the high transmission loss, a property shared by all transparent solid materials known at that time. Different materials had been studied, but among the best suited was glass with a chemical composition given by SiO_2, known as *fused silica*. But even in fused silica, light was attenuated by at least one third after a distance no longer than 1 m. This ruled out the transmission over any long distances.

Then, in 1966, Kao and Hockham of Standard Telecommunications Laboratories in London published a paper with a remarkable prediction [5]. The authors, none of them a materials expert by training, argued that the strong attenuation of glass was not really an inherent, intrinsic property but was rather caused by chemical impurities in the glass composition. They predicted the possibility of producing, by way of suitably refined procedures, glass with an attenuation no more than 20 dB/km instead of the 1000 dB/km common at the time. This would represent a reasonable value for transmission.

Here and in the following, we will make extensive use of decibel (dB) units. They are in ubiquitous use in all of electrical engineering, and it is indispensable that the reader is aware of what they mean. If you are unsure, check Chap. 13 for a thorough explanation.

In hindsight, the paper by Kao and Hockham came out at just the right time. Very quickly tremendous progress in this direction was achieved in Japan, England, and the USA. In a cooperation of Nippon Sheet Glass Co. and Nippon Electric Co., in 1969, the first gradient index fiber was made that was suitable for communication purposes. It was given the name SELFOC fiber (as in *self-focusing*), and it had an attenuation below 100 dB/km. In England, coordinated by British Post Office, a cooperation between universities and industry was launched, and in the USA, Corning Glass Works and Bell Laboratories joined forces. The latter cooperation was the first to be able to announce the attenuation factor quoted by Kao and Hockham: In 1970, Kapron and coworkers at Corning created several hundred meters of a single-mode fiber with an attenuation below 20 dB/km. The production technique involved thin layers of fused silica deposited on the inside surface of a glass tube (see Sect. 6.2). It allowed much better chemical purity than before. It also provided the possibility of adding Germanium oxide in precisely controlled concentration, so that a radial index structuring can be obtained, which is crucial for waveguiding.

Later on, both this manufacturer and others improved the attenuation to 4 dB/km by continuous fine-tuning of the procedure. At this point a limit was reached, which is indeed due to the structure of the pure fused silica itself. Nonetheless, losses could be reduced further when it was understood that the loss is wavelength-dependent (see Chap. 5). Operating with infrared light, in 1976, the milestone of 1 dB/km was reached in Japan, and it has now become commercial routine to obtain less than 0.2 dB/km, a value that is indeed very close to the limit of what is possible at all with fused silica.

As product maturity developed, so did the industrial-scale production capacity. This, in turn, had a profound effect on the cost. When fiber was first introduced in a mass market in 1981, standard fiber commanded a price around 5 $/m. Within less than 2 years, that number dropped to one tenth, and today the price may well be below 10 cents/m. The reason is simple: Of three main factors affecting the cost of a product, two are insubstantial here.

Raw material	is cheap because it is abundant. Go to the beach: How much sand is there?
Labor cost	is low because production can be almost completely automated.
Capital investment	is high, but as long as a sufficient quantity of fiber is sold, the cost per meter is low.

A first major field trial of fiber-optic transmission was performed in 1976 in Atlanta, followed by a first exploratory operation in 1978 in Chicago. Germany started in 1977 in Berlin; other countries have similar stories.

Further progress concerning fibers was linked to progress with respect to light sources. Semiconductor laser diodes had been known since the early 1960s, but the first version required cryogenic cooling and operated only in pulsed mode. In 1970, the first continuous wave laser diodes at room temperature were introduced, but their life expectancy was quite short (just a couple of hours). Today laser diodes are specified as being able to handle x gigabits per second, but in the early 1970s it was x gigabits—and that was it! Progress since then has been truly impressive, and today's laser diodes can easily reach a useful lifetime of 10^5 h (corresponding to 10 years of continuous operation) and more.

As fiber was beginning to be deployed, the need for a number of other auxiliary components arose. This includes the permanent or reconfigurable connection between fibers, which requires to maintain quite narrow tolerances in the relative positioning of the fibers. It took a while to master such tight tolerances but then the progress on the learning curve eased the transition from multimode fibers, which have more relaxed requirements, to single-mode fibers that require the highest precision.

Multimode fibers are characterized by a relatively large diameter of the light-guiding core, which is much larger than the wavelength of the light. In the most common version, the core diameter is 50 μm, embedded in a fiber of 125 μm outside diameter. In contrast, single-mode fibers have a core diameter that is larger than the wavelength only by a small factor; typical values range between 7 and 10 μm (Fig. 1.3). This does not affect the outer diameter of the fiber, which may be the same

Fig. 1.3 Multimode fibers and single-mode fibers only differ in the diameter of the light-guiding core, which is made from a glass that is doped in a slightly different way than the surrounding cladding

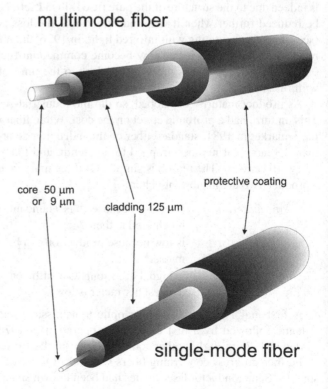

core 50 μm
or 9 μm

cladding 125 μm

protective coating

Fig. 1.4 A standard optical fiber in comparison to a match

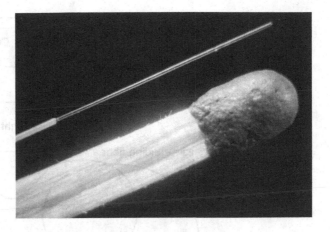

as for a multimode fiber; indeed, the outside diameter of 125 μm is the standard value for both fiber types. For a size comparison see Fig. 1.4.

In first applications, multimode fibers were used. They allow better incoupling efficiency, and there are fewer losses when connecting fibers together. However, as we shall see in Sect. 2.3, single-mode fibers allow higher data rates over longer distances. Therefore, once the connector tolerance issue was solved satisfactorily, single-mode fibers became the favored choice and are almost exclusively used today at least for the long haul. Only for short distances, in particular in local area networks between several computers inside one building, multimode fiber is still preferred because the highest data rate is not required, but some savings can be had in coupling and connecting.

At this point, we should take a look at the basic ingredients of any data transmission system (see Fig. 1.5). The information to be transmitted can originate from a person speaking into a telephone, but it might also come from a telefax machine sending data or from a computer hooked up to a line. In the case of a human speaker, the acoustic signal is first converted to an electric signal. Then it gets coded to whatever format is appropriate for the transmission.

The coded signal is then passed onto a light source to modulate it. This means that some property of the light wave, for example its amplitude or phase, is influenced by the coded message. The simplest case would be to switch the light source on and off in accordance to a digital signal representing the message.

The modulated light is then sent through the fiber and reaches the receiver where it is decoded and then converted to the required format: In the case of a telephone, this is a sound wave from the handset; for a telefax, a printout on paper.

Everything would work just fine if transmitter and receiver were sitting next to each other (back to back). The exciting part, and the reason why all this is done at all, is that one can "insert distance" between both stations. One only needs to make sure that over the distance, there are no more than minimal distortions of the signal, so that after decoding the message is still intact and transmission errors are not perceptible. The founder of information theory, Claude Shannon,

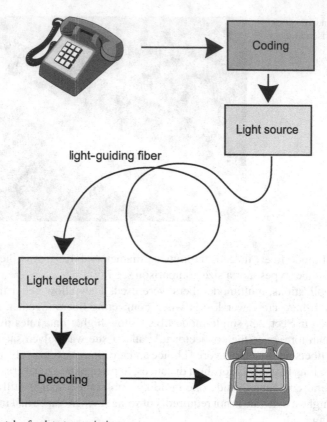

Fig. 1.5 Sketch of a data transmission

has mathematically stated the relations between the rate of data transmission, the bandwidth of the line, and the disturbances occurring on the line (see Sect. 11.1.8).

It is important for a successful transmission that the signal is not attenuated too much on the line. As mentioned above, first field trials used visible light, but soon people realized that infrared wavelengths are much better in this respect. One speaks of a *first generation* of fiber-optic systems that operated around 850 nm, a wavelength that was convenient due to the availability of very economic gallium arsenide laser diodes. This spectral region is also known as the *first window* for fiber transmission.

The *second generation* operated at a wavelength around 1300 nm (the *second window*). This wavelength was favored because the fiber's dispersion (which is the subject of Chap. 4) is particularly low there. As we shall see, this fact provides a considerable increase of both reach and data rate. The major part of all systems installed today is designed for this wavelength, although emphasis meanwhile shifts to the third window.

The *third generation* moves on to wavelengths near 1550 nm (the *third window*). This is the regime where fibers made of fused silica have their global loss minimum (see Chap. 5).

There have been numerous attempts to make fibers such that the main advantage of the second window—low dispersion—would occur at the wavelength of the third window, so that the best of both could be combined. A truly successful implementation would have allowed to phase in the third generation more rapidly. However, while it is possible in principle, the commercial success of these attempts remained limited. One of the reasons that industry preferred to hang on to the second window for a long time was that the installed base of second-generation fiber-optic systems represented a value of billions of dollars; it did not seem to make business sense to give up that legacy. A technical reason also was that fibers with dispersion optimized at 1550 nm turned out to partially lose the advantage of the lowest loss. The strategy today is that different fibers are concatenated so that dispersion is partly compensated; we will consider such systems in Sect. 11.2.3.

At this point the reader may ask: Why is it that light in fiber optics is superior to the more conventional electric current over copper cable? The answer to this is found by considering the fundamental limits to transmission losses in comparison with optical fiber and copper wire. For wire it is given by the skin effect, i.e., the phenomenon that at high frequencies almost all current is carried only in a thin surface layer of a conductor, while the volume contributes little or is even counterproductive. This effect, as described in Chap. 14, increases with frequency and eventually defeats any high data rate, long-distance transmission. Optical fibers do not suffer from this limitation and therefore have a clear advantage when it comes to transmitting high data rates over long distances.

What little loss remains in optical fibers is indeed fundamental to fused silica, as detailed in Chap. 5. There have been approaches to reduce loss even further by using other glass types. On theoretical grounds, chalcogenides, fluorides, and halides hold promise to have dramatically lower loss figures than fused silica. Unfortunately, such theoretical considerations never made it into a realization. Production of such fibers is fraught with problems arising from their chemical nature: It is extremely difficult to obtain good purity of a highly reactive substance. Today significant progress in that direction is not anticipated.

Our quick survey would not be complete without mentioning optical nonlinearity. Since the early 1980s, researchers have investigated the nonlinear properties of optical fiber. The term refers to the situation that some optical property of the fiber, such as the refractive index, depends on the intensity of the light wave passing through it. Nonlinearity does not occur in copper cables (at least not to any noticeable degree, anyway), but clearly manifests itself in fibers. This was considered a nuisance for a long time, but today it is increasingly realized that it is precisely the exploitation of nonlinear effects that allows a new generation of fiber-optic transmission systems, which turns out to be vastly superior to previous technology in its data-carrying capacity. We mention here only in passing the concept of *solitons*, special light pulses that maintain their shape not in spite of the presence of nonlinearity but due to its presence. In Chaps. 9 and 10, we will discuss nonlinearity and solitons in greater detail.

In some sense, today's fiber-optic networks have many aspects in common with the telegraph networks of earlier days: either has both attenuation and dispersion; these two represent the biggest practical problems. One can beat attenuation by inserting repeater stations into the fibers at intervals of 50 or 100 km or so. The novelty in fiber optics is that there is nonlinearity, and it causes effects unknown to electrical systems engineers. Meanwhile, however, the first commercial systems exploiting and embracing nonlinearity and solitons have taken up service, and it can be anticipated that more is yet to come.

We must also point out now that optical telecommunication is by no means the only field of application of fibers. Beyond their enormous data-carrying capacity and great reach, they represent other specific properties that make them attractive for use in data acquisition systems.

One of these properties is the enormous savings in weight, as compared to copper wire. One does not so much realize it by comparing the densities (copper: $8.9\,kg/dm^3$, fused silica: $2.2\,kg/dm^3$) because equal volumes are hardly a relevant basis for comparison. There are protective jackets around both kinds of cable, both for mechanical protection and electrical insulation. These jackets represent the lion's share of the cable's mass (bare fiber weighs in at just 30 mg/m). In a realistic comparison between, let us say, fiber-optic cable and coaxial cable for use for transmission in the megahertz regime over a few kilometers, it is a rule of thumb that 1 g of fiber cable replaces 10 kg of electrical cable. (Both reach and data rate of fiber can be scaled up much higher than that of coaxial cable, though.) This represents an immediate advantage where weight limitation is an important requirement: on board of vehicles, ships, aircraft, and spacecraft.

In connection with reduced weight, there is also reduction of space requirement. This is important in cable ducts in inner cities that are crowded already; any new installation has to find a way to squeeze in. A fiber-optic cable can replace one or several coaxial cables, upgrade the data rates, and save space at the same time. It is of course better to replace an existing cable in a duct with an upgrade than to install new ducts. Just imagine a work crew digging up Broadway in Manhattan to place more ducts—this is not an option.

As a further distinctive property, fiber-optic cables are immune to electric or magnetic field interference. This is frequently a definitive advantage in industrial installations. Even in close proximity to high-voltage installations, etc., there is no interference picked up by the fiber. This feature sets it apart from electric conductors.

Moreover, glass is chemically quite inert. As long as the fiber's protective jackets are also made of inert materials, fiber-optic cables can be deployed in chemically hostile environments where metallic parts would quickly corrode. This is attractive for applications in the chemical industry.

And, finally, a fiber-optic cable guarantees a perfect electrical insulation between transmitter and receiver. The same thing can be achieved for electric cables by other means, the so-called optocouplers, but in a manner of speaking, a fiber is an optocoupler stretched long. Different fluctuating ground potentials are therefore no longer a concern when subsystems are connected with fiber optics. This is more than a small benefit when there is a potential risk from combustible fumes that one might find, e.g., on oil-drilling rigs. The combination of these properties leads to a fiber-optic sensor technology, which will be discussed in Chap. 12.

References

1. N.N., *Abbildung und Beschreibung des Telegraphen oder der neuerfundenen Fernschreibemaschine in Paris und ihres inneren Mechanismus*, F. G. Baumgärtner, Leipzig (1795). Cited after F. Skupin, *Abhandlungen von der Telegraphie oder Signal- und Zielschreiberei in die Ferne*, Nachdruck historischer Publikationen, R. v. Decker's Verlag, G. Schenck GmbH (1986)
2. Aeschylus, *Agamemnon*, in: *Aeschylus, II, Oresteia: Agamemnon. Libation-Bearers. Eumenides*, edited and translated by Alan H. Sommerstein, Harvard University Press, Cambridge, MA (2009)
3. T. S. Chu, D. C. Hogg, *Effects of Precipitation on Propagation at 0.63, 3.5 and 1.6 Microns*, Bell System Technical Journal **47**, 723 (1968)
4. G. J. Holzmann, B. Pehrson, *Optische Telegraphen und die ersten Informationsnetze*, Spektrum der Wissenschaft März, S. 78 (1994)
5. K. C. Kao, G. A. Hockham, *Dielectric-Fibre Surface Waveguides for Optical Frequencies*, Proceedings IEE **113**, 1151 (1966)
6. R. Kompfner, *Optical Communications*, Science **150**, 149 (1965)

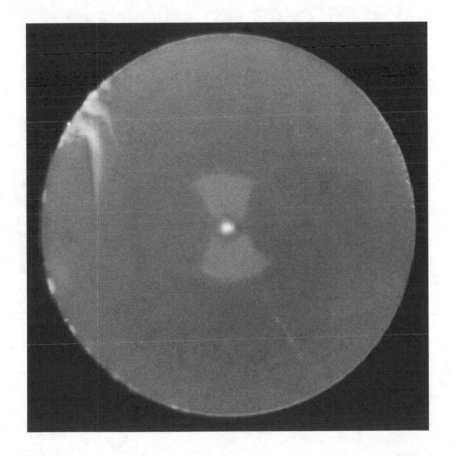

The end of a "bowtie" fiber under the microscope. The fiber's outside diameter is 125 μm. The light-guiding core is discernible as a small central bright spot. It is

surrounded by a bowtie-shaped birefringent zone that gives this fiber type its name.
For further information see Sect. 4.6.2 and Fig. 4.18 in particular.

Chapter 2
Treatment with Ray Optics

Calculations in technical optics are often done with a technique called raytracing. This is a treatment of optical systems in the framework of ray optics. It provides a particularly clear, if incomplete, view of the properties of optical systems. Strictly speaking, light propagation needs to be treated by taking the wave nature of light into account. The difference is that waves give rise to diffraction and interference phenomena which are disregarded in ray optics. Whenever the geometrical dimensions of the problem are so small as to become comparable to the wavelength of light, the ray optic treatment breaks down. This is the case in single-mode fibers.

That notwithstanding, we will first present a ray-optical consideration in order to get an idea of the phenomena to be expected. When we then proceed with a wave-optic treatment in Chap. 3, it will become apparent that in fibers the main difference consists in the fact that the direction of light propagation, which can be any direction in ray optics, is restricted to a discrete set of angles in the full picture.

2.1 Waveguiding by Total Internal Reflection

Consider a light ray impinging on some boundary to an optically less-dense medium. Less dense is optics parlance and means "having a lower index of refraction." At a suitable angle of incidence the ray will be fully reflected, instead of passing through. This process is called *total internal reflection* and is explained in any textbook on optics (see, e.g., [3, 7, 8]). Total internal reflection plays a role in many contexts: Prisms in binoculars or camera viewfinders use it, and it is the reason why to a diver the water surface appears like a mirror.

Call the angle of incidence α and the angle of refraction β (Fig. 2.1). At the boundary to the less-dense medium ($n_A < n_G$ if we think of air and glass), the inequality $\beta > \alpha$ holds. On the other hand, β cannot exceed 90°. This becomes

© Springer-Verlag Berlin Heidelberg 2016
F. Mitschke, *Fiber Optics*, DOI 10.1007/978-3-662-52764-1_2

Fig. 2.1 On the principle of total internal reflection. The ray coming in from *bottom left* at angle α strikes the boundary to the less-dense medium and is either refracted (angle β) and transmitted or, if α is too large for that as in case 3, is totally reflected towards the *bottom right*. Case 2 represents the borderline situation with a grazing angle of the outgoing beam

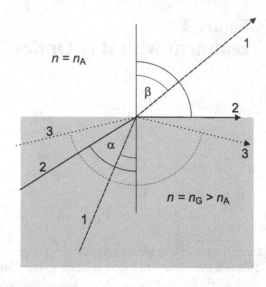

clear from an inspection of *Snell's law of refraction*

$$\frac{\sin \alpha}{\sin \beta} = \frac{n_A}{n_G}$$

when keeping in mind that $\sin \beta$ cannot exceed unity. In that limiting case,

$$\sin \alpha_{crit} = \frac{n_A}{n_G} < 1.$$

For even larger angles of incidence, the ray is reflected back into the denser medium nearly without loss. This is the meaning of the term "total internal reflection."

The same mechanism can also be used to guide light around bends. In 1870 the English scientist John Tyndall (1820–1893) during a session of the Royal Academy demonstrated an experiment which is now part of the standard repertoire of physics course demonstration experiments: A bucket of water is fitted in its lower part on one side with a small hole for the water to spit out, and on the opposing side with a window through which light from a bright lamp illuminates the hole from inside. The water falls in a parabolic curve, and this arc of water guides the light. Some part of the light is scattered off from surface irregularities so that, in a darkened lecture hall, the water column glows in the dark to spectacular effect (Fig. 2.2).[1]

The demonstration hinges on the fact that the refractive index of water exceeds that of the air surrounding it. The index of water is about $n_W = 1.33$, while that of

---

[1]Tyndall did not invent this himself. The twisted but amusing story leading up to our present-day insights about light-guiding and fibers is reported in [4].

Fig. 2.2 Total internal
reflection in water lets
illuminated fountains glow.
Submerged lamps illuminate
the fountain from below, and
the water column guides it up.
This picture was taken in
Boca Raton, Florida, USA

air is about $n_A = 1$. Most glasses have indices in the range of $n_G \approx 1.4$ to 1.8, and
therefore the same guiding effect can be had in strands or rods of glass.

And indeed, fibers exist which consist of nothing but basically a long cylinder of
glass or transparent plastic with a diameter of the order of 1 mm. They are used for
some special illumination applications, like guiding light into hard-to-reach places
inside some apparatus, and everybody has seen those decorative lamps in which a
whole bundle of such fibers is combined. Typical optical fibers for data transmission
have a somewhat more complex inner structure, though.

2.2   Step Index Fiber

A frequently used type of fiber is called *step index fiber*. Its internal structure is
as shown in Fig. 1.3: There is a core with circular cross-section, surrounded by
a cladding zone with ring-shaped cross-section. The core consists of a glass with
slightly higher refractive index than that of the cladding. Light is therefore guided
in the core (but we will need to make this statement more precise in Sect. 3.13). The
advantage of this two-layer structure over the simple version is that the fiber surface,
i.e., the boundary between cladding and the outside is no longer involved in the

Fig. 2.3 Sketch for
calculating total internal
reflection

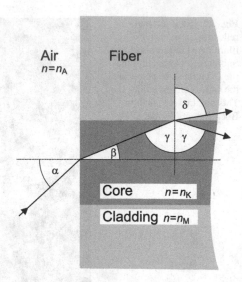

light-guiding mechanism. In the event that the fiber surface is soiled or touches other
glass, the function is not compromised. The unstructured fiber, in contrast, would
suffer from enormous loss. Just consider the case that a drop of oil or other liquid
comes into contact with the fiber surface: if it had a refractive index comparable to
that of the glass, the waveguiding by total internal reflection would break down [10].

Since the outer surface is not important for the waveguiding in step index fiber,
we may simplify its discussion by pretending that the cladding diameter is infinitely
wide so that no outside surface exists. Now we can discuss the largest angle with
respect to the fiber axis which a ray of light may take and still be guided by the fiber
(Fig. 2.3).

Near the fiber end face, we distinguish three refractive indices n, with

$$n_K > n_M \geq n_A,$$

where we used indices A for ambient air, K for core, and M for cladding.[2] In the
second relation, the equality is valid for unstructured fibers; we will concentrate on
step index fibers, though.

We apply Snell's law:

$$n_A \sin \alpha = n_K \sin \beta, \tag{2.1}$$

$$n_K \sin \gamma = n_M \sin \delta. \tag{2.2}$$

---

[2]These indices suggest the German words Kern (core) and Mantel (cladding), respectively. We
keep them from the original German edition of this book because the English words core and
cladding, as they share the same initial, do not provide a better option. Conveniently, the related
English words "kernel" and "mantel" also denote the central part of something and some kind of
enclosure, respectively.

When we assume the fiber axis to be perpendicular to the front face, $\beta + \gamma = \pi/2$ and hence $\sin \beta = \cos \gamma$. Then

$$\sin \beta = \sqrt{1 - \sin^2 \gamma}. \tag{2.3}$$

The limiting angle for total internal refraction is defined by

$$\sin \delta_{\max} = 1 \quad \Rightarrow \quad \sin \gamma_{\max} = n_M/n_K. \tag{2.4}$$

Inserting Eq. (2.4) in (2.3) and that in (2.1), we obtain

$$n_A \sin \alpha_{\max} = n_K \sqrt{1 - \frac{n_M^2}{n_K^2}} = \sqrt{n_K^2 - n_M^2}. \tag{2.5}$$

We may assume $n_A = 1$ for air. Then the limiting angle $\alpha_{\max}$ for rays to be guided is

$$\alpha_{\max} = \arcsin \sqrt{n_K^2 - n_M^2}.$$

The argument of arcsin bears a special name: it is called the *numerical aperture*, often abbreviated as NA:

$$NA = \sqrt{n_K^2 - n_M^2}.$$

(The word "aperture" derives from Latin *apertus* = open and indicates some form of opening. We also find it in "aperitif," the opening of a meal, and in "overture," the opening of an opera or a romantic affair. "Numerical" here indicates a dimensionless number.)

Clearly, the numerical aperture is a measure of the index difference between core and cladding. The largest acceptance angle for rays hitting the fiber face is given by $\sin \alpha_{\max} = NA$, *inside* the fiber by $\sin \alpha_{\max} = NA/n_K$. Using the fact that in (linear) optics ray paths can be reversed, the acceptance cone at the same time describes the exit cone of light at the other fiber end. In Fig. 2.4, this input/output cone is schematically shown.

We use the opportunity to introduce another frequently used quantity which is also a measure of the index difference between core and cladding:

$$\Delta = \frac{n_K^2 - n_M^2}{2n_K^2}. \tag{2.6}$$

The conversion between NA and Δ is given by

$$NA = n_K \sqrt{2\Delta}.$$

Fig. 2.4 Acceptance and exit
cone of a fiber are shown
schematically. In reality, the
cone is not quite as sharply
limited

Fiber

Usually the index difference is quite small (a few tenths of 1 %) so that Δ can be
simplified as

$$\Delta \approx \frac{(n_K - n_M)(n_K + n_M)}{n_K(n_K + n_M)}$$

$$\approx \frac{n_K - n_M}{n_K}.$$

This last relation justifies that Δ is called *normalized refractive index difference* or
normalized index step.

Let us consider typical realistic numbers for single-mode fibers. We assume $\Delta =$
0.3 %; with $n_K = 1.46$, this implies NA = 0.11; 0.11 rad indicates an acceptance
angle of about $\pm 7°$. Rays hitting the fiber face within a cone of this angle will
be guided in the fiber. Rays coming in at steeper angles will leave the core; they
propagate in the cladding and move away from the axis. Ultimately they are lost for
guiding: The cladding often has more loss than the core, so that part of this light is
dissipated. The rest eventually reaches the outside surface where typically a plastic
coating is applied for protection; the coating has strong optical loss. We conclude
that light rays which have left the core once are lost forever.[3]

In this chapter we have used ray optics. We have seen no reason to assume that
within the cone some angle would be preferred over any other one. In the following
chapter, a wave-optical treatment will reveal that within the continuum of angles,
only a discrete subset is physically possible. These specific angles are related to the
so-called *modes* of the light field in the fiber. The concept of modes is of central
importance for the waveguiding properties of fibers and will be studied in detail
in the next chapter. However, if many such modes exist, the continuum of angles
is approximated again, and our ray-optical approximation is the better justified
the more modes there are. In the remainder of our ray-optical treatment we will
therefore have multimode fibers in mind.

---

[3]There is one subtle exception to this otherwise reliable rule: so-called *whispering gallery modes*
will be described in Sect. 7.4.

2.3 Modal Dispersion

In this paragraph we will consider the fact that different modes, i.e., rays entering the fiber at different angles, travel different path lengths until they reach the far end of the fiber. Consequently they arrive at different times. This scatter of arrival times is known as modal dispersion. Figure 2.5 illustrates the situation.

In a fiber of length, L, let the path length of a beam propagating at an angle β with the axis be called L'. Clearly, $L' = L/\cos\beta$. Earlier we have seen that $\sin\beta$ cannot be larger than NA/n_K. In any event, $\beta \ll 1$, and therefore we may approximate $\sin\beta \approx \beta$ and $\cos\beta \approx 1 - \beta^2/2$. Then we obtain

$$L' = L\left(1 + \frac{NA^2}{2n_K^2}\right) = L\left(1 + \Delta\right).$$

Let us insert the numerical example values from above. With $\Delta = 0.3\%$ it follows that L' is 3 parts in 1000 longer than L. This implies that the path difference between the straight and the maximally inclined beam reaches one full wavelength after a distance of 333 wavelengths.

In an interference experiment, one finds a first destructive interference—and hence a mutual cancellation of both rays—after a path difference of $1/2$ wavelength; here, after a distance of 167 wavelengths which corresponds to only a fraction of a millimeter of fiber length. But of course, since both rays propagate at an angle, they do not fully cancel out but rather produce a fringe pattern of parallel bright and dark stripes across their full cross-section. Averaged over that cross-section, the interference effect cancels out.

Nonetheless, one and the same light signal coupled into the fiber may propagate along different paths so that there is a scatter of propagation times often called delay distortion. In the case of short light pulses, this causes an increased duration, i.e., a widening of the pulse. This can go so far that pulses widen to more than their separation; then neighboring pulses spill into each other. When this

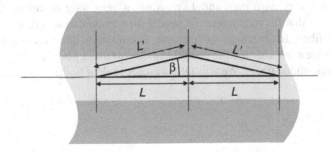

Fig. 2.5 Modal dispersion: rays propagating at an angle with the fiber axis travel a longer distance than those remaining parallel to the axis. This leads to different arrival times

intersymbol interference happens, the transmitted message is mangled and may be undecipherable.

A rough estimate will suffice to show that this is indeed a serious problem. Let us for simplicity take the velocity of light in the fiber as c/n.[4] Then, the propagation time τ for fiber length L in a step index fiber with core index n_K is $\tau = n_K L/c$. For the ray along the axis, the travel time acquires its minimal value:

$$\tau_{\min} = \frac{n_K L}{c}.$$

Meanwhile, the ray traveling at the maximal angle takes the longest time:

$$\tau_{\max} = \frac{n_K L}{c} (1 + \Delta) = \tau_{\min}(1 + \Delta).$$

In comparison, the difference $\delta\tau = \tau_{\max} - \tau_{\min}$ is

$$\delta\tau = \frac{n_K L}{c} \Delta = \tau_{\min} \Delta.$$

This shows the simple result that the relative amount of propagation time scatter is given by Δ:

$$\boxed{\frac{\delta\tau}{\tau_{\min}} = \Delta}.$$

Let us again take $\Delta = 0.3\,\%$ as a typical value. In a fiber of 1 km length, the arrival times will spread over ca. 15 ns.

This is just a rough estimate, of course: we used approximations and we have neglected that in addition to meridional rays there can also be helical rays (Fig. 2.6). Nonetheless it tells us that there is some maximum data rate above which the transit time spread will begin to deteriorate the signal integrity. The maximum rate is given by the inverse of the maximum scatter: in our example we obtain about 70 MHz.

That is not a very high rate, and 1 km is not a very long distance, either. We therefore realize that the mechanism of modal dispersion can severely hamper the usefulness of fibers for practical applications. Fortunately, there are ways to avoid the problem. One can either use the so-called *gradient index fibers* or, for the highest demands, single-mode fibers. We will take a closer look at both.

---

[4]By doing so, we momentarily ignore the distinction between phase and group velocities.

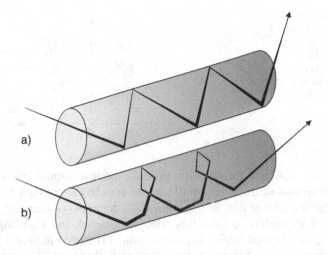

Fig. 2.6 Light guiding by total internal reflection in a fiber. There are meridional and helical rays. Meridional rays (**a**) propagate in a plane, helical rays (**b**) on a twisted path

2.4   Gradient Index Fibers

In order to avoid the scatter of arrival times, one can use a certain radial profile of the refractive index in the fiber. Instead of a step index profile, let us consider a gradient index profile where the index depends on the radial position like

$$n(r) = \begin{cases} n_K \sqrt{1 - 2\Delta \, (r/a)^\alpha} : & |r| \leq a \\ n_M & : & |r| > a, \end{cases} \tag{2.7}$$

where a denotes the core radius. The resulting profile is sketched for selected values of the profile exponent α in Fig. 2.7.

The optimum index profile is the one which minimizes the differences in transit time. In first approximation, the optimum is obtained for $\alpha = 2$; in a parabolic index profile, fiber rays follow a curved—rather than zigzag—path. While the curved path is still geometrically longer than the straight path along the axis, the detour is made up for by the lower index away from the axis so that the optical path is the same.

Now one obtains for the scatter of transit times

$$\alpha = \infty : \quad \delta\tau = \frac{n_K L}{c} \Delta \qquad \text{as above}$$

$$\alpha = 2 : \quad \delta\tau = \frac{n_K L}{c} \frac{\Delta^2}{2} \qquad \text{improvement by } \frac{\Delta}{2} \approx 10^{-3}$$

This is a considerable improvement: For a gradient index fiber with parabolic profile, the transit time spread is reduced by about three orders of magnitude. The modal dispersion is then reduced to a few tens of picosecond per kilometers.

Fig. 2.7 Some common
index profiles, as described
by Eq. (2.7). For $\alpha = 2$ the
profile is parabolic. For
$\alpha \to 1$ the profile becomes
triangular, and for $\alpha \to \infty$,
rectangular (step index
profile)

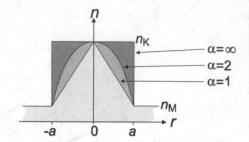

A precise calculation of the optimum profile exponent is quite involved due to the sudden transition of the index profile in the core to the constant index in the cladding. It has been found that the optimum value does not occur exactly at $\alpha = 2$, but slightly off, depending on glass type, doping material, and wavelength [2, 5]. One reason is the so-called profile dispersion, which is treated in Sect. 4.2. Simply stated, it occurs because Δ depends on wavelength, due to the fact that both n_K and n_M depend on wavelength in not exactly the same way. Moreover, the optimum exponent is not the same for meridional rays and helical rays; it thus also depends on the specific mix of excited modes [1]. For these reasons, the significance of the theoretical optimum is reduced. Unavoidable manufacturing tolerances in making the fibers make it difficult to maintain a target value with high precision anyway. Therefore, improvement over the parabolic index profile through further perfecting the index profile is only marginal.

2.5 Mode Coupling

The distribution of power over the different modes in a multimode fiber is not necessarily maintained as the light propagates down the fiber. Whenever the fiber is bent, there is coupling between modes. Any motion of the fiber on the table or lab bench, indeed just small temperature fluctuations, can and will modify the distribution of power over the modes (the "mode partition"). This has no further consequences as long as the detector at the fiber end correctly measures the sum of all partial powers. In practice, however, detectors are not necessarily uniformly sensitive across their surface; in such case some modes would register stronger than others. Then random changes of the mode partition will be reflected as random fluctuations of the received power, a phenomenon called *mode partition noise*.

As the mode partition fluctuates, the transit time scatter is mitigated to some degree. It becomes unlikely that a certain photon travels the total distance in the fastest or the slowest mode; more likely it will undergo a random walk between faster and slower modes, and experience an averaging effect. Provided that the fiber length exceeds a certain minimum called the coupling length L_{coupl}, the temporal spread does not grow in proportion to distance L, but only as $\sqrt{L}$. A typical value

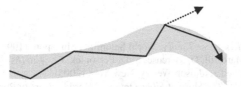

Fig. 2.8 Light guiding in a bent fiber: bends imply that rays impinge on the core–cladding interface at a different angle. Part of the light may even be lost because the maximum angle for total internal reflection is exceeded (*dashed*)

for the coupling length is on the order of 100 m for step index fibers and a few kilometers for gradient index fibers.

Mode mixing is then beneficial for reducing modal dispersion. One can even enhance this effect by enforced mode mixing. This is accomplished in mode mixers which are mechanical fixtures that deform and bend the fiber (Fig. 2.8). It is also a well-known fact that sometimes a fiber can transmit larger bandwidth when it is made from a concatenation of several pieces, rather than one single piece. One might have expected that irregularities at the joints (the fiber splices, see Sect. 8.3.2), would be detrimental, but the opposite is true!

2.6 Shortcomings of the Ray-Optical Treatment

The treatment given so far is not accurate. We have pretended that there are rays of light which are reflected at the core–cladding interface like at an ideal mirror. Of course, light is a wave phenomenon. The wave partially protrudes across the interface and reaches into the second medium down to a penetration depth on the order of a wavelength. This makes the ray path longer; equivalently one can also speak of an additional phase shift known as *Goos–Hänchen shift* [9]. We are dealing with fibers which have core diameters not a whole lot larger than the wavelength, and therefore we must expect significant corrections.

However, rather than attempting to incorporate such corrections into a ray-optical treatment, we take the high road and replace it altogether with a proper wave-optical treatment in the next chapter. As we shall see, wave optics predicts automatically that part of the light penetrates into the cladding, that the exit cone does not have a perfectly sharp boundary, and it will tell us that there is a discrete set of possible distributions of the electrical field in the fiber cross-section known as the fiber modes. This is equivalent to saying that rays cannot make any angle with the axis between zero and the maximum, but only one out of a discrete set.

References

1. P.-A. Bélanger, *Optical Fiber Theory*, World Scientific, Singapore (1993)
2. D. Gloge, E. A. J. Marcatili, D. Marcuse, S. D. Personick, Ch. 4 in [6]
3. E. Hecht, *Optics*, 4th ed., Addison Wesley, New York (2005)
4. J. Hecht, *City of Light: The Story of Fiber Optics*, Oxford University Press, Oxford (1999)
5. E. A. J. Marcatili, Ch. 2 in [6]
6. S. E. Miller, A. G. Chinoweth (Eds.), *Optical Fiber Telecommunications*, Academic Press, London (1979)
7. F. L. Pedrotti, L. M. Pedrotti, L. S. Pedrotti, *Introduction to Optics*, 3rd ed., Benjamin-Cummings, Upper Saddle River, New Jersey (2006)
8. B. E. A. Saleh, M. C. Teich, *Fundamentals of Photonics*, 2nd ed., John Wiley & Sons, New York (2007)
9. A. W. Snyder, J. D. Love, *Goos-Hänchen Shift*, Applied Optics **15**, 236 (1976)
10. A. C. S. Van Heel, *A New Method of Transporting Optical Images Without Aberrations*, Nature **173**, 39 (2 January 1954)

Chapter 3
Treatment with Wave Optics

In this chapter we will start with Maxwell's equations, derive a wave equation, apply this to the geometry of the fiber, and finally arrive at the modal structure. Closed solutions can be obtained for step index fibers and for gradient index fibers without cladding (i.e., when the gradient continues ad infinitum). We will restrict our treatment to step index fibers. For the sake of clarity, we will also use several approximations in order to emphasize important issues over detail.

3.1 Maxwell's Equations

In MKS units of measurement, Maxwell's equations are [3]

$$\nabla \cdot \vec{D} = \rho, \tag{3.1}$$

$$\nabla \cdot \vec{B} = 0, \tag{3.2}$$

$$\nabla \times \vec{H} = \vec{J} + \frac{\partial \vec{D}}{\partial t}, \tag{3.3}$$

$$\nabla \times \vec{E} = -\frac{\partial \vec{B}}{\partial t}. \tag{3.4}$$

Here,

$\vec{E}$	electric field strength	(V/m)
$\vec{H}$	magnetic field strength	(A/m)
$\vec{D}$	dielectric displacement	(As/m^2)
$\vec{B}$	magnetic induction	(Vs/m^2=T)
$\vec{J}$	current density	(A/m^2)
ρ	charge density	(As/m^3)

© Springer-Verlag Berlin Heidelberg 2016
F. Mitschke, *Fiber Optics*, DOI 10.1007/978-3-662-52764-1_3

Some textbooks simplify by considering only processes in vacuum. Of course, there is no use for us in doing so; we need to describe processes inside a material. Therefore we need to use quantities which are given by the material's properties:

$\vec{P}$ polarization
$\vec{M}$ magnetization
σ conductivity

Polarization and magnetization describe the distortion of atomic orbitals as they are produced by the influence of the electromagnetic field. Conductivity describes the transport of electric charges (as is well known, there are no magnetic charges); in the general case it takes the form of a tensor.

The following relations hold:

$$\vec{D} = \epsilon_0 \vec{E} + \vec{P}, \tag{3.5}$$

$$\vec{B} = \mu_0 (\vec{H} + \vec{M}), \tag{3.6}$$

$$\vec{J} = \sigma \vec{E}. \tag{3.7}$$

where

ϵ_0 vacuum permittivity   (Dielectric constant of free space),
μ_0 vacuum permeability   (Permeability constant of free space).

The numerical values are given by

$$\epsilon_0 = \frac{10^7}{4\pi c^2} \frac{\text{Am}}{\text{Vs}}$$

$$\approx 8.85 \times 10^{-12} \frac{\text{As}}{\text{Vm}},$$

$$\mu_0 = \frac{4\pi}{10^7} \frac{\text{Vs}}{\text{Am}}$$

$$\approx 1.26 \times 10^{-6} \frac{\text{Vs}}{\text{Am}}.$$

Two combinations have special relevance: the product

$$\mu_0 \epsilon_0 = 1/c^2,$$

where $c = 2.99792458 \times 10^8$ m/s is the speed of light in vacuum, and the ratio

$$\mu_0 / \epsilon_0 = \left(\frac{4\pi c}{10^7} \right)^2 = Z_0^2.$$

$Z_0 \approx 377\ \Omega$ is the vacuum impedance and denotes the amplitude ratio of the electric and the magnetic part of the electromagnetic wave:

$$\frac{\vec{E}}{\vec{H}} = Z_0.$$

In air and glass we may simplify as follows:

- $\rho = 0$ There are no free charges (Approximation 1)
- $\vec{J} = 0$ There are no currents (Approximation 2)
- $\vec{M} = 0$ There is no magnetization (Approximation 3)

Hence, of all properties of the material, we retain only the ones which influence the polarization. Using these approximations, Maxwell's equations are reduced to

$$\nabla \cdot \vec{D} = 0, \tag{3.8}$$

$$\nabla \cdot \vec{B} = 0, \tag{3.9}$$

$$\nabla \times \vec{B} = \mu_0 \frac{\partial \vec{D}}{\partial t}, \tag{3.10}$$

$$\nabla \times \vec{E} = -\frac{\partial \vec{B}}{\partial t}. \tag{3.11}$$

3.2 Wave Equation

Applying $\nabla \times$ to Eq. (3.11) yields

$$\nabla \times \nabla \times \vec{E} = \nabla \times \left(-\frac{\partial \vec{B}}{\partial t} \right), \tag{3.12}$$

$$\nabla(\nabla \cdot \vec{E}) - \nabla^2 \vec{E} = -\frac{\partial}{\partial t}(\nabla \times \vec{B}). \tag{3.13}$$

We rearrange the RHS using Eqs. (3.10) and (3.5) and obtain

$$\nabla(\nabla \cdot \vec{E}) - \nabla^2 \vec{E} = -\frac{\partial}{\partial t} \left(\mu_0 \frac{\partial \vec{D}}{\partial t} \right) \tag{3.14}$$

$$= -\mu_0 \frac{\partial^2}{\partial t^2} \vec{D} \tag{3.15}$$

$$= -\mu_0 \epsilon_0 \frac{\partial^2}{\partial t^2} \vec{E} - \mu_0 \frac{\partial^2}{\partial t^2} \vec{P}. \tag{3.16}$$

We thus find the wave equation

$$-\nabla(\nabla \cdot \vec{E}) + \nabla^2 \vec{E} = \frac{1}{c^2}\frac{\partial^2}{\partial t^2}\vec{E} + \mu_0 \frac{\partial^2}{\partial t^2}\vec{P}. \tag{3.17}$$

A fully analogous equation can be derived for the magnetic field.

Now we must make some statement about the relation between the polarization $\vec{P}$ and the field strength $\vec{E}$. This involves properties of the material. We will make the assumption that the polarization follows a change of field strength instantaneously, i.e., quicker than any other relevant time scale involved (Approximation 4). Then we can write the polarization as

$$\vec{P} = \epsilon_0 \left(\chi^{(1)}\vec{E} + \chi^{(2)}\vec{E}^2 + \chi^{(3)}\vec{E}^3 + \cdots \right). \tag{3.18}$$

Now we introduce a further approximation: We will assume that the polarization of the material is always parallel to the field strength (Approximation 5). This is a justified assumption: In a homogenous medium, the tensor $\chi^{(i)}$ takes the form of a scalar. It is true that certain crystals are in use in optics which are decidedly nonhomogenous, but glass is homogenous due to its structure (see Sect. 6.1.2). In a fiber, the homogeneity is only slightly perturbed due to the refractive index profile. On the other hand, wave guiding essentially occurs parallel to the axis. In this geometry one may make the paraxial approximation which plays a role in many optical arrangements. Here it means that propagation will make only small angles with the axis. Then the index change between core and cladding, which is small to begin with, is almost inconsequential because $\vec{E}$ and $\vec{H}$ are both perpendicular to the interface and are proportional to each other. (The proportionality constant is the impedance, which in free space is given by Z_0.) In this book, we will use the scalar approximation throughout, because (a) a vectorial treatment is considerably more involved and (b) the impact on the result is minimal. Below we will briefly point out the difference between the modes obtained in the scalar treatment and the so-called hybrid modes from a vectorial calculation.

We return to the wave equation, in which we can now introduce a simplification. Given that now $\vec{E}\|\vec{P}$ and thus $\vec{D}\|\vec{E}$, it follows that $\nabla \cdot \vec{D} = \nabla \cdot \vec{E} = 0$. On the LHS of the wave equation, the term with $\nabla \cdot \vec{E}$ then disappears and it remains

$$\nabla^2 \vec{E} = \frac{1}{c^2}\frac{\partial^2}{\partial t^2}\vec{E} + \mu_0 \frac{\partial^2}{\partial t^2}\vec{P}. \tag{3.19}$$

3.3   Linear and Nonlinear Refractive Index

We will now go one step further and make specific assumptions about the relation between electric field and polarization.

3.3.1 Linear Case

In many situations, it is well justified to truncate the series expansion of Eq. (3.18) after the linear term

$$\vec{P} = \epsilon_0 \chi^{(1)} \vec{E}. \tag{3.20}$$

This is the *linear approximation* (Approximation 6); it is valid for low light intensities. Due to Eq. (3.5) we then get

$$\vec{D} = \epsilon_0 \vec{E} \left(1 + \chi^{(1)}\right). \tag{3.21}$$

The expression inside the bracket is the relative dielectric constant

$$1 + \chi^{(1)} = \epsilon = \left(n + i\frac{c}{2\omega}\alpha\right)^2,$$

where n is the refractive index and α is Beer's coefficient of absorption. We are going to study propagation in extremely pure, low-loss glass. If there ever was a justification for using the low-loss approximation that $\alpha \approx 0$ (Approximation 7), this is it. Then, ϵ is real and is given by

$$\epsilon = n^2. \tag{3.22}$$

On the RHS of Eq. (3.19), we insert the relation (3.20) between E and P and then obtain

$$\boxed{\nabla^2 \vec{E} = \frac{n^2}{c^2} \frac{\partial^2}{\partial t^2} \vec{E}.} \tag{3.23}$$

This is the linear wave equation, as it is obtained directly from Maxwell's equations using Approximations 1–7. An analogous equation

$$\boxed{\nabla^2 \vec{H} = \frac{n^2}{c^2} \frac{\partial^2}{\partial t^2} \vec{H}} \tag{3.24}$$

can be found by similar procedure for the magnetic component of the wave. From now on we will drop vector symbols (arrows) for convenience.

3.3.2   Nonlinear Case

If one does not truncate the serial expansion (3.18) after the linear term, one can capture some interesting physical processes that are lost in the linear approximation, but which are experimentally observed and are of relevance for advanced applications. As soon as E is no longer so small that truncation after the linear term is justified, we enter the realm of *nonlinear optics*.

Here our main interest is for light in glass. Glass is a material which has a statistical structure, which is isotropic on average. Therefore glass has an inversion symmetry so that $\chi^{(2)} = 0$. The first nonvanishing higher-order term in the series expansion is then the one containing $\chi^{(3)}$. Even higher terms, however, can still be safely neglected except in some very special circumstances since their coefficients are small so that they become noticeable only at enormous intensities. This is why we can restrict our discussion to the impact of the $\chi^{(3)}$ term (alternative Approximation 6). It will turn out, though, that this term can make a big difference.

As before, we keep the low-loss approximation, so that the only conceivable effect is a modification of the refractive index. In the linear case we had

$$P = \epsilon_0 \chi^{(1)} E$$

and

$$n^2 = \epsilon = 1 + \chi^{(1)}.$$

In the interest of a clear distinction, we shall denote the ϵ appearing here as ϵ_{linear}. Similarly, from now on the refractive index n in this equation shall be denoted by n_0; we will call it the *small signal refractive index*. For the nonlinear case we obtain

$$P = \epsilon_0 \left\{ \chi^{(1)} + \chi^{(3)} E^2 \right\} E \tag{3.25}$$

and

$$\epsilon = n^2 = 1 + \chi^{(1)} + \chi^{(3)} E^2 = \epsilon_{\text{linear}} + \chi^{(3)} E^2. \tag{3.26}$$

This is the same as

$$\epsilon = \epsilon_{\text{linear}} \left(1 + \frac{\chi^{(3)}}{\epsilon_{\text{linear}}} E^2 \right). \tag{3.27}$$

Since the nonlinear contribution to the refractive index is a small correction, we obtain

$$n = n_0 \sqrt{1 + \frac{\chi^{(3)}}{\epsilon_{\text{linear}}} E^2} \approx n_0 \left(1 + \frac{\chi^{(3)}}{2 n_0^2} E^2 \right). \tag{3.28}$$

We rewrite this as

$$n = n_0 + \bar{n}_2 E^2$$ (3.29)

with

$$\bar{n}_2 = \frac{\chi^{(3)}}{2n_0}.$$ (3.30)

The numerical value of $\bar{n}_2$ for fused silica is slightly frequency-dependent and is also influenced by dopants. However, these dependencies are weak and we can use the typical value of $10^{-22} \, \text{m}^2/\text{V}^2$. The intensity I (power per area) of a light field is proportional to the square of the field amplitude. Therefore it is quite common to write

$$n = n_0 + n_2 I$$ (3.31)

with $I = (n_0/Z_0)E^2$ and

$$n_2 = 3 \times 10^{-20} \, \text{m}^2/\text{W}.$$ (3.32)

We see that inclusion of the $\chi^{(3)}$ term results in a modification of the refractive index: The index always depended on wavelength, but now it also depends on intensity.

Under conditions that one would consider "reasonable", this modification is tiny indeed: Even an irradiated power of 1 kW, focused down to a spot of $100 \, \mu\text{m}^2$, will result in an increase of the index of only

$$n_2 I = 3 \times 10^{-20} \, \text{m}^2/\text{W} \frac{10^3 \, \text{W}}{10^{-10} \, \text{m}^2} = 3 \times 10^{-7}.$$ (3.33)

This is a change which is much smaller than the core–cladding index difference of a fiber. As we proceed to consider the field distribution in the fiber, this term will therefore be inconsequential. Equation (3.23) remains valid—in the linear case one can equate n with n_0, but in the nonlinear case $n(I) = n_0 + n_2 I$. We will see later (beginning with Chap. 9) that this nonlinearity unfolds its impact when the phase evolution of a light wave is considered.

3.4 Separation of Coordinates

At this point, we introduce simplifications which are based on the special geometry of fiber: circular cross-section, extended in the longitudinal direction. This strongly suggests the use of cylindrical coordinates r, ϕ, and z. We take the propagation direction as the positive z direction. As is well known, the Laplacian in cylindrical coordinates reads

$$\nabla^2 E = \frac{1}{r}\frac{\partial}{\partial r}\left(r\frac{\partial}{\partial r}E\right) + \frac{1}{r^2}\frac{\partial^2}{\partial \phi^2}E + \frac{\partial^2}{\partial z^2}E. \tag{3.34}$$

We introduce the following ansatz for the optical field of the light wave:

$$E = E_0 \mathcal{N}\mathcal{Z}\mathcal{T}. \tag{3.35}$$

Here,

$$\mathcal{N} = \mathcal{N}(r,\phi)$$

is the field amplitude distribution in the plane normal to the z-axis,

$$\mathcal{Z} = \mathcal{Z}(z) = \mathrm{e}^{-i\beta z}$$

denotes a running wave with wave number β, and

$$\mathcal{T} = \mathcal{T}(t) = \mathrm{e}^{i\omega t}$$

denotes a monochromatic wave with (angular) frequency ω. Such separation is permitted due to the linearity discussed above, which makes it possible to pull out a factor of E_0, and paraxiality, which implies that both the electric and magnetic field components are basically perpendicular to the propagation direction; thus, longitudinal and transverse processes are decoupled. We write β, not k for the propagation constant; this way we admit a difference between the wave vector and its longitudinal component. This is to allow propagation in analogy to the rays that make an angle with the fiber axis (see Sect. 2.3).

Using cylindrical coordinates and this ansatz, the wave equation takes the form

$$\frac{1}{r}\frac{\partial}{\partial r}\left(r\frac{\partial}{\partial r}E_0\mathcal{N}\mathcal{Z}\mathcal{T}\right) + \frac{1}{r^2}\frac{\partial^2}{\partial \phi^2}E_0\mathcal{N}\mathcal{Z}\mathcal{T} + \frac{\partial^2}{\partial z^2}E_0\mathcal{N}\mathcal{Z}\mathcal{T} = \frac{n^2}{c^2}\frac{\partial^2}{\partial t^2}E_0\mathcal{N}\mathcal{Z}\mathcal{T}. \tag{3.36}$$

Obviously, all terms contain the constant factor E_0, which is thus cancelled out. The physical reason, again, is the linearity assumed here.

Partial derivatives act differently upon $\mathcal{N}, \mathcal{Z}$, and $\mathcal{T}$. The first term can be rewritten as

$$\frac{1}{r}\frac{\partial}{\partial r}\left(r\frac{\partial}{\partial r}\mathcal{N}\mathcal{Z}\mathcal{T}\right) = \mathcal{Z}\mathcal{T}\frac{1}{r}\frac{\partial}{\partial r}\left(r\frac{\partial}{\partial r}\mathcal{N}\right),$$

which can be simplified to yield

$$\mathcal{ZT}\left(\frac{1}{r}\frac{\partial}{\partial r}\mathcal{N} + \frac{\partial^2}{\partial r^2}\mathcal{N}\right).$$

The second term becomes

$$\mathcal{ZT}\frac{1}{r^2}\frac{\partial^2}{\partial \phi^2}\mathcal{N},$$

and the third

$$-\beta^2\mathcal{NZT}.$$

On the RHS we obtain

$$-\frac{n^2}{c^2}\omega^2\mathcal{NZT}.$$

We will denote the vacuum wave number by k_0. Inside a medium with refractive index n, we will write $k = nk_0 = n\omega/c$. Then, the RHS becomes

$$-k^2\mathcal{NZT}.$$

Now the factor $\mathcal{ZT}$ is common to all terms and is thus cancelled out, too. This is caused by the homogeneity of the problem in space (at least in propagation direction) and time. We are left with the field amplitude distribution in the plane normal to the propagation direction. As typical fibers are circular in cross-section, it is useful to perform a further factorization:

$$\mathcal{N}(r,\phi) = \mathcal{R}(r)\,\Phi(\phi). \tag{3.37}$$

When we now reinsert all terms and multiply with $r^2/\mathcal{R}\Phi$, we obtain

$$\frac{r}{\mathcal{R}}\frac{\partial}{\partial r}\mathcal{R} + \frac{r^2}{\mathcal{R}}\frac{\partial^2}{\partial r^2}\mathcal{R} + \frac{1}{\Phi}\frac{\partial^2}{\partial \phi^2}\Phi - r^2\beta^2 = -k^2r^2, \tag{3.38}$$

which after sorting of terms becomes

$$-\frac{1}{\Phi}\frac{\partial^2}{\partial \phi^2}\Phi = \frac{1}{\mathcal{R}}\left(r^2\frac{\partial^2}{\partial r^2}\mathcal{R} + r\frac{\partial}{\partial r}\mathcal{R} + r^2\left(k^2 - \beta^2\right)\mathcal{R}\right). \tag{3.39}$$

We see that now the LHS contains Φ and not $\mathcal{R}$; on the RHS it is the other way around. Therefore, both sides must be equal to some constant. We will denote this constant by m^2. Now we have two independent equations for the azimuthal and the radial parts of the field amplitude distribution:

$$\frac{\partial^2}{\partial \phi^2} \Phi + m^2 \Phi = 0 \tag{3.40}$$

and

$$r^2 \frac{\partial^2}{\partial r^2} \mathcal{R} + r \frac{\partial}{\partial r} \mathcal{R} + r^2 \left(k^2 - \beta^2 \right) \mathcal{R} = \mathcal{R} m^2,$$

which, by using the abbreviation $\kappa^2 = k^2 - \beta^2$, is written in simpler form as

$$r^2 \frac{\partial^2}{\partial r^2} \mathcal{R} + r \frac{\partial}{\partial r} \mathcal{R} + (\kappa^2 r^2 - m^2) \mathcal{R} = 0. \tag{3.41}$$

To obtain some understanding of the meaning of quantities κ, k, and β, we recall the ray-optical description where we had rays propagating at a small angle with the axis. We assign the propagation constant k to the wave. β was introduced as its component in the propagation direction. Then, one can look at κ as its transverse component.

3.5 Modes

The equation for the azimuthal structure, Eq. (3.40), has the general solution

$$\Phi = c_0 \cos(m\phi + \phi_0) \tag{3.42}$$

with c_0 and ϕ_0 constants. Surely, Φ and $\partial \Phi / \partial \phi$ must be continuous at ϕ_0 and $\phi_0 + 2\pi$. But then, m must be an integer number. This constrains the number of possible solutions of the equation for the radial structure, Eq. (3.41).

Equation (3.41) has the form

$$x^2 y'' + x y' + (\kappa^2 x^2 - m^2) y = 0$$

when one makes the identifications $y = \mathcal{R}$ and $x = r$ and interprets the prime as a derivative with respect to x. In this form, or after scaling out κ, the equation is given in mathematical tables. It is called *Bessel's differential equation*. For integer m it is solved by

$$y = c_1 J_m(\kappa x) + c_2 N_m(\kappa x), \tag{3.43}$$

whenever κx is real (i.e., $\kappa^2 x^2 \geq 0$), or by

$$y = c_3 I_m(\kappa x) + c_4 K_m(\kappa x), \tag{3.44}$$

whenever κx is imaginary (i.e., $\kappa^2 x^2 < 0$). Here, functions $J_m(\kappa x)$, $N_m(\kappa x)$, $I_m(\kappa x)$, and $K_m(\kappa x)$ are Bessel functions. Properties of Bessel functions are given in Chap. 15.

In order to find the coefficients c_1–c_4, we must finally fix the specific geometry of the fiber. Up to now we have only assumed that the geometry is cylindrical, and that the value of the refractive index difference is small. Now we further specify that we consider a step index fiber: This is both a particularly simple structure and a realistic choice. For a step index fiber we have

$$n = \begin{cases} n_K: & r \leq a \quad \text{(inside the core),} \\ n_M: & r > a \quad \text{(in the cladding).} \end{cases} \tag{3.45}$$

Of course, $n_K > n_M$ because there would be no waveguiding otherwise. In order to distinguish the refractive indices in core and cladding, and also the wave numbers, we will use indices "K" and "M" as in Sect. 2.2. Index "0" continues to denote the respective quantity in vacuum, e.g., $k_K = n_K k_0$.

In the limiting case when the wavelength is much smaller than the geometric dimensions of the fiber cross-section, we expect to recover the results from ray optics: Light is guided inside the core. Therefore we look for solutions with the dominant part of the light wave concentrated in the core. We conjecture, at the same time, that light in the cladding will have field amplitudes which decrease further away from the center and will at least not contribute dominantly to the guided wave.

But then, κr must be real in the core and imaginary in the cladding. In other words, at least in the core, $k \geq \beta$ must hold: The wave number must be larger than or equal to its longitudinal component. In the cladding we may well have the opposite situation. This corresponds to solutions to Bessel's equation with transversal standing waves in the core and radially decaying waves in the cladding.

We expect on physical reasons that the field amplitude distribution does not have singularities. This implies that for the core the coefficient in the N_m term must vanish. In the cladding, similarly, the coefficient of the I_m term must vanish. This makes good physical sense: Far away from the core we expect the field amplitude to decay at least as rapidly as $1/r$ because otherwise the integral of power over the entire transverse plane would diverge.

In order to have κr real in the core, it is required that at $r \leq a$, $(\kappa r)^2 \geq 0$, or $\left(k_K^2 - \beta^2\right) r^2 \geq 0$. In contrast, in the cladding (i.e., at $r > a$), we need to have $(\kappa r)^2 \leq 0$ and thus $\left(k_M{}^2 - \beta^2\right) r^2 \leq 0$. Taken together, this implies

$$k_K \geq \beta \geq k_M.$$

The range of possible wave numbers for propagation down the fiber is thus constrained by the requirement of waveguiding.

Once again we introduce abbreviations: The transversal components of the wave number in core and cladding are given by

$$\kappa_K^2 = k_K^2 - \beta^2,$$
$$\kappa_M^2 = -\left(k_M^2 - \beta^2\right);$$

it is customary to use the product of these quantities with the core radius a:

$$u = \kappa_K a, \tag{3.46}$$

$$w = \kappa_M a. \tag{3.47}$$

u and w are dimensionless, real positive quantities often used in the literature. To comment on their physical significance, suffice it to say that u describes the progression of phase and w the transverse decay of amplitude. One might call u the longitudinal phase constant and w the radial decay constant.

In order to establish a relation between these somewhat abstract quantities and measurable quantities, we use the following relation between u and w:

$$u^2 + w^2 = \left(k_K^2 - \beta^2\right) a^2 - \left(k_M^2 - \beta^2\right) a^2 \tag{3.48}$$

$$= \left(k_K^2 - k_M^2\right) a^2 \tag{3.49}$$

$$= k_0^2 \left(n_K^2 - n_M^2\right) a^2. \tag{3.50}$$

It is clear that $u^2 + w^2$ equals a constant. This constant is of central importance and is called *normalized frequency* or simply *V number*. It is given by

$$V^2 = k_0^2 a^2 \left(n_K^2 - n_M^2\right)$$

or

$$\boxed{\begin{aligned} V &= k_0 \, a \, \mathsf{NA} \\ &= \frac{2\pi}{\lambda_0} a \, \sqrt{n_K^2 - n_M^2} \end{aligned}} \tag{3.51}$$

and contains all relevant data of an experimental situation. A step index fiber is characterized by the core radius a and the refractive indices n_K and n_M or, alternatively, numerical aperture. Either wavelength or vacuum wave number completes the description of the experimental situation. In the following, the value of V will be the decisive criterion to establish how many modes a fiber can support.

Using u and w, the general solution of the wave equation for a step index fiber can be written as:

$$\mathcal{N}_K = C_K J_m(ur/a)\cos(m\phi + \phi_0) : \quad r \leq a,$$
$$\mathcal{N}_M = C_M K_m(wr/a)\cos(m\phi + \phi_0) : r > a. \tag{3.52}$$

We recall that from a ray-optical treatment, one would naïvely expect a rectangular field distribution: constant 100 % amplitude everywhere in the core and 0 % in the cladding. It should by now be obvious that reality is different from that.

At $r = a$ both solutions must connect in a smooth way. This means that we do not expect a discontinuity: Rather, we expect the transition to be continuous and differentiable. This is only possible when the angular dependence is identical for both solutions, which is why in Eq. (3.52) we already wrote the same ϕ_0 and m.

The conditions for smooth transition are

$$\mathcal{N}_K(r = a) = \mathcal{N}_M(r = a), \tag{3.53}$$

$$\frac{\partial}{\partial r}\mathcal{N}_K(r = a) = \frac{\partial}{\partial r}\mathcal{N}_M(r = a). \tag{3.54}$$

After inserting, we find

$$C_K J_m(u) = C_M K_m(w), \tag{3.55}$$

$$C_K \frac{\partial}{\partial r} J_m(ur/a)\Big|_{r=a} = C_M \frac{\partial}{\partial r} K_m(wr/a)\Big|_{r=a}. \tag{3.56}$$

In the second of these equations we can introduce

$$\frac{\partial}{\partial r} = \frac{u}{a}\frac{\partial}{\partial(ur/a)} \tag{3.57}$$

and thus use the argument ur/a throughout; then we can write the derivative at $r = a$ as

$$C_K \frac{u}{a} J'_m(u) = C_M \frac{w}{a} K'_m(w) \tag{3.58}$$

(the prime ($'$) denotes the derivative with respect to the argument).

For the existence of a solution, it is required that the determinant of coefficients be zero:

$$C_K C_M J_m(u) \frac{w}{a} K'_m(w) - C_K C_M K_m(w) \frac{u}{a} J'_m(u) = 0. \tag{3.59}$$

From this, $C_K C_M/a$ can be eliminated immediately. Now we use a well-known recursion relation between Bessel functions (see Chap. 15):

$$u J'_m(u) = m J_m(u) - u J_{m+1}(u), \tag{3.60}$$

$$w K'_m(w) = m K_m(w) - w K_{m+1}(w). \tag{3.61}$$

With this the derivatives can be eliminated:

$$J_m(u)\big(m K_m(w) - w K_{m+1}(w)\big) = K_m(w)\big(m J_m(u) - u J_{m+1}(u)\big)$$

$$w J_m(u) K_{m+1}(w) = u K_m(w) J_{m+1}(u)$$

or

$$\boxed{\frac{J_m(u)}{u J_{m+1}(u)} = \frac{K_m(w)}{w K_{m+1}(w)}.} \tag{3.62}$$

From this relation between u and w, we will now obtain the permitted solutions for the fundamental mode of the fiber. It is obvious that functions $K_m(w)$ on the RHS are always positive while functions $J_m(u)$ on the LHS frequently change their sign. From this, one sees certain combinations of argument values (and thus V numbers) for which solutions are possible.

Explicit solutions are best obtained numerically. However, for our present purpose, we can inspect some special cases which will give us insight without invoking a computer.

3.6 Solutions for $m = 0$

Let us first consider the case of $m = 0$, which stands for rotationally symmetric field distributions over the fiber's cross-section. Then the equation is reduced to

$$\frac{J_0(u)}{u J_1(u)} = \frac{K_0(w)}{w K_1(w)}. \tag{3.63}$$

For a survey of possible solutions, we use the following table in which LHS and RHS of Eq. (3.63) are juxtaposed:

u	J_0	J_1	J_0/uJ_1		K_0/wK_1		w	
0	1	0	∞	$\Rightarrow$	∞	$\Rightarrow$	0	
$\vdots$	+	+	+	$\Rightarrow$	+	$\Rightarrow$	+	Branch of solutions
2.405	0	+	0	$\Rightarrow$	0	$\Rightarrow$	∞	
$\vdots$	–	+	–	$\Rightarrow$	–	$\Rightarrow$	–	No solution
3.832	–	0	∞	$\Rightarrow$	∞	$\Rightarrow$	0	
$\vdots$	–	–	+	$\Rightarrow$	+	$\Rightarrow$	+	Branch of solutions
5.520	0	–	0	$\Rightarrow$	0	$\Rightarrow$	∞	
$\vdots$	+	–	–	$\Rightarrow$	–	$\Rightarrow$	–	No solution

There is an alternation of ranges with and without solutions. The table only checks the algebraic sign; we can go beyond that and actually compute the values pertaining to the branches of solutions and plot them in a diagram of the (u, w) plane. This is done in Fig. 3.1. The curves represent ratios of two Bessel functions; their shape looks similar to a plot of a tan function. This is not surprising as we state in Chap. 15 that J_0 resembles a cosine function and J_1 a sine function.

Additionally, we can plot the locus of all points pertaining to a given V number. They form segments of circles in the (u, w) plane. For a given fiber, circles with different radii correspond to experiments at different wavelengths. The other way to look at it is that for a given wavelength, different radii correspond to different core diameters.

Fig. 3.1 Solutions of the eigenvalue equation for $m = 0$ in the u–w plane. As u increases, there is an alternation between regimes with existing (e.g., $0 \leq u \leq 2.405$) or nonexisting (e.g., $2.405 \leq u \leq 3.832$) solutions. Labels at the branches denote indices mp of the LP$_{mp}$ modes, as explained in Sect. 3.9

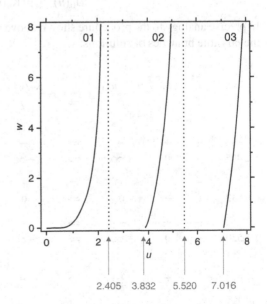

Points of intersection between the tan-like branches with the circle segments define possible solutions, i.e., combinations of u and w in a particular given situation (V fixed). Obviously the first branch of solutions exists at any V number between zero and infinity. The second branch shown here exists only above a minimum V. It takes the value of $V = 3.832$, as given by the first zero of Bessel function J_1. For all other branches shown one can make a similar statement about the minimum V which is always defined by a zero of J_1.

3.7 Solutions for $m = 1$

We might proceed by directly inserting $m = 1$ into the eigenvalue equation Eq. (3.59). This is not a problem for a computer solution. However, here we want to get a feel for the situation without recourse to a computer, and therefore it is advantageous to use an alternative recursion relation of Bessel functions which contains not $m + 1$ but $m - 1$. This way we obtain an equation for $m = 1$ which again only contains J_0 and J_1 (instead of J_1 and J_2); this makes our argument more transparent.

So let us use

$$u J'_m(u) = -m J_m(u) - u J_{m-1}(u) \text{ and}$$

$$w K'_m(w) = -m K_m(w) - w K_{m-1}(w)$$

in Eq. (3.59) to obtain

$$-\frac{J_1(u)}{u J_0(u)} = \frac{K_1(w)}{w K_0(w)}.$$

In precise analogy to the procedure shown above we can again write a table to locate the possible branches of solutions:

u	J_1	J_0	$-J_1/uJ_0$		K_1/wK_0		w	
0	0	1	0	$\Rightarrow$	0	$\Rightarrow$	$-\infty$	
								No solution
$\vdots$	$+$	$+$	$-$	$\Rightarrow$	$-$	$\Rightarrow$	$-$	
2.405	$+$	0	$-\infty$	$\Rightarrow$	$-\infty$	$\Rightarrow$	0	
$\vdots$	$+$	$-$	$+$	$\Rightarrow$	$+$	$\Rightarrow$	$+$	Branch of solutions
3.832	0	$-$	0	$\Rightarrow$	0	$\Rightarrow$	∞	
$\vdots$	$-$	$-$	$-$	$\Rightarrow$	$-$	$\Rightarrow$	$-$	No solution
5.520	$-$	0	$-\infty$	$\Rightarrow$	$-\infty$	$\Rightarrow$	0	
								Branch of solutions
$\vdots$	$-$	$+$	$+$	$\Rightarrow$	$+$	$\Rightarrow$	$+$	

Fig. 3.2 Solutions of the eigenvalue equation for $m = 0$ and $m = 1$ in the u–w plane for $m = 1$. Now there are branches of solutions where there were gaps in Fig. 3.1

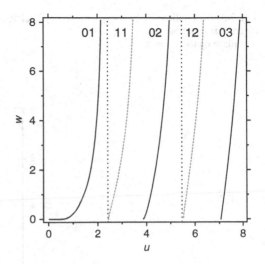

Again we find an alternation between permitted and forbidden regimes, with the changes occurring at zeroes of Bessel functions. In comparison to $m = 0$, here the regimes switch roles. This way there is a branch of solutions beginning at $V = 2.405$, constituting a second mode beyond the fundamental mode. Figure 3.2 combines all solutions found so far, i.e., for $m = 0$ and $m = 1$.

3.8 Solutions for $m > 1$

At larger m values one again finds that allowed and forbidden ranges alternate, with the transitions occurring where V equals zeroes of Bessel functions. Figure 3.3 shows all modes up to $V = 8$.

We wrap up what we have learned:

- For $V < 2.405$, there is only one branch of solution.
- For $V \geq 2.405$, there are initially two branches.
- At certain still higher V values, more branches come up.

The particular value $V = 2.405$ marks the transition from the existence of a unique solution to more than one solutions. Below, the fiber is said to be single-moded. This first mode is called the fundamental mode, the transition point is the *cutoff* of higher-order modes. For any given fiber, one can calculate the corresponding cutoff frequency or wavelength from $V = 2.405$. At lower frequencies (longer wavelengths) the fiber is single-moded. This implies that some fiber is not single-moded in an absolute sense: Such statement has meaning only in relation to a specified wavelength.

Fig. 3.3 All solutions of the eigenvalue equation in the u–v plane up to $u, w = 8$

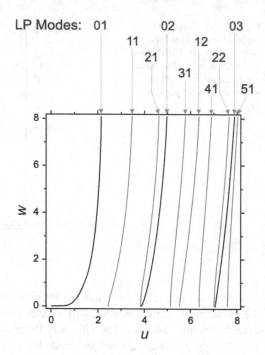

3.9 Field Amplitude Distribution of the Modes

We now see that the modes form a two-parameter family. One of the parameters is m. m indicates the angular dependence of the field distribution of the mode. For $m = 0$ the distribution is rotationally invariant, i.e., on any circular path around the center one would find a constant field amplitude (and thus intensity). For $m = 1$, the field amplitude will vary according to a sine function of the azimuthal angle. It therefore has two zeroes at mutually opposite positions; in between there are a positive and a negative branch, or lobe. Either branch contains a maximum of the intensity while the algebraic sign indicates the phase of the field. Thus, in one lobe the field oscillates in opposite phase to the other. For $m = 2$, a circular path would run through two full periods of the sine function; the intensity pattern then resembles a four-leafed clover. Again, each pair of leaves in opposite positions has the same phase while the other pair has opposite phase. When m takes even higher values, the angular dependence of the intensity has $2m$ leaves.

m also fixes which Bessel functions govern the field distribution in radial direction: We have found a combination of J_m in the core and K_m in the cladding. Since J_m oscillates (at any m), there are infinitely many ways to smoothly connect J_m to K_m even after m has been fixed. (Recall that the signs of coefficients C_M and C_K in Eq. (3.52) were arbitrary.) This set of possibilities is labeled with p, the second parameter. We adopt here the terminology of modes as introduced in 1971 by Gloge [2]: Modes are designated with "LP_{mp}" on grounds that they are essentially

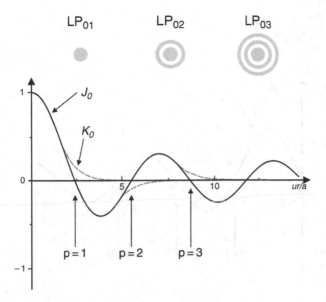

Fig. 3.4 Construction of the radial intensity distribution for modes with $m = 0$

linearly polarized. Index m designates the number of pairs of nodes in the azimuthal coordinate, and index p counts the possibilities in the radial coordinate.

We can now sketch what the intensity distribution of the modes looks like. Figures 3.4, 3.5, 3.6 show the various possibilities to make the connection between the J_m and K_m terms and give an idea about the intensity distribution.

Between $V = 0$ and $V = 2.405$, there only exists the branch of solution pertaining to the LP_{01} mode. This is called the *fundamental mode* of the fiber; it has a particularly simple shape, not unlike a bell shape. Between $V = 3.832$ and $V = 5.520$ we additionally find the LP_{02} mode. As V increases, new modes keep coming up.

This reasoning is borne out very well by experimental observation. In [5], all modes were excited separately so that the intensity pattern at the fiber end could be photographed individually. Figure 3.7 shows the result.

3.10   Numerical Example

We consider a typical numerical example to illustrate the transition from a single-mode fiber to a multi-mode fiber. Take a fiber with $a = 4\,\mu\text{m}$, $\Delta = 3 \times 10^{-3}$, and $n_K = 1.46$. From the definition of V, using the approximation $\text{NA} = n_K\sqrt{2\Delta}$, and the cutoff condition $V = 2.405$, one immediately obtains the condition for the *cutoff wavelength*:

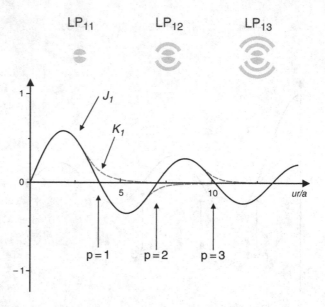

Fig. 3.5 Construction of the radial intensity distribution for modes with $m = 1$

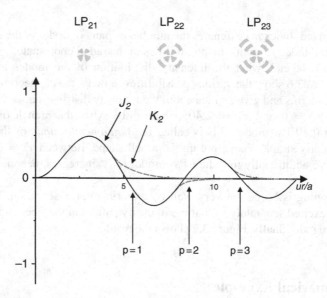

Fig. 3.6 Construction of the radial intensity distribution for modes with $m = 2$

$$\lambda_{\text{cutoff}} = \frac{2\pi \, a \, n_K \sqrt{2\Delta}}{2.405}.$$

(3.64)

Fig. 3.7 Observed intensity distribution of all modes existing at $V = 8$. For reproduction purposes the *gray scale* of the original photograph has been reduced to a binary *black* and *white*. From [5] with kind permission

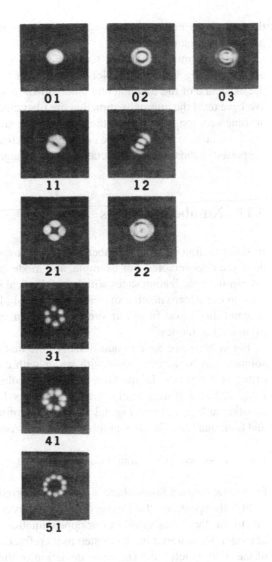

Inserting the numbers, we obtain $\lambda_{cutoff} = 1.182\,\mu m$. This fiber is then a single-mode fiber for all wavelengths longer than $1.182\,\mu m$. This includes the 1.3- and 1.55-μm range preferred in telecommunication. For wavelengths shorter than λ_{cutoff} there is more than one mode: A second mode appears at this cutoff, and at some particular even shorter wavelength yet another mode appears. This wavelength is obtained from the same condition by simply replacing $V = 2.405$ with $V = 3.832$. One obtains 742 nm.

For even larger V (even shorter wavelengths) more and more modes are added. The same fiber which supports just a single mode in the infrared will carry several modes in the visible! As V grows very large, the wavelength becomes much smaller than the core radius and we approach the multimode case. This confirms our previous heuristic assumption.

The LP_{01} mode exists all the way down to arbitrarily small V, i.e., at any arbitrarily long wavelength. None of the other modes has this property. However, the existence of the fundamental mode down to $V = 0$ must not be taken literally: We have used the approximation that the fiber cladding is infinitely wide. However, at some very long wavelength the field will penetrate far enough into the cladding to reach the outside surface. There is always a practical limit for the longest wavelength supported in the fiber, often dictated by wavelength-dependent losses.

3.11   Number of Modes

In order to find the total number of modes at any given V, we have to note that there are degeneracies. For example, any mode can exist in two distinct, mutually orthogonal polarization states which are identical in any other respect as long as we stick to our approximation that fibers are circularly symmetric. (We will later look at small deviations from this symmetry.) Then, all LP_{0p} modes are actually pairs (*doublets*) of modes.

For $m \neq 0$, we have to additionally note that the azimuthal dependence of the solution can be written either with sin or with cos; again, these are two mutually orthogonal variants. Taking both this and the polarization degeneracy into account, LP_{mp} with $m \neq 0$ are actually *quartets* of modes. Let us consider, as an example, the situation at $V = 6$: From Fig. 3.3 we see six nominal modes, two of which are pairs, and four, quartets. The total number thus is 20. Asymptotically one can approximate

$$V \to \infty \quad \Rightarrow \quad \text{total number of modes} = V^2/2 \quad \text{(step index fiber)}.$$

For gradient index fibers, there is a similar approximation with $V^2/4$.

Strictly speaking, the modes do not have precisely linear polarization. This is due to the fact—neglected in our approximation—that the fiber is not a perfectly homogenous material but has a step in its refractive index. This leads to distortions of the field which produces some deviation of the modes, mostly for higher-order modes. We can live with that because we are mostly interested in the fundamental mode.

3.12   A Remark on Microwave Waveguides

The reader may or may not be aware that discrete modes also exist in microwave waveguides. Microwave waveguides are metal pipes with conducting walls. This enforces a node of the electrical field on the boundary. In contrast, optical fibers are weakly guiding conduits. Therefore we could use here an approximation which is not valid in microwave guides whereas there one finds different types of modes and uses a different terminology [1]. Many of the modes derived here are linear combinations of metallic waveguide modes; the following table presents the correspondence:

LP modes	Microwave guide modes
LP_{01}	HE_{11}
LP_{11}	HE_{21}, EH_{01}
LP_{21}	HE_{31}, EH_{11}
LP_{02}	HE_{12}
LP_{31}	HE_{41}, EH_{21}

Nevertheless, similarities do exist between optical fibers and microwave waveguides. The biggest difference may be that they always have a minimum frequency even for the fundamental mode; below, no mode is supported at all. This can be traced directly to the conducting walls. In Sect. 4.5.2, we will present a particular case in which a fiber with special structure has a well-defined finite lower cutoff even for the fundamental mode, too.

3.13 Energy Transport

We have calculated the modal structure of fibers under the assumption of circular symmetry. Waveguiding arises from the guiding of modes by the refractive index structure. As soon as a fiber is bent, the circular symmetry is broken. And, as may be expected, then additional energy loss arises; bending losses are discussed in Sect. 5.2. Here we can already note this much:

The fiber core may be decisive for waveguiding, but it would be an oversimplification that the light power is guided in the core exclusively. We have seen that the field amplitude decays radially like an exponential function; this implies that there is always a certain fraction of power well outside the core. Different modes have different geometric field distributions; the fraction outside the core must therefore also be different for different modes.

The energy transport out of (or into) a certain volume element is described by the Poynting vector

$$\vec{S} = \vec{E} \times \vec{H}. \tag{3.65}$$

The direction of propagation is perpendicular to the plane containing $\vec{E}$ and $\vec{H}$. Disregarding anisotropic materials, all three vectors are pairwise orthogonal. The reader is reminded that by convention, the direction of polarization designates the direction of the oscillation of $\vec{E}$ (historically there was once a convention to refer to $\vec{H}$, but that has long been obsolete).

In most cases, one describes the energy transport of a wave by giving its *irradiance* or *intensity*. By this, one means the temporal average of the *instantaneous intensity*

$$I_{\text{inst}} = E(t) \, H(t).$$

I gives power per unit area and thus has units of W/m^2. Obviously I is the temporal average of the Poynting vector:

$$I = \langle |S| \rangle.$$

In the special case that the wave is harmonic, the relations between rms and peak values are $\hat{E} = \sqrt{2} \, E_{\text{rms}}$ and $\hat{H} = \sqrt{2} \, H_{\text{rms}}$. Then we find the intensity as

$$I = \frac{1}{2} \, \hat{E} \hat{H}.$$

Using the relation

$$\hat{E} = Z_0 \, \hat{H},$$

one can also write

$$I = \frac{1}{2Z_0} \hat{E}^2.$$

When the wave propagates in nonmagnetic matter with real index n, this becomes

$$I = \frac{n}{2Z_0} \hat{E}^2.$$

Power P is the integral of intensity over the cross-sectional area s:

$$P = \int_s I \, ds.$$

After a calculation which we will not show in detail here (see [2, 4]), one finds the energy fraction in the core (P_{K}) and in the cladding (P_{M}) for the fundamental mode:

$$\frac{P_{\text{K}}}{P} = \left(\frac{w}{V}\right)^2 \left[1 + \left(\frac{J_0(u)}{J_1(u)}\right)^2\right]$$

$$\frac{P_{\text{M}}}{P} = 1 - \left(\frac{w}{V}\right)^2 \left[1 + \left(\frac{J_0(u)}{J_1(u)}\right)^2\right]$$

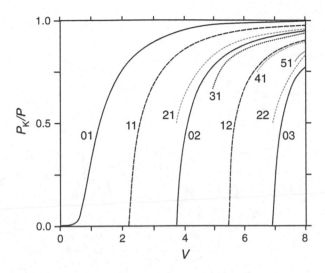

Fig. 3.8 Frequency dependence of the partition of power between core and cladding. The relative power inside the core is shown as a function of V for all modes up to $V = 8$. Modes with $m = 0, 1$ are guided when only a tiny fraction of power is in the core (at their cutoff these curves begin at zero). It is true for all modes that for large V the fraction approaches unity. After [2] with kind permission

For very large V, the mode is strongly concentrated in the core. As V decreases, the field begins to spread into the cladding. At the mode cutoff, practically all the power is in the cladding. This is precisely what causes the loss of waveguiding. For modes with $m \geq 2$, waveguiding is lost already when the power fraction outside the core exceeds $1/m$. Figure 3.8 illustrates the situation.

We have seen that fields of different modes have different cladding penetrations. For any given mode, the degree of spreading into the cladding depends on wavelength. This observation is also relevant for the fiber's dispersion, which will be treated in Chap. 4.

References

1. N. J. Cronin, *Microwave and Optical Waveguides*, Institute of Physics Publishing, Bristol (1995)
2. D. Gloge, *Weakly Guiding Fibers*, Applied Optics **10**, 2252 (1971)
3. J. D. Jackson, *Classical Electrodynamics*, 3rd edition, John Wiley & Sons, New York (1998)
4. E.-G. Neumann, *Single Mode Fibers*, Springer Series in Optical Sciences Vol. **57**, Springer-Verlag, Berlin (1988)
5. R. H. Stolen, W. N. Leibolt, *Optical Fiber Modes Using Stimulated Four Photon Mixing*, Applied Optics **15**, 239 (1976)

Chapter 4
Chromatic Dispersion

A light signal propagating in an optical fiber is subject to a variety of ways in which it can get distorted. Many of these are based on different propagation velocities for different parts of the signal. After such distortion, there is a risk that the signal arrives at the receiver in such a mangled form that it may be impossible to correctly decipher it.

We already encountered one such mechanism: modal dispersion in multimode fibers. Now we will address such distortions as they arise in single-mode fibers.

At the center of explanation is the fact that the refractive index of the fiber glass, just like that of any other material, depends on wavelength (or frequency). No light signal is ever truly monochromatic; rather, it contains Fourier components from a certain spectral interval. In other words, a light pulse of finite duration by necessity has a nonzero spectral width. Different frequency components, however, will propagate with different velocity. This gives rise to differential transit time and thus to signal distortion called delay distortion.

The wavelength dependence of the refractive index is behind three different contributions to delay distortion. They are collectively called *chromatic dispersion*. Individually, they are

Material dispersion. D_m: This contribution arises directly from the wavelength dependence of the index. Material dispersion is not specific to fibers but can be found in any bulk glass. It is independent of geometry and (given the material) depends solely on wavelength.

Waveguide dispersion. D_w: In the particular geometry of optical fibers, there is a modification to the differential propagation time. The reader is reminded that we saw in Chap. 3.13 that the signal power is partitioned between core and cladding; the splitting ratio depends on the wavelength. On the other hand, core and cladding indices are slightly different. As the wavelength is varied, we have a crossover from mostly core index to mostly cladding index. The result is a contribution to the wavelength dependence of the effective index.

Profile dispersion. D_p: Strictly speaking, the index difference between core and cladding itself, and thus Δ, is also wavelength-dependent. (Core index and

© Springer-Verlag Berlin Heidelberg 2016
F. Mitschke, *Fiber Optics*, DOI 10.1007/978-3-662-52764-1_4

cladding index do not vary "in parallel.") This gives rise to another correction which, however, is often much smaller than material and waveguide dispersion.

Another reason for dispersive distortions in single-mode fibers is related to the state of polarization of the light. As mentioned above, each mode can be decomposed into two mutually orthogonal parts. An ideal fiber has perfect circular symmetry; then both polarization states (*polarization modes*) propagate with identical velocities. However, real-world glass always has at least some residual birefringence; this implies a slightly different effective index for both states. One can argue that the term "single-mode fiber" is a misnomer: Even when it is true that only a single geometric field amplitude distribution (LP_{01}) can propagate, it still consists of two polarization modes. This is why in real fibers there is *polarization mode dispersion*. Typically it is a small contribution; we will discuss it further below.

To characterize dispersion, one normally specifies the size of the effect per distance. For both modal and polarization mode dispersion, one can write the dispersion parameter

$$D = \frac{1}{L}\delta\tau \,, \tag{4.1}$$

where $\delta\tau$ designates the difference of propagation time after distance L. Units of ps/km are commonly used. For chromatic dispersion, including material, waveguide, and profile dispersion, the following specification is common:

$$D = \frac{1}{L}\frac{d\tau}{d\lambda} \,. \tag{4.2}$$

Here, D contains the three parts just mentioned:

$$D = D_m + D_w + D_p \,. \tag{4.3}$$

This dispersion parameter indicates the propagation time difference per distance and per wavelength difference; therefore, units of ps/(nm km) are commonly used.

4.1 Material Dispersion

For any glass, the refractive index varies with wavelength. This gives rise to chromatic defects of lenses and the color-separating capability of prisms. For historical reasons, it became common practice (see, e.g., Schott glass catalog [2]) to characterize types of glasses by giving their indices at three wavelengths:

- n_D, the refractive index at wavelength 589.30 nm (yellow, Fraunhofer's D line of sodium),

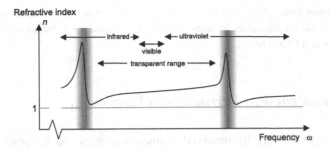

Fig. 4.1 Schematic plot of refractive index vs. wavelength. It is dominated by resonances which occur both in the ultraviolet and the infrared

- n_F at wavelength 486.13 nm (blue–green, Fraunhofer's F line of hydrogen), and
- n_C at wavelength 656.27 nm (red, Fraunhofer's C line of hydrogen).

This choice was motivated by the availability of narrowband emission lamps at these wavelengths. As a further characterization, often the Abbe number is given, defined by

$$v_D = \frac{n_D - 1}{n_F - n_C}. \tag{4.4}$$

Obviously this is a metric of the wavelength dependence of the index near the central (yellow) wavelength.

Glass catalogs often specify only n_D and v_D. It is instructive, however, to survey the variation of the index over a wide spectral range. This is schematically shown in Fig. 4.1. There are absorption bands due to electronic transitions in the ultraviolet at wavelengths on the order of 100 nm and molecular vibrations in the infrared around 10 μm. In the vast interval in between, including all of the visible and near-infrared, pure silica glass does not exhibit any resonances. This is the reason, of course, why it appears "crystal-clear" to the eye. The position of the absorption resonances is reflected in the refractive index.

Pure fused silica (SiO_2) has a refractive index of $n_D = 1.456$, decreasing slightly toward longer wavelengths. As long as one stays clear of the resonances, one can empirically describe the wavelength dependence with interpolation formulas. One of the best known such formulas is Sellmeier's equation

$$n^2(\omega) = 1 + \sum_{j=1}^{m} \frac{A_j \lambda^2}{\lambda^2 - \lambda_j^2}, \tag{4.5}$$

but there are also variants to this. Coefficients A_j denote the resonance strengths and λ_j the pertaining wavelengths. These coefficients are tabulated in the literature (for various glass types, see [2], and for fibers with various doping materials and concentrations, see e.g., [3]). In most cases three terms in the sum are considered sufficient; sometimes, five. Let us emphasize again that a Sellmeier curve is an empirical fit: the coefficients λ_j must not be construed to indicate the resonance wavelengths in a literal sense.

4.1.1 Treatment with Derivatives to Wavelength

We now turn to the actually observed propagation times and the scatter thereof. Consider a plane monochromatic wave with angular frequency ω and wave number β. It is well known that it propagates with phase velocity

$$v_{ph} = \omega/\beta , \tag{4.6}$$

whereas the propagation of a signal, like a wave packet, is governed by the group velocity

$$v_{gr} = d\omega/d\beta . \tag{4.7}$$

The group propagation time then is

$$\tau = \frac{L}{v_{gr}} = L\frac{d\beta}{d\omega} \tag{4.8}$$

$$= L\frac{d\beta}{d\lambda}\frac{d\lambda}{d\omega} . \tag{4.9}$$

Since $\beta = nk_0 = 2\pi n/\lambda$, we can rearrange the first fraction in the last line:

$$\frac{d\beta}{d\lambda} = 2\pi\frac{d}{d\lambda}\left(\frac{n}{\lambda}\right) = \frac{2\pi}{\lambda^2}\left(\lambda\frac{dn}{d\lambda} - n\right). \tag{4.10}$$

The second fraction can be rearranged using $\lambda = 2\pi c/\omega$ to yield

$$\frac{d\lambda}{d\omega} = 2\pi c\frac{d}{d\omega}\left(\frac{1}{\omega}\right) = -\frac{2\pi c}{\omega^2} . \tag{4.11}$$

Combining, we obtain

$$\tau = L\frac{2\pi}{\lambda^2}\frac{2\pi c}{\omega^2}\left(n - \lambda\frac{dn}{d\lambda}\right) \tag{4.12}$$

or, since $\lambda\omega = 2\pi c$,

$$\tau = \frac{L}{c}\left(n - \lambda\frac{dn}{d\lambda}\right). \tag{4.13}$$

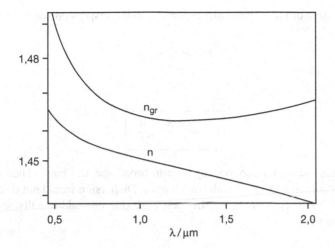

Fig. 4.2 Refractive index and group index as a function of wavelength, calculated from a three-term Sellmeier's equation

We have now expressed the wavelength dependence of the group propagation time as a function of the easily measured quantities n and λ. We emphasize that we here find the bracketed expression taking the role of the usual index n which only appears in the first term of that expression. Therefore we call the bracketed expression the *group index*:

$$n_{gr} = \left(n - \lambda \frac{dn}{d\lambda} \right) \tag{4.14}$$

Figure 4.1 shows that throughout the visible and near infrared, n decreases with increasing wavelength; thus, in this range $n_{gr} > n$. Figure 4.2 shows a comparison of both indices for fused silica.

The scatter of arrival times at the receiver is obtained from

$$\delta\tau = \frac{d\tau}{d\lambda} \delta\lambda . \tag{4.15}$$

Herein, the derivative is

$$\frac{d\tau}{d\lambda} = \frac{L}{c} \frac{d}{d\lambda} \left(n - \lambda \frac{dn}{d\lambda} \right)$$

$$= -\frac{L}{c} \lambda \frac{d^2n}{d\lambda^2} .$$

The contribution of the material dispersion D_m to the dispersion coefficient

$$D_m = \frac{1}{L}\frac{d\tau}{d\lambda} \,,$$ (4.16)

is then

$$D_m = -\frac{\lambda}{c}\frac{d^2 n}{d\lambda^2} \,.$$ (4.17)

In some cases the signal may occupy a quite broad spectral band. Then one must take into account that D varies with wavelength: Dispersion would not be described to sufficient accuracy by D alone. In such case one can additionally specify the dispersion slope:

$$S_m = \frac{dD_m}{d\lambda} \,.$$ (4.18)

4.1.2 Treatment with Derivatives to Frequency

An alternative terminology to describe dispersion arises when one takes derivatives not with respect to wavelength but with respect to frequency. One starts from a series expansion of the propagation constant β

$$\beta(\omega) = n(\omega)\frac{\omega}{c} = \beta_0 + \beta_1(\omega - \omega_0) + \frac{1}{2}\beta_2(\omega - \omega_0)^2 + \cdots$$ (4.19)

with

$$\beta_m = \frac{d^m \beta}{d\omega^m}\bigg|_{\omega=\omega_0} \,.$$ (4.20)

Let us assess the meaning of β_m:

$$\beta_0 = \beta(\omega = \omega_0) = kn$$ (4.21)

where n is the ordinary (phase) index.

$$\beta_1 = \frac{d\beta}{d\omega}\bigg|_{\omega=\omega_0}$$

$$= \frac{d}{d\omega}\left(n(\omega)\frac{\omega}{c}\right)\bigg|_{\omega=\omega_0}$$

$$= \frac{1}{c} \left(n(\omega_0) + \omega \left. \frac{dn(\omega)}{d\omega} \right|_{\omega=\omega_0} \right).$$

By inserting $\omega = 2\pi c/\lambda$, this can be written as

$$\beta_1 = \frac{1}{c} \left(n(\lambda_0) - \lambda \left. \frac{dn(\lambda)}{d\lambda} \right|_{\lambda=\lambda_0} \right). \tag{4.22}$$

The bracketed expression is the group index n_{gr} as introduced in Eq. (4.14), which implies that

$$\beta_1 L = \frac{L}{c} n_{gr} = \tau \tag{4.23}$$

holds. Then

$$\beta_1 = \frac{1}{v_{gr}} = \frac{n_{gr}}{c} . \tag{4.24}$$

β_1 is of the order of $\beta_1 \approx 5\,\text{ns/m}$.
    For β_2 we find

$$\beta_2 = \frac{d\beta_1}{d\omega} = \frac{1}{c} \left(2\frac{dn}{d\omega} + \omega\frac{d^2n}{d\omega^2} \right). \tag{4.25}$$

This quantity is called *group velocity dispersion parameter* or GVD parameter. It is commonly given in units of ps^2/km. The GVD parameter is preferred by theorists over the dispersion coefficient D which is widely used by technicians. To convert, one uses

$$D_m = \frac{d\beta_1}{d\lambda} = \frac{d\beta_1}{d\omega}\frac{d\omega}{d\lambda} = \beta_2\frac{d}{d\lambda}\left(\frac{2\pi c}{\lambda} \right) \tag{4.26}$$

and thus

$$\boxed{D_m = -\beta_2 \left(\frac{2\pi c}{\lambda^2} \right) = -\frac{\omega}{\lambda} \beta_2 .} \tag{4.27}$$

There are some cases in which higher-order dispersion terms become relevant. Then one can also specify the third-order dispersion (TOD) β_3. To convert between dispersion slope S and β_3, one can use

$$S_m = \frac{(2\pi c)^2}{\lambda^4} \beta_3 + \frac{4\pi c}{\lambda^3} \beta_2 . \tag{4.28}$$

4.2   Waveguide and Profile Dispersion

In addition to material dispersion D_m, in fibers there is waveguide dispersion D_w. Without explicitly deriving the result, we state that for step index fibers, it can be calculated from [8, 17]

$$D_w = -\frac{Vn_K\Delta}{\lambda c}\frac{d^2}{dV^2}(Vb)\,,\qquad(4.29)$$

where

$$b = \frac{\beta^2 - k_M^2}{k_K^2 - k_M^2}\,.$$

For very large V number, the dimensionless quantity b tends to $b = 1$; at the cutoff of each mode there is $b = 0$.

The reason for the waveguide contribution can be intuitively understood by noting that for increasing wavelengths, the field extends more and more into the cladding so that the light wave experiences more and more of the cladding index, rather than just the core index (Fig. 4.3).

If one takes into account that on top of this Δ is also not constant but depends slightly on wavelength, one obtains what is called profile dispersion or *differential*

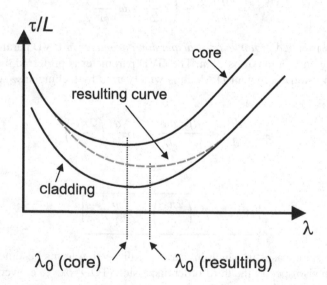

Fig. 4.3 The travel time of a signal in a fiber is obtained from taking a suitably weighted average of travel time pertaining to core and cladding material. At short wavelength, light is guided predominantly in the core; at long wavelength, overwhelmingly in the cladding. The zero-dispersion wavelength (corresponding to minimum travel time) therefore shifts toward longer wavelength, compared to the core material alone

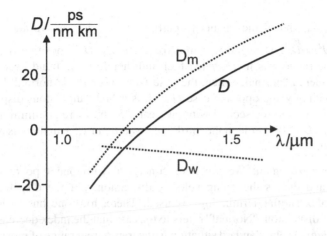

Fig. 4.4 Total dispersion D results from the material contribution of the core D_m and the waveguide contribution D_w. The zero-dispersion wavelength is shifted by D_w with respect to what would be expected from D_m alone

material dispersion D_p. This contribution is usually small and will not be further discussed here. Below we will refer to the sum $D = D_m + D_w + D_p$ (Fig. 4.4).

Specifications by fiber manufacturers quote either D (and sometimes S) or β_2 (and sometimes β_3) at specific wavelengths. The values given refer to the total; relative contributions of material, waveguide, and profile dispersion are not normally provided. Conversions (4.18), (4.27) and (4.28) remain valid when indices "m" are dropped.

4.3 Normal, Anomalous, and Zero Dispersion

Let us consider some typical numerical values. Figure 4.2 shows the refractive index for fused silica as obtained from a Sellmeier's equation. We make the following observations:

- Throughout the visible and near infrared, $dn/d\lambda < 0$. This implies that $n_{gr} = n - \lambda(dn/d\lambda) > n$. The group index is larger than the phase index. In the near infrared, n_{gr} is nearly constant at about $n_{gr} = 1.46$.
- The refractive index $n(\lambda)$ has an inflexion point at $\lambda \approx 1.27\,\mu m$. At this point, group delay is minimal and $D_m = \beta_2 = 0$, so that this point is referred to as *zero-dispersion wavelength*. Strictly speaking, this commonly used term is slightly incorrect since it is only the leading order in the series expansion of the dispersion that vanishes here. All higher-order terms still contribute. Below we will call this particular wavelength λ_0.

- For $\lambda < \lambda_0$ and in the visible in particular, $\dfrac{d^2n}{d\lambda^2}$ is positive and thus $D = -(\lambda/c)(d^2n/d\lambda^2)$ is negative, while for $\lambda > \lambda_0$, D is positive. Historically, the visible range was investigated first and therefore the trend observed there was considered "normal." Then, the case $D < 0$ is called "normal dispersion." Correspondingly, the opposite case $D > 0$ is called "anomalous dispersion." If the fiber is used in the second window near $1.3\,\mu$m, there is a minimum of the dispersion ($D \approx 0$), while in the third window around $1.5\,\mu$m there is anomalous dispersion.

Let us emphasize again: We are here concerned with one type of dispersion exclusively and that is the group velocity dispersion. For the dispersion of the refractive index similar terminology is used: There, too, one has "normal" and "anomalous" dispersion. "Normal" refers to the case that the index decreases toward longer wavelengths, the standard situation in the transparent range of most materials. The opposite only occurs near atomic resonance frequencies; that is then called anomalous dispersion (of the index). Unfortunately, some authors do not always make it entirely clear just which type of dispersion they refer to, so that occasionally confusion may arise.

The waveguide contribution to the total dispersion is negative throughout the visible and near infrared. A typical value for standard fibers is $-2\,\mathrm{ps}/(\mathrm{nm}\ \mathrm{km})$. At long wavelengths, it acts opposite to the material dispersion. Consequently, the zero-dispersion wavelength in standard fibers is slightly shifted with respect to bulk fused silica, toward longer wavelengths by typically about 20–30 nm. According to a CCITT[1] standard in effect since 1984, the dispersion of fibers for telecommunication purposes shall be bounded as follows:

$|D| \leq 3.5\,\mathrm{ps}/\mathrm{nm\,km}$ for $1285\,\mathrm{nm} \leq \lambda \leq 1330\,\mathrm{nm}$,
$|D| \leq 20\,\mathrm{ps}/\mathrm{nm\,km}$ at $\lambda = 1550\,\mathrm{nm}$.

Near $1.55\,\mu$m, a value of $D = 18\,\mathrm{ps}/\mathrm{nm\,km}$ is typical. (According to Eq. (4.27) this corresponds to $\beta_2 = -23\,\mathrm{ps}^2/\mathrm{km}$.) This value will generate a propagation time difference between two wavelength components that are 1 nm apart, which after a distance of $L = 10\,\mathrm{km}$ reaches $\delta\tau = 180\,\mathrm{ps}$. The zero-dispersion wavelength near 1300 nm provides minimal spread in propagation time in the second window. Third-order dispersion varies not as much with wavelength as the second-order term. Typical numbers are $S(\lambda_0) = 0.085\,\mathrm{ps}/\mathrm{nm}^2\,\mathrm{km}$, corresponding to $\beta_3(\lambda_0) = -0.08\,\mathrm{ps}^3/\mathrm{km}$.

As we will see shortly, the shift of the zero-dispersion wavelength by the waveguide dispersion can be intentionally increased. This allows to make fibers with custom-designed zero-dispersion wavelength in the infrared at wavelengths beyond the zero of pure fused silica.

---

[1]Comité Consultatif International Télégraphique et Téléphonique. This committee is now called ITU-T, a subunit of the International Telecommunication Union, which is a United Nations agency for information and communication technology issues.

4.4 Impact of Dispersion

Consider the propagation of a light pulse which we think of as being generated by taking a monochromatic oscillation of the electric field

$$\hat{E}\cos(\omega t - \beta z)$$

and multiply (*modulate*) it with an *envelope* function. For the latter a reasonable choice is a Gaussian:

$$e^{-\frac{t^2}{2T_0^2}} . \tag{4.30}$$

The temporal profile of the intensity (irradiance) or power of a pulse so generated is

$$I(t) = I_0\, e^{-(t/T_0)^2} . \tag{4.31}$$

Here I_0 is the peak value of the intensity and T_0 the pulse duration. Note that there is not a unique way to specify pulse duration: T_0 refers to the time interval between the peak and the point where the intensity has dropped to $1/e$ of the maximum. However, experimentalists and technicians often prefer the use of the half width of the pulse, i.e., the time elapsed between the points, where the power or intensity takes $1/2$ of the peak value. This half width is often denoted by "FWHM" (*full width at half maximum*); we will designate it by τ. The conversion for a Gaussian is $\tau = 2\sqrt{\ln 2}\,T_0$.

After propagation over fiber length L, both pulse duration and peak power are modified. One can show that the pulse duration now is

$$T_L = T_0\sqrt{1 + \left(\frac{L}{L_D}\right)^2}, \tag{4.32}$$

where

$$L_D = \frac{T_0^2}{|\beta_2|} \tag{4.33}$$

is a characteristic length called the *dispersion length*. After distance L_D, the pulse duration has increased by $\sqrt{2}$. After considerably longer distance, the pulse duration grows in proportion to distance as

$$L \gg L_D \quad \Rightarrow \quad T_L = |\beta_2|L/T_0 . \tag{4.34}$$

The shorter initially, the longer in the end! We point out that there is a very close analogy with diffraction, the transverse spreading of a narrow fan of light rays, and

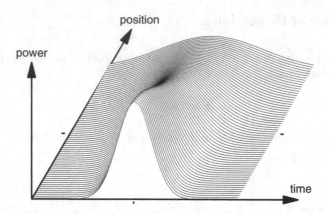

Fig. 4.5 Dispersive broadening of a Gaussian pulse. The figure shows a pulse with initial width (FWHM) of 0.5 ps as it propagates over a distance of 21 m and widens in the process. Its peak height is reduced because energy is conserved. Fiber dispersion was chosen here as $\beta_2 = -18\,\text{ps}^2/\text{km}$ and $\beta_3 = 0$

dispersion, the longitudinal spreading of a short light pulse. Far-field diffraction (Fraunhofer diffraction) is the most transparent case: The spread increases in proportion to distance, i.e., at a constant angle of divergence. The functional form of the fan of rays is given by the Fourier transform of the initial shape and remains unaltered; only scale factors evolve. In the near field (Fresnel diffraction) the situation is more involved, but for a Gaussian it is true that its shape is maintained except for scale factors (Fig. 4.5). Gaussians display a particularly simple behavior under this transformation, which is of course why we considered this special shape.

This close analogy becomes especially clear when we replace the Gaussian envelope of Eq. (4.30) with a rectangular envelope

$$
I(t) = \begin{cases} 0: & t < -\dfrac{T_0}{2}\,, \\[2mm] 1: & -\dfrac{T_0}{2} \le t \le +\dfrac{T_0}{2}\,, \\[2mm] 0: & t > \dfrac{T_0}{2}\,. \end{cases}
$$

This is certainly not a realistic proposition, but it comes closest to resemble diffraction at a slit. Initially certain undulations are generated near the steep slopes; as propagation proceeds, they spread out. After some wiggling and interfering, the pulse shape eventually approaches the functional form of $(\sin(x)/x)^2$ (see Fig. 4.6). The close relation to diffraction at a slit, and the transition from near field to far field, is quite obvious here.

In Eq. (4.33) we used β_2 and T_0. However, experimentalists and technicians often prefer to use the dispersion parameter D and the full width at half maximum τ. Using the conversions $\tau_0 = \tau(L = 0) = 2\sqrt{\ln 2}\, T_0$ and $|\beta_2| = |D|\lambda^2/(2\pi c)$ as

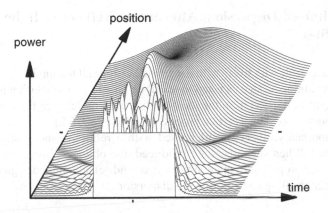

Fig. 4.6 Dispersive broadening of a rectangular pulse. This case is purely academic, but it shows in particular clarity how steep slopes of the initial shape are deformed strongly by dispersion

given above, we can write the relevant term in Eq. (4.32) as follows:

$$\frac{L}{L_D} = \frac{L|\beta_2|}{T_0^2} = \frac{L|D|\lambda^2}{\tau_0^2}\frac{2\ln 2}{\pi c}. \tag{4.35}$$

The first fraction on the RHS specifies the fiber(L, D) and the light signal (λ, τ_0). The second fraction contains only constants and is thus independent of the specific situation. Its value equals 1.4709×10^{-9} s/m. If one now inserts L in km, D in ps/nm km, λ in μm, and τ_0 in ps, units combine to give an additional numerical factor of 10^9, and we can write

$$\tau_L = \tau_0 \sqrt{1 + \left(\frac{1.47L|D|\lambda^2}{\tau_0^2}\right)^2}. \tag{4.36}$$

This equation contains directly measurable quantities in technically common units and is thus of practical value. The "magic number" 1.47 is valid for Gaussian pulses; for other shapes somewhat different values apply. For example, the hyperbolic secant squared shape (sech^2) often encountered for solitons requires a value of 1.87.

Dispersive broadening limits the information-carrying capacity because pulses must be kept at sufficient temporal distance from each other. The highest capacity would be obtained at the lowest dispersion, which in turn is found at the zero-dispersion wavelength. This is why a large fraction of all installed fibers is designed for operation in the second window near 1.3 μm. However, this apparently obvious conclusion was arrived at in the framework of the *linear* approximation, i.e., at sufficiently small powers or intensities. Nonlinear effects (Chap. 9 ff.) will modify the result and maximize capacity at a different condition.

4.5   Optimized Dispersion: Alternative Refractive Index Profiles

So far we have dealt with fibers with a step index profile. One might note that there never is such a thing as an exact step index fiber. Due to manufacturing limitations, there are slight deviations from the ideal profile, e.g., quite often there is a central dip of the index caused by a certain process step (see Sect. 6.2).

More importantly, fibers are often used with a refractive index profile that is more complex. When such fibers are produced, the objectives are to (a) maintain the single-mode property, (b) maintain low loss, and (c) add more design degrees of freedom for controlling and tailoring the dispersion.

4.5.1   Gradient Index Fibers

In the context of multimode fibers, we have already mentioned a radial dependence of the index according to

$$n(r) = n_K \sqrt{1 - 2\Delta \left(\frac{r}{a}\right)^\alpha} \; ; \tag{4.37}$$

single-mode fibers can be endowed with a similar gradient index profile. Ray optics fails to provide a good interpretation in this case. A wave-optic calculation yields the following:

$\alpha = \infty$:   This limiting case is the step index profile (SI profile). The cutoff of the second (LP_{11}) mode is at $V = 2.405$.

$\alpha = 2$:   For a parabolic profile (Fig. 4.7) the cutoff of the second mode shifts to $V = 3.518$.

$\alpha = 1$:   This is a triangular profile, hence the name "T fiber" (as in *triangular*). Here the cutoff of the second mode is even higher. As a rough approximation, the cutoff occurs at $V \approx 2.405\sqrt{1 + 2/\alpha}$.

Fig. 4.7 Pseudo-3D rendering of the refractive index profile. Across the circular section the index is plotted in vertical direction. From [14]

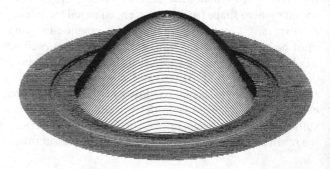

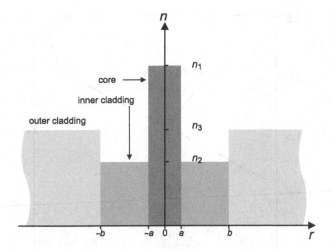

Fig. 4.8 Schematic shape of the index profile of a W fiber (*depressed-index cladding profile*). There are three indices for core, inner, and outer cladding, labeled here as n_1, n_2, and n_3, respectively

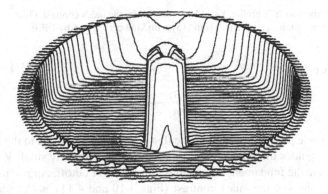

Fig. 4.9 Pseudo-3D rendering of the refractive index profile of a W fiber (*depressed-index cladding profile*). From [14]

4.5.2   W Fibers

There may be an additional zone between core and cladding having its own lower refractive index (Fig. 4.8). Then a cross-sectional index profile roughly resembles the letter W; hence the name "W fiber" (Fig. 4.9). An alternative name is "DIC fiber" for *depressed-index cladding fiber*. This profile provides ample freedom for designing the dispersion variation.

For this profile, we define a V number

$$V = \frac{2\pi}{\lambda} a \sqrt{n_1^2 - n_3^2} \qquad (4.38)$$

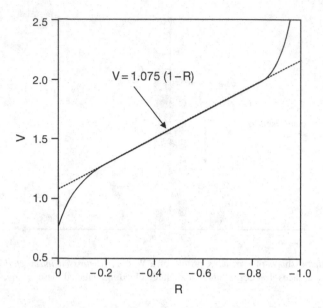

Fig. 4.10 For fibers with W profile, V can be controlled by the index contrast. Over a certain range a linear approximation is appropriate. From [15] with kind permission by IEEE

and an index contrast

$$R = \frac{n_2 - n_3}{n_1 - n_3} \, . \tag{4.39}$$

It is a remarkable property of this profile that—in marked contrast to the step index profile which guides the fundamental mode down to arbitrarily small V at least in principle—here the fundamental mode has a finite lower cutoff. Approximately and for medium values of the index contrast (Figs. 4.10 and 4.11), at the fundamental mode cutoff one has

$$V_0 \approx 1.075 \, (1 - R) \, . \tag{4.40}$$

Note that in the limit $n_2 \rightarrow n_3$ which reproduces the step index profile, this simple linear trend is not maintained, and V goes to zero in accord with our earlier result for step index fibers.

4.5.3 T Fibers

T fibers or *triangular* fibers are popular because the dispersion trend is more favorable than in step index fibers, while losses are, if anything, even lower. The

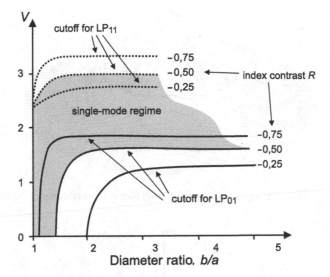

Fig. 4.11 For fibers with W profile, the cutoff behavior can be controlled through the ratio of radii b/a. There is even a cutoff for the fundamental mode LP_{01} as soon as b/a is sufficiently larger than unity. For example, at $R = -0.5$ and at $b/a = 3$, the fiber is single mode only in the interval $1.8 \leq V \leq 3.0$. For $V \geq 3.0$ there is the additional LP_{11} mode, and for $V \leq 1.8$ there is no guided mode at all. After [15] with kind permission by IEEE

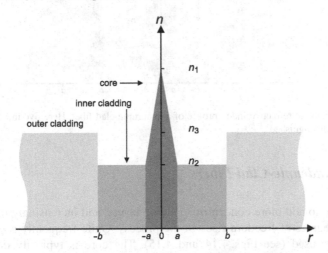

Fig. 4.12 Schematic refractive index profile of a fiber with triangular core profile, shown with a depressed inner cladding. Again, three indices n_1, n_2, and n_3 need to be distinguished

latter can be traced back to the interface between core and cladding: for the sudden transition of glass composition there is an enhanced chance of mechanical stress which is mitigated by a more gradual transition. Figures 4.12 and 4.13 show a modified T profile which is really a combination of T and W profiles.

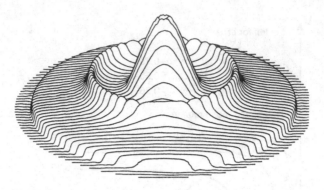

Fig. 4.13 Pseudo-3D rendering of a triangular profile, here with more complex cladding composition. From [14]

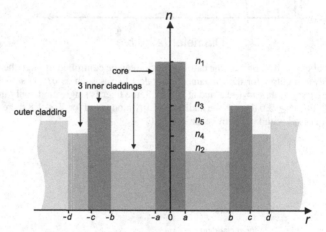

Fig. 4.14 Schematic refractive index profile of a quadruple-clad fiber. Here five indices and four radii must be distinguished

4.5.4 Quadruple-Clad Fibers

It is possible to add more concentric cladding layers, and increasingly the number of design degrees of freedom rises in the process. Quite frequently a *quadruple-clad fiber* is used (see Figs. 4.14 and 4.15). The core is typically doped with germanium and thus has a raised refractive index. The first cladding zone can be doped with phosphorus and fluorine and has lowered index. In the second and third cladding zones germanium and phosphorus/fluorine are repeated with suitable concentrations. The outermost cladding can then remain undoped fused silica.

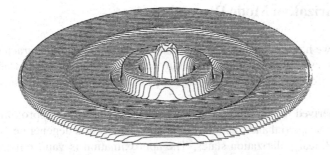

Fig. 4.15 Pseudo-3D rendering of a quadruple-clad fiber profile. From [14]

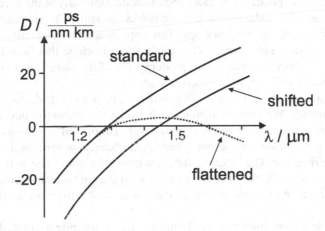

Fig. 4.16 Tailoring of the dispersion curve through choice of suitable index profiles: dispersion-shifted and dispersion-flattened fiber in comparison to a standard step index fiber

4.5.5 Dispersion-Shifted or Dispersion-Flattened?

There is an important distinction between *dispersion-shifted* and *dispersion-flattened* fibers. In comparison to a step index fiber, by using a triangular core profile with a depressed cladding zone as in Fig. 4.12, one can achieve a shift of the dispersion curve toward longer wavelengths (Fig. 4.16). The zero-dispersion wavelength can thus be moved all the way to 1550 nm if desired. Using a quadruple-clad design one can even achieve a very low dispersion simultaneously at both 1300 and 1550 nm by bending the dispersion curve flat.

The motivation to tailor the dispersion curve is to get the best of two worlds: the minimal dispersion of the second window combined with the minimal loss in the third window. Dispersion flattened fibers have low dispersion at both the second and third windows at the same time; such fibers can be used as direct replacement for older fiber designed for the second window but provide the added benefit of also performing well in the third window.

4.6 Polarization Mode Dispersion

Up to now we have disregarded the fact that a light field is fully characterized only when one takes its state of polarization into account. Different states of polarization propagate differently, thus arises a special type of dispersion which we now examine closer.

As we derived the modal profiles in Chap. 3, we used the approximation of a homogeneous material and found that all modes are twofold degenerate into distinct orthogonal linear polarization states. The approximation is valid only, of course, when there is weak guiding (Δ very small). In the more general case the modes are not exactly linearly polarized, because the index discontinuity at the core–cladding interface distorts the modal structure. Nonetheless, the approximation can be useful.

Moreover, we had assumed isotropy. This implies that the orientation of the two planes of polarization is arbitrary. One might then conclude that launching linear polarized light at any orientation will produce light of the same linear polarization at the fiber end. This is not what experience shows.

In any real fiber there is some—potentially very weak—deviation from ideal circular symmetry. We have to distinguish between (a) geometric deviations, like when the core is asymmetric or not centered well; (b) optical deviations like when the material index is not homogenous; and (c) mechanical deviations due to stress-induced birefringence. The latter contribution may arise either due to tension built into the fiber—after all, the fiber is rapidly cooled during its manufacturing which may well introduce tension—or due to tension created during use as the fiber is being bent.

All these deviations from perfect circular symmetry conspire to create differences for the propagation of the two polarization modes. This gives rise to polarization mode dispersion. We will now consider how it manifests itself and how to avoid it.

4.6.1 Quantifying Polarization Mode Dispersion

Rather than a single propagation constant β we now need to use two, β_x and β_y, to describe conditions for the two polarization states linearly polarized in x and y directions. As soon as $\beta_x \neq \beta_y$, the two light waves polarized in parallel to x and y will propagate differently; hence, they will be alternatingly in phase and out of phase with each other. This occurs with spatial period

$$\Lambda = \frac{2\pi}{\beta_x - \beta_y} \, , \tag{4.41}$$

which defines the *beat length* Λ. Alternatively, some authors use the modal birefringence

$$B = \frac{\lambda}{\Lambda} = \frac{\lambda}{2\pi}(\beta_x - \beta_y) = \frac{\beta_x - \beta_y}{k_0} = n_x - n_y \, . \tag{4.42}$$

B is only weakly wavelength-dependent, while Λ is essentially proportional to wavelength. For a standard fiber B is of the order 10^{-7} to 10^{-8}, with random orientation of the axes. Then, the beat length is $10^7 \lambda$ to $10^8 \lambda$, which is on the order of a couple of meters. If the fiber is strongly bent (coiled on a spool!), birefringence can reach $B = 10^{-5}$, with correspondingly shorter beat length.

The propagation time difference for an arbitrarily polarized light signal (decomposed into the two polarization states) is when one considers phase velocity:

$$\Delta t = \frac{L}{c} B \, . \tag{4.43}$$

This translates into a dispersion of

$$\frac{\Delta t}{L} = \frac{B}{c} \approx \frac{10^{-7}}{3 \times 10^8} \frac{\text{s}}{\text{m}} = 0.3 \, \text{ps/km} \, . \tag{4.44}$$

If instead, and more correctly, one considers group velocity, one finds

$$\Delta t = \left| \frac{L}{v_{\text{gr},x}} - \frac{L}{v_{\text{gr},y}} \right| = L \left| \frac{d\beta_x}{d\omega} - \frac{d\beta_y}{d\omega} \right| \tag{4.45}$$

from which one obtains the dispersion contribution through $\Delta t/L$. A typical value for standard fibers is 0.1 ps/km, a small difference indeed—and yet, consequential in some contexts. Polarization mode dispersion is now the largest obstacle to further increase of the data-carrying capacity of fibers.

4.6.2 Avoiding Polarization Mode Dispersion

The state of polarization is not maintained in standard fiber. In order to render a fiber polarization-maintaining, one might try to reduce its residual birefringence. However, this is a tedious task: Even when the built-in tensions could be eliminated in a modified manufacturing progress, the manufacturer has no control over bending of the fiber by the user.

In 1982 two people came up with the same surprising idea practically simultaneously: R.H. Stolen then at AT&T Bell Laboratories in the USA and D.N. Payne at the University of Southampton in England. They drew the surprising conclusion that when it is not possible to reduce birefringence to negligible levels, one can achieve

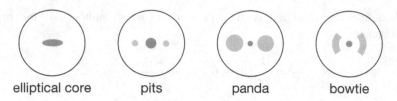

elliptical core   pits   panda   bowtie

Fig. 4.17 Several polarization-maintaining structures. In each case the circular symmetry is broken

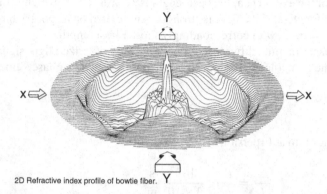

2D Refractive index profile of bowtie fiber.

Fig. 4.18 Pseudo-3D rendering of the refractive index profile for a bowtie fiber. With kind permission from Fibercore [1]

the same ends by making it intentionally much larger! To do this is easy: One can either make the core elliptic, or one can insert additional structural elements that break the circular symmetry. In most cases the symmetry is broken by the insertion of elements with a slightly different thermal expansion coefficient, so that during the cooling of the glass at the end of the fiber manufacturing process mechanical stress is built into the fiber.

Figure 4.17 shows some popular versions with elliptic core, so-called *pits*, PANDA geometry (the latter named after the facial expression of a cutie in the zoo), and bowtie geometry (Fig. 4.18).

Such geometries allow $B = 3$ to 8×10^{-4} corresponding to $\Lambda = 1$ to 3 mm. The beat length is thus reduced by three orders of magnitude and is now shorter than the tightest possible bend radii. Therefore, the built-in birefringence due to this structure overwhelms the random birefringence including any that may occur during operation due to bending. Why is this, then, a polarization-maintaining fiber?

If one launches light which is linearly polarized along the direction of one of the two axes of the elliptical structure, this state of polarization will be maintained. If, however, the light is polarized at an angle with the axes, one can mentally decompose it into the two parts along the axes: These will propagate with different velocity because they experience a different refractive index. The state of polarization will then cyclically evolve through linear $\rightarrow$ elliptic $\rightarrow$ circular $\rightarrow$ elliptic again $\rightarrow$ linear again, etc. This evolution can be exploited to measure the

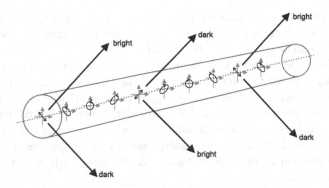

Fig. 4.19 Sketch to illustrate the beat length. The state of polarization evolves from linear through elliptical to circular, etc. Depending on the state, emission into a given direction away from the fiber axis is more or less efficient. From the periodic appearance of bright and dark zones, one can immediately read the beat length

Fig. 4.20 Measurement of the beat length under a microscope. The periodic changes of brightness of the light scattered off the core are plainly visible. In this case the beat length was measured as 0.41 mm

beat length in a particularly simple experiment: The weak scattered light exiting the fiber sideways appears modulated with a spatial period because dipoles do not radiate energy in the direction along their own axis (see Figs. 4.19 and 4.20).

How well does a polarization-maintaining fiber actually maintain the state of polarization? This is quantified by the extinction ratio E, defined as

$$E = -10 \log_{10} \frac{P_s}{P_p + P_s} \cdot \qquad (4.46)$$

Here, indices p and s distinguish the fractions of power P which are polarized parallel (p) or perpendicular (s, as in German *senkrecht*) to the initial polarization plane.

For short fibers, say, less than 20 m, $E \approx 40$ dB would be considered normal. When the fiber is a kilometer in length, this value will degrade to typically 20 dB, and if the fiber is tightly bent or is squeezed, it may go down to 15 dB.

Sometimes the holding parameter h is specified; it is defined by

$$h = \left(\frac{P_s}{P_p + P_s} \right) / L \cdot \qquad (4.47)$$

It corresponds to the extinction ratio after 1 m of fiber, expressed in linear units rather than decibel. A typical value for a polarization-maintaining fiber is $h = 10^{-5}$ to 10^{-6}, sometimes $h = 10^{-7}$ is reached.

It is unfortunate that the manufacturing process for polarization-maintaining fibers is more complex than for standard fibers and that losses tend to be slightly larger. Polarization-maintaining fibers never became standard, but are used only in applications where polarization has particular importance (and where fiber length is limited anyway). This is often the case in metrological applications (see Chap. 12). In long-haul transmission, such as over transoceanic distances, polarization-maintaining fibers are not used even though polarization mode dispersion poses a challenge.

4.7 Microstructured Fibers

In recent years a very different type of fibers has emerged[5, 11, 18]. These novel fibers consist of a cylindrical glass body just like ordinary fibers; however, in the cladding zone there are voids running along the entire length of the fiber so that a pattern of holes appears in the cross section (see Fig. 4.21). Manufacturing of these fibers differs from that of conventional fibers; details are presented in Sect. 6.2.

The array of holes in the cladding gives rise to the (somewhat tongue-in-cheek) name of "holey" fibers. The air-filled holes reduce the effective index of the cladding so that dopants to raise the core index are not normally applied: The index difference

Fig. 4.21 A schematic view of a typical holey fiber. Tubular channels run along the entire length of the fiber parallel to the core which they surround in a certain geometric arrangement. The hole diameter d and pattern pitch Λ are indicated. In the case shown here, the core is at the position of the "missing" channel at the center

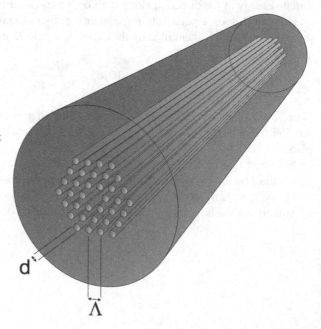

can easily exceed the 1 % limit maximally obtainable with dopants by far. The regularity of the hole pattern is not crucial in this type of microstructured fibers; in fact, even fibers with random hole arrangements have been demonstrated [16].

On the other hand, a strictly periodic arrangement with a pitch not much different from the wavelength of the light can give rise to resonant reflectivity when a certain Bragg condition is met; this is very reminiscent to effects with X-rays passing through crystals (actually, this is how we know the size of crystal cells) and thus gives rise to the name of *photonic crystal fibers*. There is an actual distinction between these two fiber types, but as of this writing, the names are not used very consistently in the community. We will here adopt the following definitions:

Holey fiber   designates a microstructured fiber with hollow channels surrounding the core which itself, however, is massive.

Photonic crystal fiber   designates a microstructured fiber with hollow channels surrounding the core which is also hollow.

Figure 4.22 shows examples of both types in comparison. We will briefly outline both types and point out their remarkable properties which cannot be had from conventional (massive) fiber and which open up exciting possibilities for novel applications. See also Fig. 4.23.

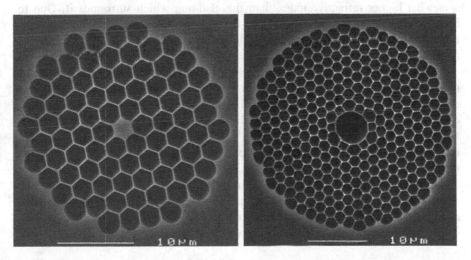

Fig. 4.22 Comparison of two basic types of microstructured fibers. In both cases a central region around the core is shown. *Left*: Holey fiber (Type NL-24-800). *Right*: Photonic crystal fiber (Type HC-633-01). Both fibers are manufactured by Crystal Fibre AS. With kind permission by NKT Photonics, Birkerød, Denmark

Fig. 4.23 A holey fiber seen in an electron microscope together with the eye of an ant. Insects are endowed by nature with a regularly patterned eye. The structure of a holey fiber follows a similar design. The author thanks Toralf Ziems for his assistance in taking this picture

4.7.1   Holey Fibers

In conventional fibers, there is a core which by way of suitable doping has a somewhat higher refractive index than the cladding which surrounds it. Due to constraints in the chemistry, the index difference can be no larger than about 1 %. In holey fibers, the cladding has a sizeable air fraction so that its effective index is lowered. These fibers are also known as solid-core photonic crystal fibers.

The light wave experiences an index which has contributions from both the air holes and the remaining glass in between. In the case of a regular hole pattern, one can distinguish two quantities: the hole diameter d and the pitch Λ. These are often combined with the wavelength of the light into the normalized hole diameter d/Λ and the normalized spatial frequency Λ/λ. A precise calculation of the effective index and the modal structure requires numerical procedures which can be quite involved. It is straightforward, though, to see the following.

The void content lowers the effective index with respect to that of the glass. The effective cladding index is therefore a function of the air fill fraction **AFF**, the ratio of air channel volume to total cladding volume. For the hexagonal geometry shown in Fig. 4.21 it is calculated by straightforward geometry as

$$\mathsf{AFF} = \frac{\pi}{2\sqrt{3}} \left(\frac{d}{\Lambda} \right)^2 . \tag{4.48}$$

This expression, by the way, tells us that as the holes get bigger to the point that the glass walls in between vanish at $d = \Lambda$, the air-filling fraction is bounded by

$$\mathsf{AFF}_{\max} = \frac{\pi}{2\sqrt{3}} = 0.9069 .$$

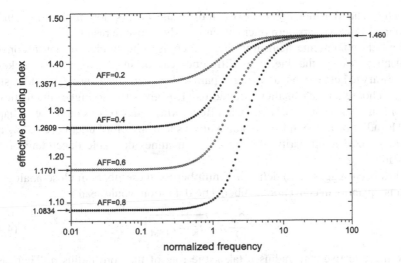

Fig. 4.24 The effective cladding index for an infinite triangular array as a function of frequency for different hole sizes (parameter: air fill fraction). For simplicity, the index of air $n_{air} = 1$ and glass $n_{glass} = 1.46000$ were considered constant. The author thanks Christoph Mahnke for the calculations for this figure

When the wavelength is shorter than both the pattern pitch and the hole size, light will be guided primarily by the glass bridges between the holes. This suggests that in the limit of $\lambda \rightarrow 0$ the effective index tends to that of the glass alone. When on the other hand the wavelength is much larger than the structural dimensions, the light field cannot 'feel' the holes and interstitial glass separately. The effective index may then be expected to be some suitably weighted average of the indices of glass and air. This is the reason for a strong dependence of effective cladding index on wavelength [13].

The situation is demonstrated in Fig. 4.24 which shows the effective cladding index for an infinite triangular array as a function of normalized optical frequency $\bar{\nu} = \nu\Lambda/c$ for four different air fill fractions. The normalization was chosen so that conveniently at $\bar{\nu} = 1$ the (vacuum) wavelength coincides with the pattern pitch. The glass index was here assumed to be at a constant $n_{glass} = 1.4600$, and that of air at $n_{air} = 1$, to avoid a complication of the present discussion by issues of material dispersion. Data points were calculated using the freely available software described in [10].

For very high frequencies the effective index tends to that of the glass as expected (arrow at right axis). For very low frequencies the effective index tends towards a weighted average of glass and air index. By interpolation between glass and air indices according to air fill fraction using the Lorentz–Lorenz equation [6, 9] one obtains the values indicated by arrows on the left.[2] The agreement could hardly be

---

[2]In [7] a linear interpolation is suggested, but does not fit quite as well.

any better. The transition regime between the limits occurs close to $\bar{\nu} = 1$ when the wavelength equals the pattern pitch, an eminently plausible result.

The light-guiding mechanism in these fibers is quite similar to that in conventional fibers, except that the index difference can be much larger. That makes it unnecessary to bother about applying dopants. However, given the bigger index step, one can choose a much smaller core radius. This gives rise to higher intensities in the core and thus to stronger nonlinear effects which may be desirable (see Chap. 9 ff.). The additional design freedom also allows to design for larger core radius and thus minimized nonlinearity; this, too, is sometimes desirable depending on the application.

It has been suggested to define a V number for these fibers in close analogy to the same quantity in conventional fiber. The definition would read

$$V = \frac{2\pi}{\lambda} \rho \sqrt{n_K^2 - n_{FSM}^2} , \qquad (4.49)$$

where the effective core radius ρ takes the role of the core radius a. There is a certain ambiguity how to define ρ in terms of the pattern pitch Λ (one can, e.g., identify Λ with the core radius [4]), and thus the numerical value of V at cutoff may be different from that in conventional fibers. n_{FSM} is the effective cladding index. The name derives from the fundamental space-filling mode, i.e., the fundamental mode that would occupy an infinitely extended cladding pattern without the central defect which is the core.

As pointed out above, n_{FSM} tends to the material index as frequency increases. This partially cancels the λ term in the denominator so that V is not proportional to frequency but rather becomes almost constant at $\Lambda/\lambda \gg 1$, with the specific value depending on d/Λ (see Fig. 4.25).

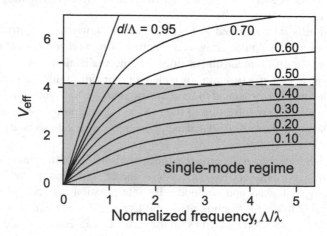

Fig. 4.25 V number in a holey fiber as a function of normalized frequency Λ/λ. The parameter is d/Λ. Toward high frequencies, V_{eff} tends to some constant. It never crosses the cutoff (marked by *horizontal dotted line*) as long as $d/\Lambda < 0.4$. Then the fiber supports only one mode at any frequency. After [5]

Through this effect one obtains a remarkable property which is called the *endlessly single-mode* property [4]. It has been found that for d/Λ small enough (more precisely: $d/\Lambda < 0.406$ [12]), there is only one mode supported at any frequency. This means that within the practical limits set by wavelength-dependent loss, the fiber is always a single-mode fiber. This has been verified over a tremendous frequency range of efficient waveguiding, which may run from ultraviolet to infrared—a factor of 4 [4].

Holey fibers have a second very remarkable property, and this regards their dispersion behavior. For small air-filling fractions, the influence of the holes is small and the wavelength dependence of group velocity dispersion can be expected to closely follow the material dispersion. This is indeed the case. However, as the hole size increases, there is a growing contribution from waveguide dispersion which can reach the point of overwhelming the material dispersion. Since the waveguide contribution can be anomalous at short wavelengths, the zero-dispersion wavelength can shift toward *shorter* wavelengths. The reader will recall that this is not possible with conventional fiber. At about $d/\Lambda = 0.30$, dispersion is flattened over a sizable interval. The precise wavelength range of this interval can be shifted by adjusting the pitch size Λ. Fibers are being offered commercially where this range begins at about 1 μm or even 800 nm. Finally, by judicious choice of pitch and hole size, the dispersion can be even made to have a maximum in this spectral range so that there can be two zero-dispersion wavelengths, similar to the case of the dispersion-flattened conventional fiber shown in Fig. 4.16 but at a wavelength considerably shorter than the material's zero-dispersion wavelength (see Fig. 4.26). Applications have been found for these specialty fibers, but this topic is beyond our present scope.

4.7.2 Photonic Crystal Fibers

This type of microstructured fiber is also known as photonic-bandgap photonic crystal fiber. Its most remarkable feature is that the core is also a hollow channel, giving rise to the alternative name of hollow-core photonic crystal fiber. A hollow (i.e., air-filled) core implies that the light-guiding property certainly cannot rely on the index step between core and cladding—this step goes the wrong way. Instead, it is now the regularity of the hole pattern: the periodic array of holes forms what is called a photonic crystal [20]; for the right wavelengths, there is a photonic band gap which keeps the light from leaving the core through a coherent scattering effect. This is related to periodic structures in nature which selectively reflect certain wavelengths and, e.g., give butterfly wings their fancy colors. We repeat that for solid core holey fibers, the periodicity of the hole pattern is not decisive, whereas for hollow core photonic crystal fibers it is of crucial importance. By the details of the pattern, an interval of wavelengths, typically 50–150 nm wide, becomes the guiding range of the fiber. The value of the pitch Λ is particularly important: By and large, the range of guidance is shifted proportionally when Λ is varied.

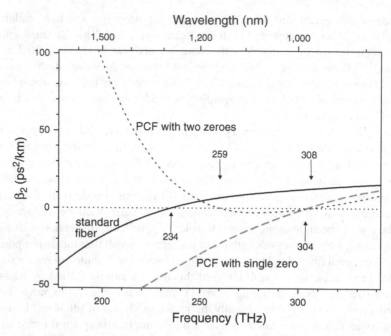

Fig. 4.26 Comparison of dispersion of a standard single-mode fiber (compare Fig. 4.4) with that of two different types of holey fibers (both obtained from a commercial vendor). Note that in this figure a linear frequency scale is used, and that dispersion is given as β_2. The fiber with a single zero-dispersion wavelength has a 3-μm core diameter, the one with two zero-dispersion points, 2.3 μm. In either case, the zeroes are shifted toward higher frequencies with respect to the standard fiber. The measured zero-dispersion frequencies are indicated in terahertz (*arrows*); manufacturer's specifications provide only approximate values

In an interesting departure from standard practice, a hollow core photonic crystal fiber with radially varying pitch of the hole pattern was demonstrated in [19] to allow more freedom in designing the dispersion properties.

4.7.3 New Possibilities

Microstructured fibers offer a variability in almost all fiber parameters, which is unattainable with conventional fibers. The zero-dispersion wavelength can be shifted to much shorter wavelengths, the strength of nonlinear effects can be enhanced or reduced, and the single-mode regime can be greatly enlarged. This is why these fibers will find a whole range of applications which were not possible with conventional fibers. On the downside, they are a lot more difficult to manufacture and therefore quite expensive. They also have much higher loss than conventional fibers, and mechanically they are not nearly as robust. It is therefore not anticipated that they will replace conventional fibers in applications like long-haul transmission.

References

1. Fibercore Limited, Fibercore House, Chilworth Science Park, SO16 7QQ (UK). See www. fibercore.com.
2. Schott AG, Mainz (Germany): Optical Glass Catalog. Available for download at www. schott.com/advanced_optics/english/download/schott-optical-glass-collection-datasheets-july-2015-eng.pdf
3. E. E. Basch (Hrsg.), *Optical-Fiber Transmission*, Howard W. Sams & Co., Indianapolis, IN (1987)
4. T. A. Birks, J. C. Knight, P. J. St. Russell, *Endlessly Single-Mode Photonic Crystal Fiber*, Optics. Letters **22**, 961 (1997)
5. A. Bjarklev, J. Broeng, A. Sanchez Bjarklev, *Photonic Crystal Fibers*, Kluwer Academic Publishers, Dordrecht (2003)
6. M. Born, E. Wolf, *Principles of Optics*, 7th ed., Cambridge University Press, Cambridge (1999)
7. J. M. Fini, *Microstructure Fibres for Optical Sensing in Gases and Liquids*, Measurement Science Technology **15**, 1120 (2004)
8. D. Gloge, *Weakly Guiding Fibers*, Applied Optics **10**, 2252 (1971)
9. J. D. Jackson, *Classical Electrodynamics*, 3rd edition, John Wiley & Sons, New York (1998)
10. S. G. Johnson, J. D. Joannopoulos, *Block-Iterative Frequency-Domain Methods for Maxwell's Equations in a Planewave Basis*, Optics Express **8**, 173 (2001)
11. J. C. Knight, *Photonic Crystal Fibres*, Nature **424**, 847 (2003)
12. Boris T. Kuhlmey, Ross C. McPhedran, C. Martijn de Sterke, *Modal Cutoff in Microstructured Optical Fibers*, Optical Letters **27**, 1684 (2002)
13. Ch. Mahnke, F. Mitschke, *A useful approximation for the cladding index of holey fibers*, Applied Physics B **99**, 241 (2010)
14. S. E. Miller, A. G. Chinoweth (Eds.), *Optical Fiber Telecommunications II*, Academic Press, London (1988)
15. M. Monerie, *Propagation in Doubly-Clad Single-Mode Fibers*, IEEE Transactions in Quantum Electronics **QE-18**, 535 (1982)
16. T. M. Monro, P. J. Bennett, N. G. R. Broderick, D. J. Richardson, *Holey Fibers with Random Cladding Distribution*, Optics Letters **25**, 206 (2000)
17. E.-G. Neumann, *Single Mode Fibers*, Springer Series in Optical Sciences Vol. **57**, Springer-Verlag, Berlin (1988)
18. P. J. St. Russell, *Photonic Crystal Fibers*, Science **299**, 358 (2003)
19. J. S. Skibina, R. Iliew, J. Bethge, M. Bock, D. Fischer, V. I. Beloglasov, R. Wedell, G. Steinmeyer, *A Chirped Photonic-Crystal Fibre*, Nature Photonics **2**, 679 (2008)
20. E. Yablonovitch, T. J. Gmitter, K. M. Leung, *Photonic Band Structure: The Face-Centered-Cubic Case Employing Nonspherical Atoms*, Physical Review Letters **67**, 2295 (1991)

Chapter 5
Losses

Charly Kao, then with Standard Telecommunications Labs in England, proposed in 1966 that it should be possible to produce fibers with loss below 20 dB/km [5]. He had arrived at this conclusion after noting that losses were not an intrinsic property of the glass itself, but rather were due to impurities. His remarkable insight earned him the Nobel prize in physics in 2009.

At the time one could make glass with about 1 dB/m, this was an improvement over glass of ancient Egypt by four orders of magnitude. Then, in less than 20 years, another improvement of four orders of magnitude was reached. The loss came down to 0.2 dB/km, a figure now routinely obtained at 1.5 μm. Part of this progress stems from longer wavelengths now being used: In the visible, the best glass had and has several dB/km loss. Once loss contributions due to impurities had been almost completely eliminated, a fundamental limit was reached, which is defined by the structure of the glass itself.

5.1 Loss Mechanisms in Glass

This fundamental limit is determined by three factors: (a) the long-wave tail of material resonances in the ultraviolet (electronic transitions), (b) the short-wave tail of material resonances in the infrared (molecular vibrations), and (c) Rayleigh scattering due to the statistical structure of the glass. Rayleigh scattering is the same mechanism that makes the sky blue and the sun yellowish (and contributes to its orange to reddish appearance just prior to sunset). The reason is its strong wavelength dependence: The efficiency of Rayleigh scattering scales with the negative fourth power of wavelength.

At the short wavelength end of the visible (in the blue and violet), the contribution from ultraviolet resonances is the leading factor. In much of the visible spectral range and also in the first and second window for telecommunication, Rayleigh scattering dominates while infrared resonances are irrelevant. The third window

© Springer-Verlag Berlin Heidelberg 2016

F. Mitschke, *Fiber Optics*, DOI 10.1007/978-3-662-52764-1_5

finally is in the transition regime: Beginning around 1.6 μm the contribution from infrared resonances overtakes the Rayleigh contribution.

It is not a trivial task to determine with any precision what the theoretical limit for the lowest possible loss is. Taking clues from data taken on bulk glass, a minimum of 0.114 dB/km was derived; this figure corresponds to an energy loss of 2.6 %/km. In 1986, researchers at the Sumitomo company succeeded in making a single piece of fiber with a confirmed loss as low as 0.154 dB/km. This record was not repeated, much less improved upon, for a long time. Only in 2002, a fiber loss of 0.151 dB/km was reported, and this newer record was then quickly bested to 0.1484 dB/km [6]. In mass production of fibers, industry has routinely obtained 0.2 dB/km since the late 1980s, and occasionally 0.18 dB/km. The reader might think that bargaining for the last few percentage points might be of little relevance, but that would be a wrong conclusion: Implications are enormous. Even a minor reduction in loss allows to use longer spans for transmission. On any long-haul distance, a number of intermediate amplifiers or signal conditioners is required; their number can be reduced as soon as the loss goes down. The consequence of any loss reduction by, say, a few percent then translates into savings of potentially millions of dollars.

As can be seen in Fig. 5.1, there is a local loss peak between the second and third windows, at about 1.39 μm. It is caused by impurity molecules in the glass. Optical materials must be exceedingly pure to be transparent at all at this wavelength; water vapor in the atmosphere is particularly detrimental. The loss peak is caused by molecular vibrations of water molecules embedded in the glass. There is a characteristic strong vibrational resonance of the OH bond at 2.8 μm. Water is always present in our environment and so ambient air is opaque at 2.8 μm. However,

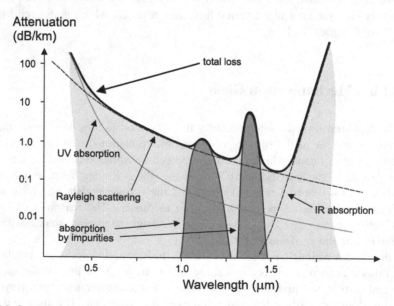

Fig. 5.1 Spectral dependence of the contributions to energy loss in a fiber

these oscillations are not purely harmonic, but have overtones (anharmonicity). The first overtone (the second harmonic) at twice the frequency, i.e., half the wavelength, is the peak that we see in Fig. 5.1. To be sure, this is a weak overtone, and the OH concentration in the fiber is low, but the resonance stands out conspicuously because the Rayleigh background is so low. Since 1998, manufacturers have succeeded to produce fibers with a much reduced OH content so that the peak disappears into the background (e.g., "AllWave Fiber" by Lucent Technologies [1]).

Besides water, numerous other impurities can contribute to loss. Among these are the transition metals Fe, Cu, Co, Cr, Ni, and Mn. To appreciate the required purity, consider this: At 800 nm, one ppb of Cu produces an absorption of several tenths of dB/km. (One ppb, or *part per billion*, indicates a concentration of 10^{-9}.)

5.2 Bend Loss

We have seen that the composition of the glass gives rise to loss mechanisms. On top of all that, there are further losses when the fiber is deployed for use, in particular when it is being bent. This may be intuitively plausible since in a bend the cylindrical symmetry is broken (Sect. 3.4). One has to distinguish two contributions to bending loss known as *macro-bending loss* and *micro-bending loss*.

Macro-bending loss occurs when fibers are bent with a "macroscopic" radius of curvature, i.e., in the range of centimeters. If one falls back to a ray optic view (which is applicable only to multimode fibers, of course), one realizes that the critical angle for total internal reflection can be exceeded in a bent portion. More physically correct, in a wave-optic picture we have seen that the field distribution of any mode is not restricted to the core, but extends into the cladding. As the fiber is being bent, there must be a certain distance from the fiber's axis toward the outside of the curve where the propagation velocity (as determined by the effective index for that mode) begins to exceed the maximum possible velocity in the cladding (given its index). Of course, the velocity is not actually exceeded. Rather, the phase fronts cease to be plane and fall behind. There is then a component of the Poynting vector pointing radially outward; this implies energy radiated away from the guided mode (see Fig. 5.2).

Strictly speaking, there is mechanical tension in the bent fiber so that the inside is compressed, the outside expanded. This creates deviations of the refractive index, effectively lowering it on the outside. The mechanism just described is counteracted by this, but it is not compensated.

It should be clear that the critical radius at which radiative loss begins must be proportional to the bend radius of the fiber. On the other hand, the modal field radially decays exponentially. Together, it follows that bend loss decreases exponentially when the bend radius increases. It is also implied that the stronger the cladding penetration, the higher is the bend loss. Cladding penetration is high when the index difference between core and cladding is small; a large value of Δ is thus beneficial in this context. Also, cladding penetration grows with increased

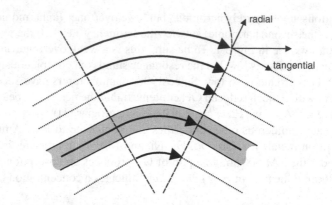

Fig. 5.2 Bending a fiber creates additional loss. On the outside of the bend, the wave cannot keep up due to its limited velocity. The wavefront gets distorted so that a radial component of radiation is created

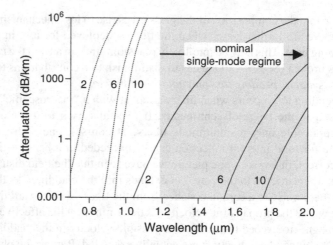

Fig. 5.3 Macro-bend loss for the LP_{01} fundamental mode (*solid lines*) and the LP_{11} mode (*dotted lines*), for a loop with radius as indicated in centimeter. In the limit of infinite bend radius the values for the LP_{11} mode constitute the limit of the single mode regime. At finite radius, the cutoff is shifted toward shorter wavelengths. At the same time, loss in the single-mode regime increases. After [7] with kind permission

wavelength; therefore the highest reasonable V number is also beneficial. It is therefore good practice to operate close to the cutoff wavelength of higher-order modes. Higher-order modes also extend farther into the cladding and are more attenuated by bending than the fundamental mode. Bending thus shifts the effective cutoff toward shorter wavelengths (Fig. 5.3). This must be observed when the cutoff wavelength is measured, see Sect. 7.4.

It is a standard laboratory trick to shift a fiber's cutoff from, say, 1200 to 1000 nm by winding it 20 turns on some bobbin with 20 mm diameter.

While macro-bending loss is reasonably understood, micro-bending loss is considerably more complicated. At least so much is clear: The effect is dominated by the statistics of the deviation of the fiber from a straight line. A roughness of 100 nm to 1 μm plays a major role. This raises the issue of the surface material of spools. It is well known in the trade that styrofoam, for example, is particularly bad, probably due to its structure which is best described as air bubbles separated by thin walls. It is also known that micro-bending in multimode fibers creates only a small, wavelength-independent loss contribution while in single-mode fibers there is a sharp loss onset at large wavelength (Fig. 5.4). This tends to shift the wavelength regime of lowest loss, in principle centered around $\approx$ 1600 nm, toward 1550 nm.

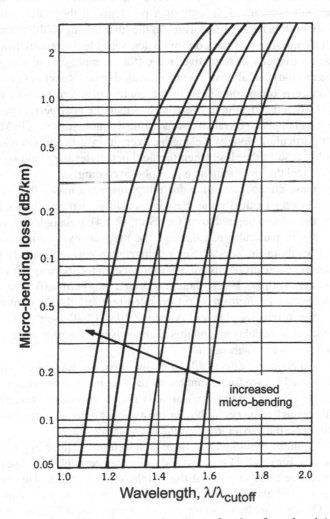

Fig. 5.4 Micro-bending loss of the fundamental mode as a function of wavelength (normalized to cutoff wavelength). From [3] with kind permission

At V numbers close to 2.4 (i.e., close to the cutoff of higher-order modes), this contribution is usually negligible. Since fibers are normally used at $2.1 < V < 2.4$ for reasons described in the previous paragraph, the problem is often avoided.

5.3 Other Losses

Quite a number of influences may give rise, at least in principle, to further loss contributions and must therefore be considered in the design of fibers. These are, of course, taken care of by the fiber's manufacturer, and the application engineer does not normally deal with such problems. It therefore suffices when we only briefly outline here: Irregularities in the fiber arising in the manufacturing process such as variations of the core diameter, variable deviations from circular symmetry, and variations in refractive index (for example due to fluctuations in dopant concentration) can all give rise to increased loss. (Roughness of the core–cladding interface is often considered in the context of micro-bend loss.) Loss is also increased when the cladding glass is of less than the near-perfect purity of the core glass as may happen in certain manufacturing technologies (e.g., in MCVD, see Sect. 6.2). At particularly long wavelengths, even an insufficient outside cladding diameter could cause further loss when the field penetrates the cladding so deeply that it begins to feel the environment, e.g., a plastic coating.

On top of these effects from manufacturing, there are losses that occur when during operation very gradually the glass composition changes. This can happen through the action of ionizing radiation (see Sect. 12.2.4). Beta radiation (electrons) and gamma rays in particular can damage the material by creating dislocations of nuclei and bonds in the crystal. Such damage can partially, slowly heal after irradiation ceases. These problems are of major concern for space applications, such as aboard spacecraft. In the 1980s, there has been a long-term exposure test called LDEF (*long-duration exposure facility*) operated by the USA. It went on several years longer than originally planned: After the 1986 Challenger disaster, further space shuttle starts were delayed, and this also involved the return vehicle. Still, the damage found was not too substantial.

There are also chemical effects affecting fiber loss. Substances can diffuse into the glass; this is well known for helium but there is not much helium out there. More relevant is the case of oxygen that is ubiquitous in the atmosphere. Nevertheless, the effect is subtle enough to be negligible for most applications; in particularly critical cases, one can deposit a barrier layer on the fiber surface.

Finally, it should be remarked that there is indeed one case of impurity that can even *lower* losses. Embedded OH^- groups, while they certainly increase loss in the infrared, can reduce them in parts of the visible and ultraviolet. This happens on a high background of about 1 dB/m and is thus irrelevant for telecommunications applications. On the other hand, sometimes short wavelength light must be guided over short distances; this is the case, e.g., in laser surgery. In such cases, one can exploit this curious fact and use glass with intentionally high OH^- content.

5.4   Ultimate Reach and Possible Alternative Constructions

At the wavelength of the third window where Rayleigh loss meets the tail of infrared absorption, fibers have their global loss minimum. Does this low loss allow a transoceanic line? Taking a realistic figure of 0.2 dB/km and a typical transoceanic distance of 5000 km, total loss comes to 1000 dB. This is too much by any standard. Consider this:

A total of 1000 dB imply a power attenuation by 100 orders of magnitude. We must certainly demand that at the very least, a single photon must arrive at the detector during the time slot reserved for a single bit. Then, we would have to launch 10^{100} photons. Since the energy of a single photon is given by $E = h\nu \approx 10^{-19}$ J, the launch energy would have to be 10^{81} J. Even when we allow a full second for the time slot, 10^{81} W are about 80 orders of magnitude more than is realistic since, beginning at several watts continuous power, one starts to damage the fiber front face. (In radio engineering a single photon has much smaller energy; this allows an attenuation of 150 dB from transmitter to receiver without ever reaching a quantum limit. For radio there is a possibility of worldwide reception, e.g., on short wave; in optics, this is not possible.)

5.4.1   Heavy Molecules

A transoceanic link with 0.2 dB/km is thus not viable without several intermediate amplifiers. How about replacing the silicon dioxide in the glass with heavier molecules? This would shift the infrared resonance toward longer wavelength and postpone the onset of the corresponding loss toward longer wavelengths. One could then move to longer wavelengths and enjoy the benefit that Rayleigh scattering loss is reduced dramatically (according to the negative fourth power of wavelength) as shown in Fig. 5.5. This is certainly a very appealing idea: If one could change the fiber material so that one could go to wavelengths between 3 and 4 μm, one could reduce loss by a factor of 30. Then the whole 5000 km length of the span could be taken in one go, without the need for amplification with its associated technical complexity.

Therefore, researchers have tried for many years to come up with suitable material. In 1978–1979, three groups suggested virtually at the same time that fibers made from fluorides, chalcogenides, or halides would have dramatically lower damping, in principle down to 0.001 dB/km. There is only one problem with all these materials: They offer fantastic perspectives, but so far no one has ever succeeded in making them so that the loss would compete with existing fused silica fibers. This is due to the increased chemical reactivity of all these materials. Also, here are indications that mechanical properties of existing fibers cannot be matched by these more exotic materials: They tend to be brittle and break easily.

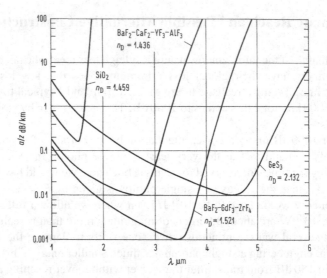

Fig. 5.5 Theoretical loss of infrared fibers made from various materials, in comparison to SiO$_2$. Infrared absorption sets on at longer wavelengths so that the λ^{-4} trend of Rayleigh scattering can be exploited toward longer wavelengths. This leads to considerably weaker damping. Unfortunately, these data remain theoretical. After [8] with permission

Fluorides. Among fluorides, the most frequently used substance is fluorozir-
conate or ZBLAN (pronounced *zee-blan*). The acronym indicates Zr (zirconium), Ba (barium), La (lanthanum), Al (aluminum), and Na (sodium) as constituent elements. In the manufacturing process, the crystalline substances are molten in a crucible and then poured into a rapidly spinning casting mould. The spectral range of reasonably low transmission is between 500 nm and 3.5 μm, with the lowest losses between 1.5 and 2.7 μm. About 15 dB/km at 2.5 μm are obtained commercially. The refractive index is similar to that of fused silica ($n \approx 1.5$), dispersion is lower. At 2.8 μm, there is a strong OH$^-$ absorption that gives rise to considerable loss; the figure can vary even among fibers by the same manufacturer from 30 to 80 dB/km. Only multimode fibers with core diameters up to 250 μm are available. Critical mechanical tension is quoted at a very low 0.6 MPa, the minimum bend radius at 10 mm. Temperatures above 150 °C pose a problem. Contact with water causes chemical change; coatings need to be employed as barriers.

Chalcogenides. There are very few glasses with good transmission between 3 and 11 μm, including the wavelength of the CO$_2$ laser at 10.6 μm. Chalcogenides composed of arsenic, germanium, and antimony in combination with sulfur, selenium, or tellurium make it possible. The ingredients are mixed, molten, homogenized, and cooled down inside fused silica ampoules under vacuum. There are practical difficulties with trapped bubbles, inclusions, and crystallites. In the visible, these fibers are basically opaque. Chemically they are reasonably stable; mechanically at 0.1–0.17 GPa critical tension not very much so. Also,

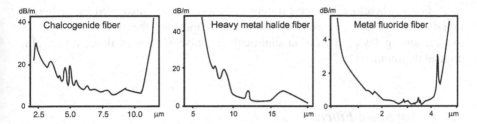

Fig. 5.6 Typical spectral transmission curves of some infrared fibers

elevated temperatures above 150 °C create difficulties. In most cases, these fibers are made without distinction between core and cladding just a cylindrical body. Diameters range from 150 to 500 μm, and there is a plastic coating for protection. The refractive index is about 2.8; therefore coupling both in and out suffers from large Fresnel loss. Overall, loss is much higher than for fluorides, but chalcogenide fibers can be used out to much longer infrared wavelengths.

In view of all the difficulties mentioned, efforts to develop such fibers for long-distance use have been reduced. Meanwhile, development has taken an entirely different route: Conventional intermediate amplifiers become unnecessary when fibers themselves are transformed into amplifiers (see Sect. 8.8.1).

There are, however, applications when a flexible light guide is used for a short distance on the order of 1 m. This is true in laser material processing laser surgery is, after all, a particular variant thereof. In this kind of application, these fibers compete with silica fibers and with sapphire fibers (Fig. 5.6). Silica fibers for these applications are usually made without cladding with 400–1000 μm diameter; common wavelengths are 1064 and 532 nm for Nd:YAG Lasers and 514 nm for Argon ion lasers. The advent of the Ho:YAG-Lasers operating at 2.1 μm created the requirement of silica fibers with less than 5 ppm OH^-.

5.4.2 Hollow Core Fibers

For use with CO_2 lasers with their high power at a wavelength of 10.6 μm, one uses a different approach: Here, fibers with hollow core (capillaries) are employed. Hollow core fibers have high loss, but that is of minor importance for guiding high power over short distance in comparison. All-important is a high damage threshold. Obviously, in a hollow core, the refractive index is close to 1. Light-guiding takes place only because, in wavelength regimes of strongly anomalous dispersion, the index of the cladding may actually fall below 1. In some doped silica glasses and in sapphire, this occurs around the wavelength of the CO_2 laser.

Hollow core fibers are also used in reverse: We will discuss fiber-optic sensors in Chap. 12 but here we jump ahead and mention that the peak wavelength of black body radiation around room temperature is within the transmission range of these fibers. Therefore, the temperature of objects can be measured by gathering radiation with a hollow core fiber without even touching the object, and guide it to a measuring device. This is an admittedly expensive, but sometimes very useful clinical thermometer!

5.4.3 Sapphire Fibers

Sapphire is a chemically stable, nontoxic material with reasonable mechanical strength. It can be worked into fibers by growing from a solution of Al_2O_3. Sapphire fibers transmit from the visible range to about 3 μm. Bend radii are limited to a few centimeter, and loss is around 1 dB/m. Sapphire fibers, too, are fabricated without cladding and with 100–500 μm diameter. They have good damage threshold and high melting point, which makes them attractive for transmission of high-power laser light, including laser delivery in surgery and dentistry.

Like hollow core fibers mentioned in the previous paragraph, sapphire fibers can be used as a part of an infrared thermometer. Their good heat resistance and chemical stability are favorable for applications in chemically harsh and/or high-temperature environments, like inside chemical reactors where sapphire fibers may offer the best way to measure temperature.

5.4.4 Plastic Fibers

Also suitable only for short-distance transmission are plastic fibers, usually called POF (*plastic, or polymeric, optical fiber*) for short. However, they are used in an entirely different field of application. Plastic is a low-cost material and can be shaped very easily into fibers (or any other shape). In most widespread use is polymethyl methacrylate (PMMA), a.k.a. acrylic glass, perspex, or plexiglas; PMMA was used for the first commercially available POF as early as 1963. Polycarbonate and polystyrene are other options.

In comparison to fused silica, the optical loss in POFs is enormous and is measured in dB/m rather than dB/km (see Fig. 5.7). While Rayleigh scattering and absorption from electronic transitions and from contaminants also exist in POFs, the dominant loss mechanism seems to be the absorption at harmonics of the CH bond, which is ubiquitous in plastic materials [4]. A successful approach to reducing this problem is to replace some of the hydrogen with heavier molecules such as fluorine, to shift the resonance. However, fluorination is an expensive process, so that the low-cost advantage is somewhat reduced.

Nevertheless, in terms of low loss, POFs cannot match silica-based fiber. Their maximum transmission distance is therefore defined by loss, not by dispersion. It is thus not a problem that realistically only multimode fibers with large numerical aperture can be made. Core diameters are often 1 mm or even more. A cladding can

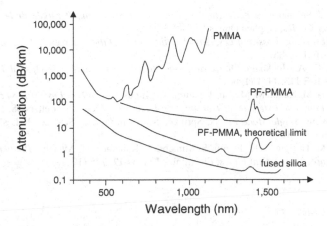

Fig. 5.7 Typical spectral transmission curves of polymethyl methacrylate (PMMA) light-guiding fibers. Data are taken from gradient index fibers. "PF" refers to perfluorinated material. After [2] with kind permission

be made from fluorinated PMMA if internal guiding is desired. A typical numerical aperture can be 0.3.

Certainly, POFs have several quite favorable aspects: Handling is easier than for fused silica, incoupling efficiency is good, and coupling between fibers is quite simple. All this adds to the low-cost aspect. One drawback is that these fibers can be damaged by high-power lasers. Thus the applications are outlined: Plastic fibers are useful for short-distance data transmission where cost limitations are stringent. Local area computer networks within a building or on premises are an example. Also, in some stereo equipment, there is optical transmission between digital audio components such as CD players, DVD recorders, etc. European car makers have been using POFs since about 1998 because interesting savings of weight are obtained. Various fiber-sensors are deployed and connected (see Chap. 12), and an on-board POF network provides the car driver and all passengers with individual access to a range of entertainment media. 10 years after the first use of fibers in cars, a series 7 BMW car or similar model by other brands contained more than 100 m of POF. Quite naturally, the aviation industry is also increasingly using POF. Increasingly data-hungry applications may, however, eventually make the use of silica-based fibers in vehicles attractive.

References

1. http://fiber-optic-catalog.ofsoptics.com/category/single-mode-optical--fibers
2. H. P. A. van den Boom, W. Li, P. K. van Bennekom, I. Tafur Monroy, G.-D. Khoe, *High-Capacity Transmission Over Polymer Optical Fiber*, IEEE Journal of Selected Topics in Quantum Electronics **7**, 461 (2001)

3. K. Furuya, Y. Suematsu, *Random-Bend Loss in Single-Mode and Parabolic-Index Multimode Optical-Fiber Cables*, Applied Optics **19**, 1493 (1980)
4. T. Kaino, *Absorption Losses of Low Loss Plastic Optical Fibers*, Japanese Journal of Applied Physics **24**, 1661 (1985)
5. K. C. Kao, G. A. Hockham, *Dielectric-Fibre Surface Waveguides for Optical Frequencies*, Proceedings IEE **113**, 1151 (1966)
6. K. Nagayama, M. Kakui, M. Matsui, T. Saitoh, Y. Chigusa, *Ultra Low Loss (0.1484 dB/km) Pure Silica Core Fibre and Extension of Transmission Distance*, Electronics Letters **38**, 1168 (2002)
7. E.-G. Neumann, *Single Mode Fibers*, Springer Series in Optical Sciences Vol. **57**, Springer-Verlag, Berlin (1988)
8. S. Shibata, M. Horiguchi, K. Jinguji, S. Mitachi, T. Kanamori, T. Manabe, *Prediction of Loss Minima in Infra-Red Optical Fibres*, Electronics Letters **17**, 775 (1981)

Part III
Technical Conditions for Fiber Technology

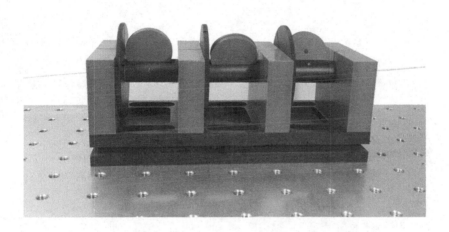

A variable "wave plate" in all-fiber technology used to adjust the state of polarization. It consists of three rotatable fiber loops and permits to translate any given state of polarization into any desired state. These components are described in Sect. 8.5.1.

Chapter 6
Manufacturing and Mechanical Properties

6.1 Glass as a Material

Blood is a quite peculiar juice. Thus spoke Mephistopheles in the tragedy "Faust" by J.W. von Goethe.[1] Glass, too, is a quite peculiar juice: There was a stone age, an iron age, and a bronze age. Glass, however, is the only artificial material that has been in use uninterruptedly for seven millennia or more without giving its name to an epoch.

Like the word "crystal," "glass" refers not to a chemical but to a physical property. Unlike a crystal, its structure is not neatly ordered but quite irregular. Glass is a liquid usually mistaken for a solid! But let us start at the beginning.

6.1.1 Historical Issues

The oldest finds date back about 7000 years before Christ, at the end of the younger stone age. They hail from the Mideast: Egypt and Mesopotamia, present-day Iraq. Independently the art of making glass was also developed in Mykenae (Greece), China, and North Tyrol.

Making glass is closely related to pottery, which has existed in Egypt more than 8000 years ago. Maybe by chance people had discovered that a glazing develops when sand with lime content is exposed to fierce heat together with soda ash. Beginning about 1500 BCE, glass was made without ceramic substrate. Blowing glass dates back to ca. 200 BCE in Sidon and Babylon. In the Roman Empire, glass articles were coveted luxury objects.

---

[1] Johann Wolfgang von Goethe (1749–1832) was a German politician and amateur scientist who is best remembered for his prolific writings of poetry and drama. Indeed, in Germany today he is widely regarded as one of the finest literary figures ever in that country.

© Springer-Verlag Berlin Heidelberg 2016
F. Mitschke, *Fiber Optics*, DOI 10.1007/978-3-662-52764-1_6

In the middle ages, Venice was an important center of the art of glass blowing. Up to 8000 people worked there. Further north in Central Europe, glass was mainly made in remote forested areas such as the German Spessart, the Thuringian and Bavarian Forests, and the Erzgebirge ("ore mountains" on the German–Czech border) because there both potash and fire wood were in abundant supply. (Potash, or potassium carbonate K_2CO_3, is the main constituent of wood ashes: All plants contain potassium salts.) Until the seventeenth century, in part the eighteenth century, there were traveling glass makers. To our day the glass industry in Central Europe is still concentrated to a good part near major forests such as the Bavarian Forest.

Modern glass technology was basically started by two Germans, Otto Schott (1851–1935) and Ernst Abbe (1840–1905) (Fig. 6.1). Schott, son of a glass maker's family from Lothringia, conducted systematic experiments with almost all chemical element to determine which influence their addition to the melt would have on the properties of the final glass.

Abbe was a professor at the German university of Jena, and he was a co-owner of the Carl Zeiss company. Zeiss needed high-quality glass in order to build optical instruments.

After many tries, Schott finally found the suitable glass recipe; this prompted a cooperation that then led to the start of "Jenaer Glaswerk Schott und Genossen" (Jena Glass Works Schott and Co.), which acquired some fame. For many specialized purposes, they developed just the right glass. The company did well, and we remark in passing that such fairly revolutionary social novelties as an 8-h work day and participation of employees in the company's profits were introduced.

After the second World War, Americans moved specialists from Jena into what was to become Western Germany. This gave rise to the new location of Schott Glass Works in Mainz.

Carl Zeiß (1816–1888)   Ernst Abbe (1840–1905)   Otto Schott (1851–1935)

Fig. 6.1 Carl Zeiss, Ernst Abbe, and Otto Schott are the founders of modern optics in which scientific methods and industrial processing are closely interwoven. From [6] with kind permission

6.1.2 Structure

As mentioned above, both words "glass" and "crystal" do not refer to a specific chemical composition but to a particular spatial arrangement of molecules. In a crystal, molecules are arranged in a repetitive, periodic pattern. In glass, by comparison, they are arranged in a disorderly fashion: glass is amorphous. Correspondingly, in glass the molecules are not densely packed (Fig. 6.2). Many substances have glassy states; the table gives a few examples:

Substance	Glass temperature (K)
Natural rubber	200
PVC	347
Water	140
Glucose	305
Selenium	303
Beryllium fluoride	570
Germanium dioxide	800
Silicon dioxide	1350

Only a limited selection of glassy substances is useful in the optics industry. We will deal almost exclusively with glass of silicon dioxide. Silicon, after oxygen, is the second most frequent element in the earth's crust (28 %). It is found in the form of silicates, silicic acids, and as anhydride SiO_2 (and of course, these days, in elementary form inside of computers). Silicic acid is a name for the oxygenic acids of Si, that is, $SiO_2(n \cdot H_2O)$ where the case of $n = 0$ (SiO_2 in its various forms) is sometimes included. In particular, ortho silicic acid H_4SiO_4 occurs frequently. Over geological time spans, it may shed water and go through intermediate stages like $H_2Si_2O_5$ until finally it becomes the anhydride SiO_2.

The most important crystalline form of SiO_2 is quartz. It constitutes the most important part of silicate rocks; there are also feldspar ($KAlSi_3O_8$), mica ($KAl_2[AlSi_3O_{10}](OH)_2$), and salts of polysilicic acids containing Mg^{++} and Ca^{++}.

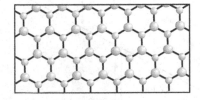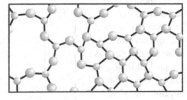

Fig. 6.2 Comparison of crystalline and glassy structures for the example of silicon dioxide. Symbolically shown are silicon ions (Si^{4-}, small spheres), oxygen ions (O^{2-}, larger spheres), and their electronic bonds. Due to the two-dimensional nature of the sketch, silicon's tetrahedron configurations with four bonds are depicted with three, rather than four bonds; the missing fourth bond may be imagined out of plane. In the crystal, there is high packing density; in glass, irregularities lead to lesser density

Among these quartz is the hardest. During geological time scales, quartz is ground down and destroyed, and small fragments remain: gravel or sand. Traces of soluble silicic acid in the waters of rivers and the sea are incorporated by plants and animals for mechanical hardness.

Depending on its crystalline modification, quartz has a melting point of 1500–1700 °C and a density of 2.3–2.6 g/cm^3. Geologically it is found in transparent crystals up to a meter in size. In form of smaller crystals, quartz is contained in all primary rocks such as granite, porphyry, and gneiss. Amorphous SiO_2, often colored by other substances, is the basis of semiprecious stones such as agate, chalcedony, and opal.

Glass is obtained for melting together quartz sand, soda ash, potash, and metallic oxides. Soda ash means sodium carbonate (Na_2CO_3) and potash is potassium carbonate (K_2CO_3). If one uses colorless metallic oxide such as CaO and white (iron-free) quartz sand, one obtains clear glass; addition of other elements can change the properties. Green or brown bottle glass is made from ordinary yellow sand (containing iron); other additives are common to control the color. Window glass consists of Na_2O CaO $6SiO_2$. Lead glass, used extensively in optics for its high refractive index, consists of K_2O PbO $6SiO_2$. Laboratory glass is similar to window glass, except that there are additions of 8 % Al_2O_3, 5 % B_2O_3, and 4 % BaO. If a lot of Al_2O_3 is mixed in, SiO_2 and Al_2O_3 will no longer mix and the substance turns turbid in the oven, stays white, and is not transparent. This is called porcelain, or China.

The main characteristic of glass, its structural irregularity, is reflected in its heat conductivity, which is about one order of magnitude lower than for the corresponding crystal. With this structure, glass is in a local, but not a global minimum of free energy (Fig. 6.3). Consequently a deglassing, a growing of crystalline structure, may occur—albeit on very long time scales due to the enormous viscosity of glass. Over

Fig. 6.3 Glass is a stiffened undercooled liquid. Starting from a crystal, by raising the temperature it will melt. When the temperature is lowered again, there can be undercooled melt rather than recrystallization. This melt then stiffens and becomes glass. In comparison, glass is less densely packed than crystal, thus occupies a larger volume. In principle, glass can recrystallize over long periods of time

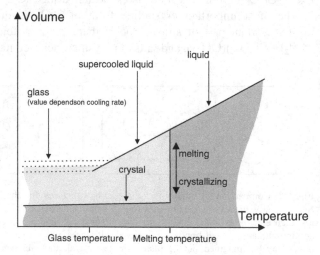

historic time spans, glass can be affected to a measurable amount: very old glass turns turbid and brittle.

Why would one use glassy, not crystalline material for light-guiding fibers? Crystals can never be made entirely without defects and dislocations, but these act as efficient scatterers of light so that losses are higher. Moreover, crystals tend to be brittle so that glass comes out as the better choice. The reader is referred to [3] for more detail about defects in silica glass.

6.1.3 How Glass Breaks

When it comes to glass, the layperson does not necessarily think of great elasticity. Rather, the common perception is that glass breaks quite easily. This may be why it is so fascinating to experience the perfect flexibility of optical fibers. It is true, though, that cracks can occur in glass, which suddenly, precipitously cause it to break (Fig. 6.4). On closer inspection, what happens is that fractures propagate across the material at a speed of hundreds of meters per second (approaching half the velocity of sound). But that is not true for all cracks: Quite to the contrary, may tiny cracks advance only at an unperceptibly slow speed. Growth rates of 10^{-14} m/s have been verified; this corresponds to one snapping atomic bond per hour, or 3 years until the crack has proceeded by 1 μm. Then, damage will become visible only after years, after a false sense of safety has developed.

If pristine glass pieces are tested under high vacuum, they withstand tensions of more than 10 GPa, about ten times the value for many metal alloys. Surface defects

Fig. 6.4 Glass breaks from cracks that grow at their tip

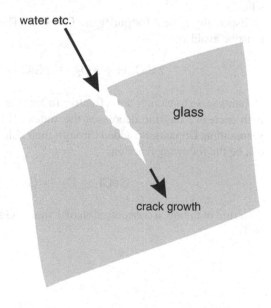

water etc.

glass

crack growth

or contact with abrasives produces microscopic cracks, though, which are the point of attack for chemicals. The cracks then grow and widen. The gravest concern is about water because it is so ubiquitous. It acts at the tip of the crack. As is well known, window panes are cut to size not with a saw as one would do with wood, but by making a scratch with a diamond, wet it with water or even saliva, and then break it. This is the same mechanism.

Atoms at the glass surface have fewer bonds than those inside the volume. Therefore they are at an elevated energetic state. Making the surface larger then requires an energy supply. When the mechanical energy stored in the material is larger than this additional surface energy, the crack will grow. Chemical reactions between the silica and intruding water reduce the required energy from 3.2 to 0.19 eV per bond.

A lot of mechanical stress builds up, at the tip of the crack in particular. In cracks that are often about 0.4 nm, water molecules of 0.26 nm diameter and similarly sized ammonia molecules can intrude, but methanol (molecular diameter 0.36 nm) has a lot less consequence. Even larger molecules do not affect crack growth appreciably.

It took a number of years until industry had learned to master the making of glass with the chemical purity required for fibers. Experience from semiconductor industry was a valuable guide in the process. In that industry, gaseous silicon chloride ($SiCl_4$), purified by distillation, is widely used as a starting material. It is now also being used for making glass, according to the reaction formula

$$SiCl_4 + O_2 \rightarrow SiO_2 + 2\,Cl_2. \tag{6.1}$$

Gaseous chlorine evaporates; solid silicon dioxide condenses as an amorphous substance on cool surfaces and can form glass at suitable temperatures ("fused silica").

Especially critical for purity are OH ions. The following reaction, for example, must be avoided:

$$2\,SiCl_3H + 3\,O_2 \rightarrow 2\,SiO_2 + 3\,Cl_2 + 2\,OH. \tag{6.2}$$

Dopants serve to modify the refractive index (Fig. 6.5). Germanium and phosphorus both increase; fluorine decreases the index. The most frequently used dopant is germanium. Dopants are added through their chlorides to the reaction gas, and there can be the following reaction:

$$GeCl_4 + O_2 \rightarrow GeO_2 + 2\,Cl_2. \tag{6.3}$$

As a rule of thumb, a concentration of 1 mol % GeO_2 in fused silica raises the index by 0.1 %.

Fig. 6.5 Influence of dopants on the refractive index at 1.0 and 1.5 μm. Shown are data for phosphorus (P$_2$O$_5$), germanium (GeO$_2$), boron (B$_2$O$_5$), and fluorine. Calculated after [7]

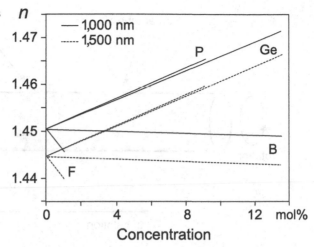

6.2 Manufacturing of Fibers

The manufacturing of optical fiber is performed in two steps: First a *preform* is made. The term refers to a rod of glass, typically ca. 1 m long, with a diameter of 10–50 mm, and with the refractive index profile already built into it. In the second step, this preform is then softened by heating and stretched out by pulling so that the final fiber is obtained.

Both process steps will now be described in some more detail. For making of the preform there are several alternative ways. All of them have certain advantages and disadvantages; each manufacturer tends to advertise the advantages, indeed the superiority of their particular proprietary technique.

6.2.1 Making a Preform

6.2.1.1 OVD

Outside vapor deposition, a technique also known as *soot process*, was the first process to achieve the reduction of losses to 20 dB/km in 1973 (Fig. 6.6). It was developed by Corning Glass Works; it is still used at Corning and, through joint ventures, at other manufacturers.

Glass is deposited on the outside of a massive cylindrical rod of aluminum oxide. It is generated when the gaseous chemicals are fed into the flame of a burner so that submicroscopic glass particles condense on the surface. All the time the rod is rotated and translated so that a uniform layer is formed. The layer is porous at first ("soot"), but during its deposition concentrations of dopants are adjusted, and

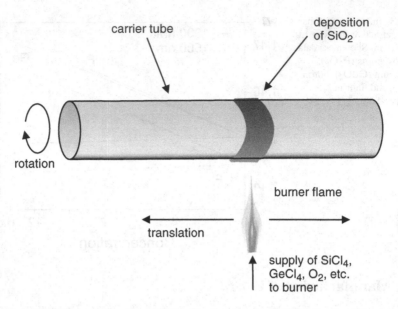

Fig. 6.6 In *outside vapor deposition* (OVD), the glass is deposited from the reaction of gaseous chemicals on the outside surface of a carrier rod

it therefore already contains the dopant profile as required for the refractive index profile of the finished fiber.

Then this rig is heated to allow evaporation of trapped gases and humidity. Next, heat is turned up to a higher temperature of 1400–1500 °C so that in a sintering process the porosity is removed. Once the glass is compact, one can pull out the ceramic carrier rod; finally, the preform is "collapsed" to fill the central hole and generate a massive object: a scale model of the fiber.

6.2.1.2 MCVD

Modified chemical vapor deposition was developed ca. 1974 at Bell Laboratories and is now in widespread use (Fig. 6.7). As compared to OVD, the inside is turned out: One starts with a glass tube (which will later become part of the cladding) and passes the gaseous reactants through the bore. Just as in OVD, a burner is moved along and around the tube; in the heated zone porous glass is deposited. The difference is that here no residual gas and no water vapor is trapped because there is the wall of the tube between the burner flame and the reaction zone, acting as a barrier. Again, in the next step, the porous glass is sintered. The resulting hollow tube is then collapsed to a massive rod.

The disadvantage to be mentioned here is that the glass tube needs to be of very high purity and uniformity. Also, during the collapsing step, some of the dopant used

Fig. 6.7 In *modified chemical vapor deposition* (MCVD), the glass is formed on the inside surface of a glass carrier tube

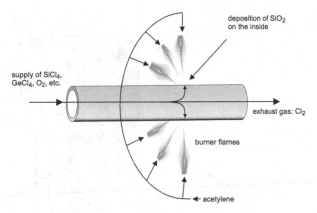

Fig. 6.8 In *plasma chemical vapor deposition* (PCVD), the glass is deposited inside a carrier tube as in MCVD; however, heat supply is quite different and relies on microwave heating

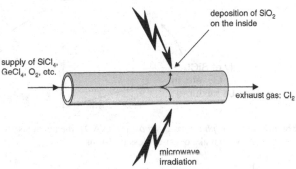

in the last, innermost layer escapes. This is why the finished fiber often exhibits a central dip in the refractive index profile.

6.2.1.3 PCVD

Plasma chemical vapor deposition goes back to Philips Research Laboratories in 1975 (Fig. 6.8). This is a variant of MCVD where not a gas burner is used for heating but a microwave generator (3 GHz, several hundreds of watts). Meanwhile, the temperature of the rod is kept at ca. 1000 °C in order to minimize mechanical tensions between tube and deposited layers during heating cycles. The plasma is uniform enough that constant turning of the tube is not required. Also, sintering is unnecessary because the glass is deposited free of pores. Moreover, the process is fast since thermal cycling is much reduced. These items combine into a distinct advantage when very many very thin layers must be deposited for the most precise control over the refractive index profile. It is not at all unusual to deposit 2000 layers. However, this method also suffers from the central index dip.

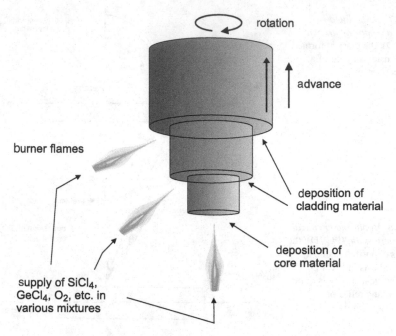

Fig. 6.9 In *vapor phase axial deposition* (VAD), the new glass forms right on the end of a carrier rod; this allows to make quasi-endless preforms

6.2.1.4   VAD

Vapor phase axial deposition was developed in ca. 1977 in Japan and is used in that country to this day, and through joint ventures elsewhere, too (Fig. 6.9). This technique is quite different from the ones described earlier in that the glass is formed at the end of a rod. One starts at the section of a seed rod, deposits glass, and lets the structure grow longitudinally. The refractive index profile is obtained through an elaborate geometry of burner flames and positions of the nozzles that bring in the reactants. Constant turning helps secure rotational symmetry.

Here, too, the glass is initially porous. One needs to sinter the soot into solid glass by pulling the entire rig through a suitably heated zone. On the other hand, no collapsing is required here.

The unique advantage is that the preform can be made to any length; in effect, endless. This allows to produce very long lengths of fiber in one piece.

6.2.1.5   Noncircularly Symmetric Fibers

We have seen in Sect. 4.6.2 that polarization-maintaining fibers intentionally deviate from a rotationally symmetric structure. To make such fibers, obviously some process step must be introduced that breaks the circular symmetry. Several approaches

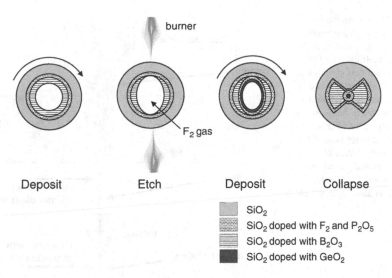

burner

F_2 gas

Deposit   Etch   Deposit   Collapse

SiO$_2$
SiO$_2$ doped with F$_2$ and P$_2$O$_5$
SiO$_2$ doped with B$_2$O$_3$
SiO$_2$ doped with GeO$_2$

Fig. 6.10 To make a nonsymmetric bowtie preform, an intermediate step is etching with heating on two opposite sides (rather than uniformly all around). All layers deposited after this step will then grow with a broken symmetry. When the fiber is pulled from the preform, the characteristic bowtie shape is obtained. After [2] with kind permission

have been explored to accomplish this, including mechanically milling a preform to generate an elliptic cross-section. It is more elegant to introduce a highly reactive gas and have it etch away some material on two opposite sides, rather than all around. As is well known, the rate of chemical reactions exponentially depends on temperature (Arrhenius factor). Figure 6.10 shows the procedure, introduced 1982 in Southampton, in the making of a bowtie fiber.

6.2.2 Pulling Fibers from the Preform

The preceding paragraphs discussed how a preform can be made. Think of a preform as a short (typically 1 m), fat (typical diameter 10–50 mm) version of fibers, complete with all internal structure. In a machine called a draw tower, the preform is heated to the temperature where glass softens and begins to melt, i.e, around 1950–2250 °C (Fig. 6.11). One can then catch a thread of glass and pull it into a fiber with a diameter of 70–250 μm, but most frequently 125 μm. In the process, the diameter is reduced some 200-fold; therefore the length increases by 40,000:1 to about 40 km. At a typical, certainly not particularly rapid, speed of advancing the preform into the heating zone of 200 μm/s, one winds up with a fiber (pun intended) at 8 m/s. At 5000 s or one and a half hour later, 1 m of preform has been spent, and 40 km of fiber has been made.

Fig. 6.11 Schematic depiction of a draw tower. The preform is heated to the onset of melting. The fiber thus formed gets coated with plastic in an extruder and is led over pulleys and onto a receiving drum. Noncontact measurement of fiber and coating diameter is used for closed-loop control of parameters such as advance speed. A test for tensile strength is also applied right here

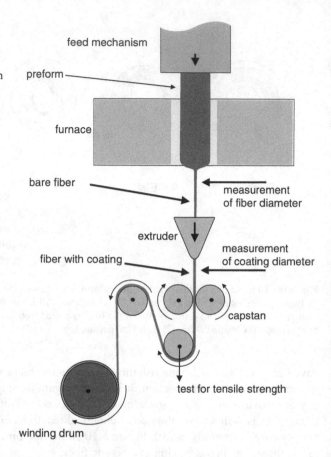

This may all sound very simple, but the technical reality is quite involved. Draw towers are two floors high. Temperature and advance speed must be maintained to the most exacting tolerance demands. Online measurements of fiber diameter and other properties are used for elaborate closed-loop control of parameters. As a result, one can maintain the fiber diameter to within $0.1\,\mu$m.

Immediately after cooling, a plastic coating is applied by way of an extruder. This is important because it protects the fibers from mechanical factors such as abrasive contact and from chemical influences by water. At the same time, it contributes to minimize micro-bending loss. Frequently the coating consists of two layers: an inner layer is soft, pliable; an outer layer, hard, abrasive-resistant. Epoxides and polyimides are used, also acrylates and silicones. Occasionally, a barrier layer is applied first to keep water out; it can consist of either amorphous carbon or metal like aluminum or gold. Before the fiber is coiled on a spool, a test for tensile strength is conducted.

In Sect. 4.7 microstructured fibers were introduced. They differ from conventional fibers in that they have air holes running the length of the fiber. To make a preform for such a fiber one starts from glass tubes which are stacked together to form a bundle. One is at liberty to replace individual tubes with massive rods, or to use tubes with various wall thicknesses. Frequently the tube at the center is replaced by a massive rod; this will end up as the massive core. The stack is then fused together in a draw tower. In most cases the result is not yet the finished preform but an intermediate product known as a cane. The cane is then stuck into the void of a wider tube. A final fusing in a draw tower then produces a fiber which is massive in the outer parts of the cladding; this provides some mechanical strength. In the center part where there was the cane there is a regular array of holes, often in a hexagonal array (compare Figs. 4.21, 4.2 and 4.23).

The technique is called *stack-and-draw*. It requires extra care to assure that during fusing of a 'holey' structure the holes do not collapse; process parameters are modified to proceed at a somewhat slower speed and at lower temperature. Also, one can pressurize the holes during the process with compressed air or inert gas.

6.3 Mechanical Properties of Fibers

6.3.1 Pristine Glass

Contrary to a widely held opinion glass is a material that can withstand quite some mechanical stress. Let us consider tensile strength. Under applied tension, there is deformation and eventually breakage.

At low stress, most materials deform elastically and stretch in proportion to tension (Hooke' law): The relative length change, the strain $\Delta l/l$, is given by

$$\frac{\Delta l}{l} = \frac{1}{E}\frac{F}{A}. \tag{6.4}$$

Here F is the applied force (measured in Newton) and A the cross-sectional area (measured in meter square). F/A is then the tension (similar to a negative pressure) and has units of $N/m^2 = Pa$ (Pascal). The proportionality constant $1/E$ contains Young's modulus of elasticity E; E is also measured in Pa.

There is a certain critical value of tension $(F/A)_{crit}$, which is called the elastic limit. Beyond this level many materials, depending on their ductility, will undergo plastic deformation. Steel wire deforms plastically, and so does copper wire: it can be stretched 20 % longer without snapping. At even higher tension, the ultimate limit is reached and the specimen is destroyed by rupturing. Glass fiber, in contrast to ductile metals, exhibits no plastic deformation but breaks immediately once the critical tension is exceeded. The following table gives some representative values of elastic properties.

The table indicates that the tensile strength of glass is much less in normal specimens than in an idealized situation. In the ideal case (absolutely pure glass without the smallest microscopic scratches in its surface), glass fibers can almost be as strong as steel. At a tension of 20 GPa over a cross-sectional area of $A = \frac{\pi}{4}(125\,\mu m)^2$, this implies a critical tension of 245 N, corresponding to a fiber suspending a weight of 25 kg. Of course, such ideal circumstances never occur in practice and so the critical strain is at a few percent. This has been studied extensively.

As may be expected, the plastic coating contributes negligibly to the overall tensile strength. The critical tension is strongly influenced by the depth of surface scratches. According to a theory by Griffith, the value stands in inverse proportion to the square root of scratch size. When in practice ca. 5 GPa is obtained, one can conclude that scratches of a few tenths of nanometers are responsible. In this context the plastic coating, applied immediately after drawing the fiber for a good reason, is all-important because it prevents scratches. It also limits the access of water. On the bottom line, the coating is important after all: not by bearing load directly, but by maintaining the initial tensile strength as intact as possible (Table 6.1).

The tension for a 1 % strain (0.7 GPa) is obtained from

$$EA \times 1\% = 8.6\,N, \tag{6.5}$$

so that it is reached at about 880 g load. Fibers are routinely tested for tensile strength right at the draw tower. Standard values are 0.35 GPa, corresponding to 0.5 % strain for fibers intended for terrestrial use, and 1.38 GPa, corresponding to 2 % strain, for fibers destined for undersea applications. It should be clear that the latter must withstand greater mechanical stress, in particular in the fiber-laying process.

After deployment, during the intended use, fibers are rarely stressed by more than one fifth of the test level; this gives a safety margin for a life expectancy of several

Table 6.1 Selected values of Young's modulus, critical tension, and critical strain

Material	E (GPa)	$(F/A)_{crit}$ (GPa)	$(\Delta l/l)_{crit}$
Various substances			
Steel	210	20	10 %
Aluminum	70	0.3	20 %
Copper	120	0.06	20 %
Glass	70	0.05	1 %
Lead	16		
Wood	10		
Optical fibers			
Ideal	70	20[4]	30 %
Real	70	2 . . . 5	3 % . . . 7 %
Coating materials			
Acrylate (inner; soft)	0.001		
Acrylate (outer; hard)	1		
Polyimide	2.5		
Polyamide (Nylon)	2	0.05	

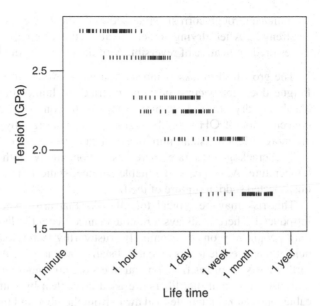

Fig. 6.12 Reliability test of optical fibers: lifetime under tensile stress. With increasing stress, life expectancy is reduced exponentially. After [5]

decades. As it is difficult to simulate long-term behavior in a short time one takes resort to statistical extrapolation (Fig. 6.12).

6.3.2 Reduction of Structural Stability

Life expectancy of fibers is dominated by all environmental changes that affect the growth of microscopic cracks. In vacuum or in a chemically inert atmosphere, fibers live longer! Unfortunately, this is of little use for technical applications. There are several ways in which glass can break.

Static fatigue. At constant tensile stress below the critical limit, some fatigue is observed that can create nasty surprises after a while.

Dynamic fatigue. With tensile stress rising linearly with time, one finds critical limits that are lower than described earlier. This phenomenon gives rise to a certain standardized measurement procedure.

Cyclic fatigue. In principle there is no material fatigue in glass as long as the temperature remains sufficiently far below the softening temperature. Therefore this process is of minor importance for fibers whereas for many other materials it is quite relevant. On the other hand, during each cycle of tension there is dynamic fatigue.

Zero stress aging. This is a case of unclear, in part contradictory evidence. If glass with a roughened surface is immersed in water at room temperature, it may even happen that the tensile strength is *increased* (30 % have been observed). In an attempt to explain, researchers have conjectured that the tip of cracks would be rounded through corrosive interaction. In most cases, however, humidity reduces strength. After drying under vacuum, the original strength can be partially restored, indicative of reversibility of the processes involved.

The growth of cracks is mostly caused by water from the environment. Static fatigue does not occur when one operates at liquid nitrogen temperature, or in absolutely dry atmosphere or in vacuum. In contrast, when there is an elevated concentration of OH ions, the cracks grow more rapidly. Different types of glass are more or less resistant; pure fused silica turns out to be the best.

The crack growth rate increases exponentially both with tension and with temperature. After years of reliable service, static fatigue can lead to an entirely unsuspected sudden rupture of the fiber.

This risk may be typical for all risks that arise when new technologies are introduced. There is always a remote chance that a hidden flaw goes undiscovered until people rely on the seemingly trustworthy technology. The risk can only be held at manageable levels by careful statistical analysis. After some early mishaps, further nasty surprises from optical fibers are no longer anticipated. In this context, it is also important that fibers are used as cables; by suitable construction of the cable, one can keep tensile load away from the fiber and thus increase reliability.

References

1. E. E. Basch (Hrsg.), *Optical-Fiber Transmission*, Howard W. Sams & Co., Indianapolis, IN (1987)
2. R. D. Birch, D. N. Payne, M. P. Varnham, *Fabrication of Polarization-maintaining Fibers Using Gas-Phase Etching*, Electronics Letters **18**, 1036 (1982)
3. D. L. Griscom, *Optical Properties and Structure of Defects in Silica Glass*, Journal of the Ceramic Society of Japan, International Edition, **99**, 899 (1991)
4. S. E. Miller, A. G. Chinoweth (Eds.), *Optical Fiber Telecommunications*, Academic Press, London (1979)
5. T. T. Wang, H. M. Zupko, *Long-Term Mechanical Behavior of Optical Fibers Coated with a UV-Curable Epoxy Acrylate*, Journal of Materials Science **13**, 2241 (1978)
6. Leaflet by Optical Museum, Ernst Abbe Foundation Jena, Germany: *Riskieren Sie einen Blick in die Geschichte der Optik!*
7. Table of Sellmeier constants in [1] p. 39. In the entry for "pure silica, annealed", the number for λ_1 is given as 0.068043; it must be 0.0684043.

Chapter 7
How to Measure Important Fiber Characteristics

It takes some special procedures to characterize an optical fiber through measurement of its relevant characteristics. These procedures, often developed along with the fibers, are presented in this chapter.

7.1 Loss

It is not trivial to measure fiber loss because the value is low, and some precautions and a very good resolution are required to obtain a meaningful value with any degree of precision. At values of a few tenths of dB/km, both resolution and accuracy should be at least a few hundredths of dB/km. Remembering that $0.01\,\mathrm{dB} = 0.23\,\%$, this means that better than one part in thousand is asked for, always a challenge for analog quantities. Here, however, there is one particular obstacle.

The naïve way to do this measurement would be to send light from some source (a lamp, say, or a laser) into the fiber with the help of some suitable focusing lens, then measure power right after and right before the fiber, and compare. However, that strategy fails because the result contains incoupling loss. This loss is mostly due to the fact that only a fraction of the incoupled light ends up in the guided mode (or modes). The rest is lost to the cladding from where it is scattered out. Additionally, there are Fresnel losses at the front and rear fiber face. With utmost care one may reduce these losses to below 10 %, but 30 % are more realistic in a typical laboratory setting. It is the *uncertainty* of this value that masks the propagation loss.

© Springer-Verlag Berlin Heidelberg 2016
F. Mitschke, *Fiber Optics*, DOI 10.1007/978-3-662-52764-1_7

This is why one does not use the power before the fiber as a point of reference, but the power shortly after the fiber input end. This requires to first measure the throughput of a very long fiber (length L preferably several kilometers), then cut it after $L_0 \approx 1$–2 m, and repeat the power measurement with the short piece. Provided that the incoupling loss did not change during the procedure, one finds the loss of the piece $L - L_0$. Fresnel losses are also cancelled out. This is known as the *cutback* technique and is the standard procedure. Still, many sources of error remain. Here are some:

- Lack of constancy of the light source.
- Lack of constancy of the incoupling.
- Lack of constancy of the detector sensitivity (either due to temperature fluctuations or by inhomogeneity of the detector surface).
- Too short L_0. Light that is not guided in the mode can travel a short distance in the cladding before it is completely scattered out. Part of it may enter the measurement.
- Macro- and micro-bending loss.

Around 1980, it became apparent that repeatable loss measurements were a necessity. Throughout the 1980s, several round-robin tests were conducted in various countries in which pieces of fiber were sent around among several institutions for loss measurement and comparison of results. In one such test in 1983/1984, 16 European laboratories in ten countries were involved. Initially, there was a spread in the results of more than 0.2 dB/km, amounting to more than 100 % of the value, even though the participants were the best labs from industry and government agencies. It took considerable effort to reach a satisfactory state of affairs. To achieve constancy of a halogen lamp, for example, it does not suffice to have a constant current run through it: It is also important to have a specific value of that current which is lamp type-dependent but often around 90 % of nominal current. At this current, the temporal change of light output is minimized. It is also important to observe the spatial orientation of the filament: For reasons of heat distribution, it makes a difference whether it hangs horizontally or vertically. If all precautions are scrupulously observed, one may achieve a constant light output within 0.1 %. Photodetectors (photodiodes) must be selected for homogeneity of sensitivity across their light-sensitive surface and must be thermostatized to within 0.1 °C.

With these and further steps, data can be taken on long fibers with a resolution of one part in thousand of 1 dB/km, and a repeatability of one part in hundred. This makes it possible to obtain data as shown in Fig. 7.1 where the Rayleigh scattering background has been subtracted out so that minute features like impurity absorption bands become visible. It turns out that such data are like fingerprints: It is possible to discern otherwise similar fibers and to identify a particular brand or type of fiber [4].

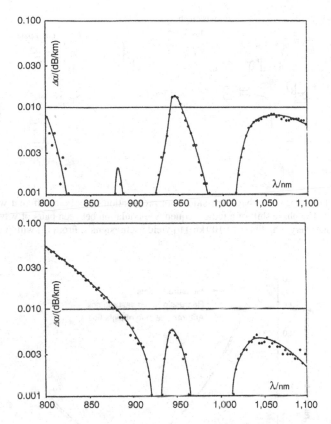

Fig. 7.1 Measurement of fiber loss. A term proportional to λ^{-4} (Rayleigh scattering) was first fitted to the data and then subtracted out; plotted are the residuals. The figure shows two very similar fibers from different manufacturers. Absorption bands due to OH groups and other impurities look quite different due to slightly different manufacturing processes. From [4] with kind permission

7.2 Dispersion

To measure fiber dispersion usually implies a measurement of the propagation time. There are two avenues one can use: One is the standard procedure in industry but requires a very long piece of fiber, like 100 km. At this length propagation time differences at different wavelengths become directly measurable: To obtain a resolution of 0.1 ps/(nm km), one needs to measure a timing differential of 10 ps/nm. With two light sources 10 nm apart, this is feasible because fast photodiodes and sampling oscilloscopes easily resolve times well below 100 ps. An example is given in Figs. 7.2 and 7.3.

A difficulty to keep in mind is that propagation time also varies due to other causes: The coefficient of thermal expansion of fiber is of the order of $10^{-5}/K$ [6, 11]; then, a minute temperature fluctuation of 0.01 °C creates a change of

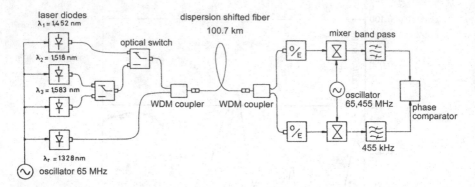

Fig. 7.2 Scheme to measure fiber dispersion from propagation time. Four different wavelengths can be selected. The phase shift of a radio frequency modulation between pairs of wavelengths is measured. It takes very long fibers (≈100 km) to yield useful signals. From [3] with permission

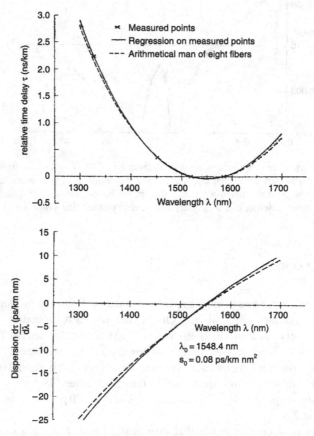

Fig. 7.3 Data taken with the setup described in Fig. 7.2. From the propagation time differentials (*top*), one finds dispersion through taking the derivative. From [3] with permission

length by 1 cm in 100 km fiber and gives rise to 50 ps change of propagation time. Therefore procedures are chosen that measure propagation time differentials between several wavelengths simultaneously. This requires several light sources and has thus higher hardware requirements. On the other hand it provides more direct, uncomplicated data acquisition that reduces manpower requirements. This is a strategy well suited to the needs of fiber manufacturers who always have access to the full-length fiber.

For the other route, it suffices to have a much shorter segment of the fiber, about 1–2 m. The fiber is inserted in one arm of an interferometer; Mach-Zehnder interferometers are the most common arrangement. The reference arm contains either some other fiber with precisely known dispersion or an air path.

Now one can tune (continuously or stepwise) the wavelength of the light source and find, at each wavelength, that path length which maximizes interference contrast. The change of this path length with wavelength leads directly to the fiber's dispersion (Figs. 7.4, 7.5, and 7.6). Alternatively, one can use broadband light (white light) and record the interferometric fringe pattern as the path length is scanned; the Fourier transform of that fringe pattern provides the phase information from which dispersion can be calculated [8]. This procedure provides the full wavelength dependence in a single step.

It is also not a trivial task to assess the refractive index profile and the core radius. The direct route is to measure refractive index in an interferometer, but since high spatial resolution is required, a setup involving a microscope is required. One expects refractive index differences of a few 10^{-3}; for a unique determination, the fiber length must therefore be a few hundred wavelengths; for visible light, this means a fiber length of no more than ca. 100 µm! We conclude that one would have to prepare a thin slice of fiber by polishing which is time-consuming and cumbersome.

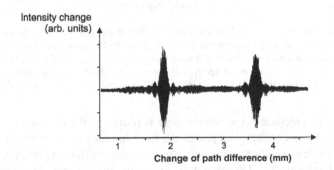

Fig. 7.4 Measurement of fiber dispersion in a Mach–Zehnder interferometer. One obtains fringe patterns (interferograms) like the one shown here. Individual fringes are spaced by one half wavelength and are thus too narrow to be resolved on the scale of the figure. In the case shown, a polarization-maintaining fiber was studied; due to its considerable birefringence, there are two clearly distinct groups of fringes

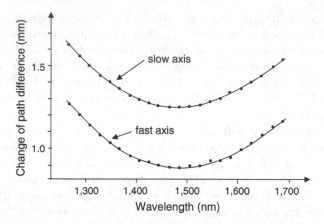

Fig. 7.5 Path differences obtained from the interferogram of Fig. 7.4, shown for both axes of the birefringent fiber. From this, the propagation time differences are obtained by division with c

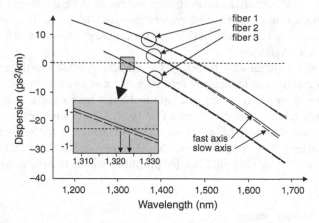

Fig. 7.6 Dispersion values β_2 obtained from propagation times as in Fig. 7.5. Shown are results for three different polarization-maintaining fibers; data for the fast axis (*solid*) and the slow axis (*dashed*) are nearly on top of each other on this scale. For fiber specimen 3, the range near the zero-dispersion wavelength is shown magnified (*inset*): on that scale both curves are clearly separated. Zero-dispersion wavelength for the fast axis here is 1324 nm and for the slow axis 1321 nm

There are also methods in which the fiber is illuminated from the side. One first removes the plastic coating, then places the fiber in index-matching gel and shines light through sideways. On a screen one captures a pattern that, in principle, contains the required information. Unfortunately the evaluation is cumbersome again (it requires integral equations) and error-prone. Similarly, one can place the fiber with transverse illumination into an arm of an interferometer. Again one can obtain the information in principle, but only after quite involved evaluation.

All told, it is a lot easier to assess the internal structure of the fiber before drawing it, i.e., from the preform. Precision is much improved because fine detail of the index profile can be seen clearly while in the finished fiber the same dimensions may be

Fig. 7.7 Core profile of a step index fiber, measured from the final fiber. Due to diffraction effects this can show no detail finer than about one wavelength. From [2] with kind permission

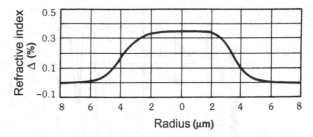

Fig. 7.8 Core profile of a triangular fiber, measured from the preform. The preform is much bigger, so much finer detail becomes visible. From [9] with kind permission

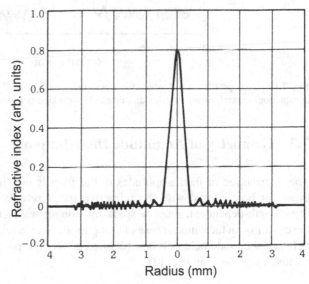

Fig. 7.9 Core profile of a typical gradient index fiber, measured from the preform. Note the central refractive index dip

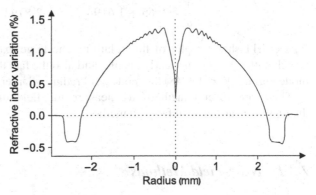

obscured by diffraction once they are smaller than one wavelength and thus below the resolution limit (see Fig. 7.7). Details like a nonperfect index step or a central index dip (see Sect. 6.2.1) can be seen much better in the preform (see Figs. 7.8, 7.9, and 7.10).

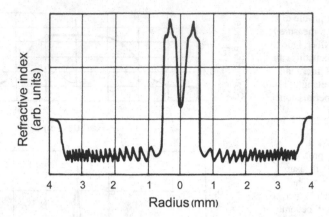

Fig. 7.10 Core profile of a double-clad fiber, measured from the preform. Again there is a conspicuous central refractive index dip. From [1] with kind permission

7.3   Geometry of Amplitude Distribution

The distribution of field amplitudes in the fiber is not limited to the core. It also must not be confused with the refractive index profile. The field distribution is wavelength-dependent; roughly speaking, longer wavelengths extend farther into the cladding. In the simplest case of a step index fiber, a relation has been formulated between the mode field radius w (defined as the $1/e$ point of amplitude), the core radius a, and the V number [7]:

$$\frac{w}{a} = 0.65 + 1.619\,V^{-3/2} + 2.879\,V^{-6} . \tag{7.1}$$

Figure 7.11 shows a plot of this relation. The large extent of the mode at long wavelengths ($V \to 0$) is clearly visible, and also the fact that throughout the single-mode regime ($V < 2.4048$) the mode field radius is larger than the core radius.

There are several methods to measure the field distribution, and one can distinguish near-field and far-field methods.

7.3.1   Near-Field Methods

To correctly identify the field distribution of the guiding mode it is essential that only light from that mode emerges from the fiber end, and that cladding light has died down. Steps to reduce cladding light may therefore be important.

Near-field methods create an image of the mode field distribution in the plane of the fiber face directly. If you now think that you only need to look at the fiber end with a microscope, consider this: We need to distinguish between conventional (or

Fig. 7.11 Plot of Eq. (7.1).
The mode field radius w is
normalized to the core radius
a and is plotted as a function
of the V number. Throughout
the single-mode regime
$V < 2.4048$, $w > a$ holds

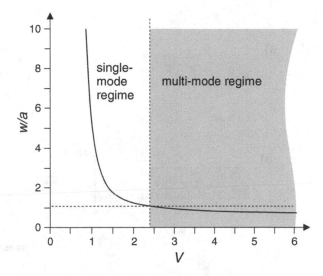

far-field) microscopes and near-field microscopes. Far-field microscopes catch the
light diffracted out from the object and transform it back to form an image. They are
subject to Abbe's theory of diffraction and do not resolve detail much smaller than
a wavelength. This limits the usefulness of the measurement. There are aggravating
facts like aberrations and other errors in the imaging; for example, the scale factor is
normally affected by the exact position of the focal plane but enters the final result
proportionally so that any uncertainty ends up in the result. Far-field microscope
techniques are therefore generally considered not very precise.

A near-field microscope works very differently. It exploits the fact that a very
small aperture, possibly much smaller than the wavelength, still transmits light,
if with strong attenuation. On the other hand, this small aperture allows to map
out an intensity pattern when it is scanned in the plane perpendicular to the
light propagation direction. Then the resolution is not limited by wavelength,
but basically by the mechanical resolution of the scanning fixture. Near-field
microscopes with atomic resolution have been built.

In practice, one uses a second fiber as a probe. The probe is scanned across
the fiber tip in very close proximity (less than $10\,\mu\mathrm{m}$) to map out the power
distribution (Fig. 7.12). This is why this is called the transverse offset method. In
the limiting case that the probe fiber has a much smaller mode field diameter than
the fiber under test, it should be clear that the desired mode structure is obtained
directly. Unfortunately this case is unlikely; more typically, both mode fields are of
comparable size. Then one obtains the convolution of both distributions; from this
the desired shape can be calculated only if the other is well known.

For typical single-mode fibers and at V numbers not too far from $V = 2.4048$,
the intensity distribution is somewhat similar to a Gaussian:

$$I(w) = I_0 \exp(-2(w/w_0)^2). \tag{7.2}$$

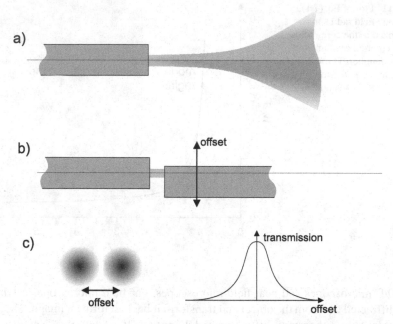

Fig. 7.12 Schematic representation of the transverse offset method to measure mode field diameter. (**a**) The fiber's exit cone (and the acceptance cone likewise) does not appreciably change its diameter over the first few micrometers. Therefore one can bring two fiber tips closely together. (**b**) Then one can measure how much power is coupled from one fiber to the other, while the transverse offset is scanned. (**c**) From mapping out the transmission as a function of position, one can draw conclusions about the mode field diameter

The mode field radius is taken, by convention, as that radius w at which the field *amplitude* is down to $1/e \approx 37\%$ of the central maximum. At this point, the intensity is down by $1/e^2 \approx 13.5\%$ of the maximum. (In old literature sometimes other definitions are found, this can create much confusion.)

The convolution of two Gaussians is a Gaussian again, which makes the Gaussian approximation convenient. The situation is particularly clear when both fibers have the same mode profile, because they are pieces of the same fiber. Then the convolution has the $\sqrt{2}$-fold radius of the mode profile of each fiber individually (see Chap. 18), and w_0 is obtained by reading the radius where the intensity is down to $1/e$ of the maximum.

The longitudinal separation of the fibers must be small enough that only the near field, not the divergent part of the exit cone is measured because otherwise one would find systematically too large values.

7.3.2   Far-Field Methods

In a very different approach, one allows light to exit from the fiber and propagate in free space until the far field is reached. This is the case when the distance z is at least

$$\frac{z}{\lambda} \gg \left(\frac{w}{\lambda}\right)^2,$$

where w is the mode-field radius (which one tries to determine, but has reasonable guesses about). This condition is easily met already after a few millimeters; in practice, one would prefer a couple of centimeters. At this distance, one can observe the far field, e.g., on a screen. If the distance were increased even more, the pattern on the screen would not change: It would just scale linearly in diameter. It is therefore appropriate to measure positions in the pattern as angles from the fiber tip. The aperture angle of the exit cone is read at the intensity $1/e^2 = -8.69\,\mathrm{dB}$ referred to the on-axis maximum.

Instead of a screen, one may use a photographic plate or an electronic camera. It is more common, though, to move a single photodetector on segments of circles around the fiber tip as shown in Fig. 7.13 and map out the far-field pattern this way. At large angles with the axis there is only weak intensity, and it is of utmost importance to safeguard against stray light.

If one again applies the Gaussian approximation mentioned in the preceding paragraph, one obtains w_0 from the condition

$$w_0 = \frac{\lambda}{\pi\theta}. \tag{7.3}$$

It is much more precise, though, not to make any approximations of that kind. The full information about the field distribution in the fiber is contained in the far-field distribution because the laws of diffraction are unique and they are known. A full measurement of the far-field amplitude distribution everywhere on the screen would yield the mode profile unambiguously. (If it were guaranteed that the fiber is

Fig. 7.13 Setup for a far-field measurement: in sufficient distance from the fiber, a detector is moved on a circular path (or over a spherical surface) centered in the fiber tip. The intensity is recorded as a function of angular position

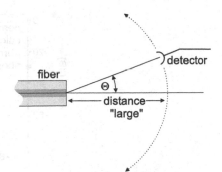

circularly symmetric, it would suffice to measure on a diameter instead of the full area.) One would have to apply a Hankel transform, which is a Fourier transform in cylindrical coordinates, using Bessel functions in place of sin and cos.

But there is a catch: Unfortunately one never measures an amplitude distribution, only an intensity distribution (Fig. 7.14). To make matters worse, one also does not measure the entire distribution in a 2π solid angle because that is difficult to do both geometrically and due to the strongly attenuated intensity at large angles from the axis. Both stray light that finds its way to the detector and the detector's own noise easily swamp data at large angles.

At large angles to the axis, the intensity goes down so rapidly that cameras cannot easily cope with the required dynamic range. It should be clear that excellent linearity is required throughout the dynamic range. This can certainly not be achieved with conventional photographic means, but electronic cameras are challenged, too. With a good photodiode and possibly with lock-in technology for the weak signals, it is easier to obtain a good dynamic range, limited only by stray light. With due care, 60 dB can be obtained, but even this usually means that at angles larger than 30° there is no useful signal.

But back to the amplitudes, rather than intensities: It does not really help to take the square root of all measured intensities because the field amplitude may have zeroes, i.e., nodal lines at which the sign of the amplitude changes. They give rise to dips, or notches, in the far-field profile at certain angles. For conventional step index fibers, the first dip typically occurs at 0.2 rad so that it can be seen only when the dynamic range exceeds 50 dB. The only option is to apply the sign change by

Fig. 7.14 Determination of the mode profile from the far field. *Top*: Measurement of far-field intensity as a function of angle shows obvious deviations from a Gaussian at large angles. *Bottom*: The result of a Hankel transform of the far field is the near field, which is the mode profile in the fiber

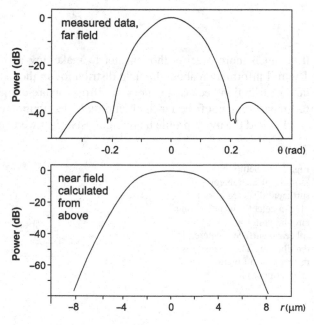

hand, but such manipulation should only be performed with utmost care and critical inspection. Sometimes what looks like a null is really only an unresolved minimum. It is precisely the far-field information at large angles that contributes most to the fine structure of the near-field result. As Abbe's diffraction theory asserts, it is the large angle information that carries the high spatial frequency content and is thus responsible for the "sharpness" of the reconstructed near field.

7.4   Cutoff Wavelength

If one determines the mode-field radius as described in the preceding section and repeats the procedure for several different wavelengths, one expects to find a trend as shown in Fig. 7.15. There is a characteristic step at the cutoff wavelength because the higher-order mode has a wider field distribution. If one takes this approach to measure the cutoff wavelength, one should observe a few subtle points:

We pointed out in the context of bend loss in Sect. 5.2 that the theoretical cutoff value at

$$\lambda_{\text{cutoff}} = 2\pi a \text{NA} / 2.4048 \qquad (7.4)$$

is only found in fibers that are stretched out straight and infinitely long. A definition better adapted to practical requirements therefore identifies the cutoff as that wavelength where the loss for the LP_{11} mode exceeds the loss of the fundamental mode by 20 dB. Strictly speaking, one would have to measure the modes individually to apply this criterion.

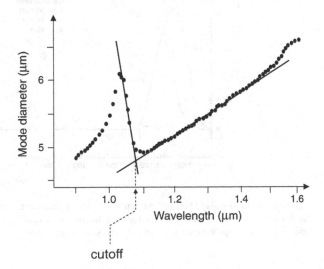

Fig. 7.15 The mode-field diameter displays a characteristic step at the cutoff

For practical use, it is helpful to think this through a little more. For short fibers, the higher-order mode shows up already at longer wavelengths where it is not an allowed mode, but its loss has not diverged yet. On the other hand, bends move the effective cutoff toward shorter wavelengths. If one judiciously selects both fiber length and bend radius, the opposing trends more or less cancel each other out, and one approaches the ideal situation. There is the standard procedure to use a fiber of 2 m length, bent to a loop of 28 cm diameter. Then the cutoff is read from the intersection of the asymptotes as shown in Fig. 7.15.

An alternative procedure is a little less involved. One measures the transmitted power as a function of wavelength and repeats with different bend radii. Changing the bend radius shifts the loss mostly for the higher-order mode (see Fig. 5.3). From the ratio of spectral transmission with and without bend, one can read the cutoff wavelength (Fig. 7.16). The standard procedure is to identify that wavelength at which the transmission differs by 0.1 dB from that in the plateau above the cutoff.

Both this and the previous method occasionally suffer from a special complication. Sometimes the characteristic step is not as clear as shown here; instead, right in the relevant range there are oscillations in the curve so that a clear reading is not possible. This is caused by the so-called whispering galley modes. These are modes that can propagate in a curved fiber in the cladding; this involves reflections at the outside surface of the fiber. To safeguard against them, in effect one removes the outside surface by stripping the plastic coating and placing the bare fiber in index-matching gel. When the indices are indeed well matched, cladding light will exit from the fiber after a very short distance and the problem is solved [10].

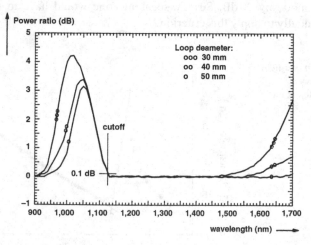

Fig. 7.16 Bend loss also shows a characteristic step at the cutoff, because higher-order modes are much more sensitive to bending. This allows to find the cutoff. Here the additional loss arising from tight fiber loops was used. From [5] with kind permission

7.5 Optical Time Domain Reflectometry (OTDR)

Fiber technology has given rise to a special tool that can be used to easily assess many properties of fibers, both in the lab and in the field. It is called *optical time domain reflectometry* or OTDR (Fig. 7.17). It is very similar in spirit to radar: A signal is launched into the fiber; whatever light is reflected or scattered back is collected and evaluated. Pulsed laser diodes are employed as light sources and photodiodes to detect the backscattered light.

The time until an echo is registered is calculated from

$$\tau_{echo} = 2nL/c, \qquad (7.5)$$

where n is the effective index for the mode and L is the length. The factor of 2 arises because light must travel forth and back before it is registered. Echo strength provides information about the type of condition that causes the echo: Rayleigh scattering gives a continuous background that gently goes down with increasing distance; localized conditions like fiber joints or breakage give sharp peaks.

There is always a certain crosstalk of transmitter light to the receiver, so that the receiver is overloaded for a short initial moment. This creates a dead zone in the short range. However, some devices are constructed to minimize the dead zone and measure even on the shortest distances (millimeters, in some cases even less).

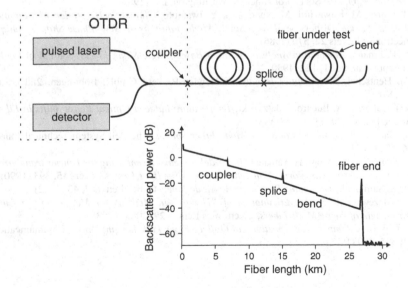

Fig. 7.17 Optical time domain reflectometry (OTDR). *Top*: Setup. A light pulse is launched into the fiber under test; the reflected light is recorded as a function of time. Time can be converted to position in the fiber. *Bottom*: The obtained data, shown here schematically, provide information about various fiber conditions

OTDR equipment is offered by several manufacturers and allows to assess a fiber over many kilometers with access only to one end. This makes OTDR a valuable tool for a wide range of tasks, notably to analyze

- fiber loss and its spatial allocation;
- loss at fiber joints like connectors or splices;
- loss at other localized conditions, e.g., sharp bends or damage;
- the location of each of these conditions;
- fiber length; and
- fiber end reflection.

In commercial installations OTDR devices are therefore indispensable in spite of their cost. Some manufacturers offer plug-in cards for computers with complete OTDR hardware; this reduces the cost because the computer does both the number crunching and the displaying.

References

1. W. T. Anderson et al., *Thermally Induced Refractive-Index Changes in a Single-Mode Optical-Fiber Preform*, in *Proc. Optical Fiber Conference*, Optical Society of America, Washington, DC (1984)
2. G. E. Berkey, *Single-Mode Fibers by the OVD Process*, in *Proceedings of the Optical Fiber Conference*, Optical Society of America, Washington, DC (1982)
3. M. Fujise, M. Kuwazuru, M. Nunokawa, Y. Iwamoto, *Chromatic Dispersion Measurement Over a 100 km Dispersion-Shifted Single-Mode Fibre by a New Phase-Shift Technique*, Electronics Letters **22**, 570 (1986)
4. W. Heitmann, *Attenuation Analysis of Silica-Based Single-Mode Fibers*, Journal of Optical Communications **11**, 122 (1990)
5. Ch. Hentschel, *Fiber Optics Handbook*, Hewlett-Packard GmbH, Böblingen, 2nd edition (1988)
6. N. Lagakos, J. A. Bucaro, J. Jatzynski, *Temperature-Induced Optical Phase Shifts in Fibers*, Applied Optics **20**, 2305 (1981)
7. D. Marcuse, *Loss Analysis of Single-Mode Fiber Splices*, The Bell System Technical Journal **56**, 703 (1977)
8. K. Naganuma, K. Mogi, H. Yamada, *Group-delay measurement using the Fourier transform of an interferometric cross correlation generated by white light*, Optics Letters **15**, 393 (1990)
9. M. A. Saifi et al., *Triangular-Profile Single-Mode Fiber*, Optics Letters **7**, 43 (1982)
10. F. Wilczewski, F. Krahn, *Elimination of "Humps" in Methods for Measuring the Cutoff Wavelength of Single Mode Fibers*, Electronics Letters **29**, 2063 (1993)
11. S. J. Wilson, *Temperature Sensitivity of Optical Fiber Path Length*, Optics Communications **71**, 345–350 (1989)

Chapter 8
Components for Fiber Technology

The best car would be good for nothing if there were no streets and no gasoline. Any technology relies on an interplay of various components. Therefore, optical fiber does not do anything useful without additional components and supporting technologies. In this chapter we introduce that "periphery."

8.1 Cable Structure

Optical fiber cables are in use for telephone data since 1980. Initially multimode fibers were used in cables of 60–144 individual fibers. At the operating wavelength of 825 nm, loss amounted to 3–3.5 db/km; therefore every 6 km an in-line amplifier or *repeater* was required. Data were transmitted at a rate of 45 Mb/s.[1] One year later, the first operation in the second window near 1300 nm was started. Initially cables for this wavelength had half as many fibers. Losses were lower, around 1 dB/km, and thus repeaters could be placed every 18 km. Data rates were 90 Mb/s. All these cables were buried in existing conduits.

Beginning in 1983, single-mode fibers were used and are now unrivalled for medium and long distances. Multimode fibers are still in use in short-range links (*local area networks* or LANs) connecting computers on-premises or within the same building. The first generation of single-mode fiber technology operated at 1310 nm, had losses around 0.5 dB/km, required repeater distances of 30 km, and could transmit 400–600 Mb/s.

The fiber count in these cables was around 20–30. The cables were no longer placed in existing ducts, because these did not provide sufficient protection from lightning flashes and from rodents.

---

[1]Date transmission rates are measured in bits per second. Mb/s stands for megabits per second.

© Springer-Verlag Berlin Heidelberg 2016
F. Mitschke, *Fiber Optics*, DOI 10.1007/978-3-662-52764-1_8

The USA has the largest domestic telecommunications market worldwide. In this market there was a profound change in 1983 which we must mention here. Before, American Telephone and Telegraph, or AT&T, had an absolutely dominant market position. In 1983 courts passed a landmark decision referred to as *divestiture*, which forced AT&T to give competitors more access. In effect the company was split into a central segment and several regional operating companies. Right after divestiture there were not as many cables as telephone service providers so that sometimes the same fiber in the same cable was used in time-sharing agreements by several competitors. Maybe that is why cables with 96 fibers were then laid.

A couple of years later, loss of 0.4 dB/km, repeater distances of 40 km, and data rates of 2 Gb/s became routine. This corresponds to 1,500,000 simultaneous telephone calls. See Chap. 11 for methods to put many calls onto the same fiber without mutual interference and Sect. 11.4 for further development.

When a cable incorporating optical fibers is manufactured, there are a couple of things to observe. Fibers must be protected from adverse environmental influences. In the interest of a long lifetime of the cable, fibers must not experience tensile load even while the cable is bent and pulled. Also, both macro- and micro-bend losses must be avoided in the deployed fiber. Several cable designs are in use to meet these objectives; Fig. 8.1 shows examples. There is always a strength member to take care of the tensile load; it may me made of fiberglass, Kevlar fiber, or steel wire. (Fiberglass is what the poles for pole vault are often made of; Kevlar is the fiber used for bulletproof vests.) Typically, fibers are individually placed in tiny tubes where they have some slack and can accommodate some extra length. If the cable is then pulled, the stress is kept away from the fibers. The tubes are filled with a gel which prevents the intrusion of water; it also damps vibrations and movement of the fiber. Sometimes a group of fibers sits in a common, slightly larger tube, again filled with gel. There are also "ribbon" constructions where several fibers are connected in a flat side-by-side structure similar to an electric flat ribbon cable. Ribbons allow to make connections of several fibers efficiently by automated machinery. All fibers

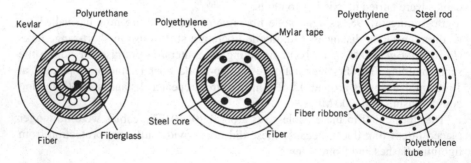

Fig. 8.1 Schematic cross-section of different cable types. *Left*: A single fiber sits loosely in a structure which is stabilized by fiberglass and Kevlar. *Center*: Several fibers are placed around a central steel wire acting as strength member. *Right*: Several fibers are combined into ribbons. Shown is a cable with several such ribbons; the structure is stabilized by steel wires. From [1]

in a ribbon can be spliced to another ribbon in one go, rather than handling each fiber individually.

There are several options for laying the cables. On long distances, they are dug into the ground, and in cities they are placed in ducts. In some countries including the USA, the cheap method is preferred in rural and suburban areas: the cables are suspended from utility poles. This, of course, is susceptible to interruptions.

The most frequent sources of damage are by humans (digging, vandalism) and natural causes such as lightning strokes and—down to 2 m below ground—rodents. In the USA, damage by gunshot occurs. Sometimes deployed fibers are subject to temperature extremes: For suspended fibers on poles, one calculates with $-25\,°C$ to $+65\,°C$ for most of the continental USA; in some areas, one has to design for $-40\,°C$ to $+75\,°C$. In the ground this range is limited to $0\,°C$ to $+30\,°C$. In this one respect, undersea cables are in a most benign environment: On the sea floor the temperature is quite constant around $10\,°C$.

8.2   Preparation of Fiber Ends

Before fibers can be used for anything at all, first the fiber end faces must be prepared (Fig. 8.2). It is mandatory that the end face, after the fiber has been cut or cleaved, is perfectly smooth and of optical quality. This is not possible by bending the fiber till it breaks, or by cutting it with scissors. The simplest way for controlled fracture is to scratch the fiber surface manually with a diamond, a tungsten carbide blade, or some other extremely hard material, and then to apply mechanical tension. With some routine one can obtain reasonably good surfaces most of the time: The reliability falls short of $100\,\%$ but in a pinch may be acceptable, but it is a good idea to check the fiber end with a microscope.

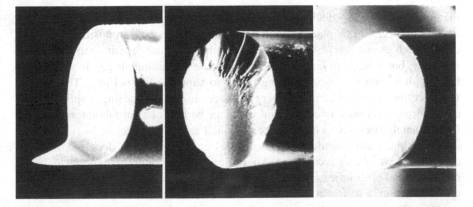

Fig. 8.2   Fiber end faces. *Left*: Here an edge remains. *Center*: An irregular surface called a hackle zone. Either is a sign of a bad preparation. *Right*: A good preparation results in a face smooth as a mirror

It is much better to use specialized equipment; the cost lies anywhere between a few hundred and several thousand euros or dollars. Fiber-breaking devices apply a well-defined longitudinal tension to the fiber while scoring it with a blade which may vibrate at ultrasonic frequency. This results in end faces which are perpendicular to the fiber axis within close tolerances and are smooth every time.

When fibers are inserted in connectors, it is important that the front face is in the same plane as the connector front. If the fiber sticks out, it will suffer from damage; if it is recessed, there will be no good match to the other fiber. One cannot obtain the cut in the exact position with the gear just described. Instead, one inserts the fiber so that it sticks out a bit, then polishes it down on special polishing pads with very fine abrasive until it fits exactly. A problem can be that the grinding and polishing exerts shear forces on the glass so that, in a thin layer just beneath the surface, the glass structure may be modified. Local changes of the refractive index to $n = 1.6$ have been observed [5]; in such cases there will be extra losses. By using a judiciously chosen sequence of initially coarse, then progressively finer abrasives one can mitigate or even eliminate the problem. There are commercial fiber-polishing machines, which can even prepare several connectors simultaneously.

8.3 Connections

Connections between two fibers can be of either one of two basic types: permanent and nonpermanent.

8.3.1 Nonpermanent Connections

Fixtures are available, which have a V-shaped groove in an otherwise smooth metal surface. A fiber can be placed in the groove where it is held in position by some clamp. Such groove can be used to bring two fibers in close proximity to each other manually, but it helps to have a steady hand. The remaining air gap is sometimes filled with a drop of index-matching liquid to suppress Fresnel loss. This way a viable connection between two fibers is made; it is called a finger splice. Such connections are easily opened again and can be useful in a laboratory setting. Unfortunately, they have a loss between one half and one decibel.

When fibers are installed for a technical application, one does not want to deal with such finicky techniques. There are various connector types which are reminiscent of electronic connectors and almost as trouble-free. They are the result of a development which first had to deal with issues of geometric tolerances. To maintain the required precision even after multiple cycles of opening and closing, the connection was a challenge initially, in particular for single-mode fibers with their extremely small mode-field radii.

Fig. 8.3 A typical fiber connector. At the center of the ferrule, one can see the fiber either as a *dark* or a *bright* spot, depending on lighting conditions

Today one can purchase such connectors for a few euros/dollars from a variety of vendors. Several connector styles are common (Fig. 8.3). Coupling loss can result from a variety of causes:

1. Both fibers have different mode field shape and diameters.
2. Between both fibers a distance (air gap) remains.
3. Both fibers are positioned with a transverse offset.
4. Both fibers are positioned with an angular offset.
5. There are surface (Fresnel) reflections.

Losses due to these factors were studied in [7]; Fig. 8.4 shows the result. It should be clear that quite close tolerances must be maintained. If the fibers to be connected are a given, the loss from (1) is unavoidable, while the loss from (2)–(4) arises from lack of precision in the connection and can be minimized.

In case of actual physical contact of both fibers the contribution from (5) would vanish, but such contact is problematic because abrasion might damage the fibers in repeated operation. Therefore Fresnel losses are usually accepted. The reflection at an interface between a medium with index n_1 and a medium with index n_2 for perpendicular incidence is given by

$$r = \frac{n_1 - n_2}{n_1 + n_2},$$

$$R = \left(\frac{n_1 - n_2}{n_1 + n_2}\right)^2,$$

where r is the reflectivity for the field amplitude, i.e., the reflected amplitude normalized to that of the incident wave. $R = r^2$ is the reflected power fraction. For fused silica in the visible and near infrared with $n \approx 1.46$, one finds $r = 0.19$ and $R = 3.5\%$. In a connection between two fibers not in physical contact, we consider two such interfaces: fiber–air–fiber. Naïvely one may expect twice the loss from an individual air–glass interface or 7 %. Unfortunately the situation is slightly more complicated than that.

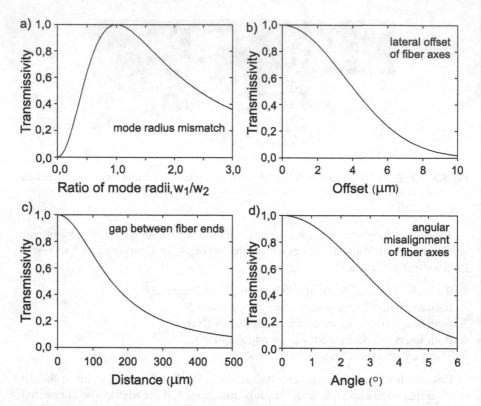

Fig. 8.4 Theoretical coupling loss between two fibers, after [7]. Shown is the expected transmissivity (Fresnel loss not considered) if (**a**) there are unequal mode-field radii, (**b**) there is transverse offset, (**c**) there is a gap, and (**d**) there is an angular misalignment. A mode-field radius of $a = 5\,\mu\text{m}$, a cladding refractive index $n_M = 1.46$, and a wavelength $\lambda = 1.5\,\mu\text{m}$ are assumed

In the case of coherent light the loss may be more or less than 7 % because both reflections may add in phase or in opposite phase. Both reflecting surfaces are nearly parallel, and light can bounce back and forth between them. Depending on the gap width-to-wavelength ratio, a resonance condition may been fulfilled (round trip path equals integer multiple of wavelength). The total reflection can vary accordingly between zero and four times the individual reflection or 14 %. In effect, one has a Fabry–Perot interferometer (see Fig. 8.5). If the light is not perfectly coherent and the gap is wider than the coherence length, resonances are washed out and eventually the naïvely expected value is approached. The coherence length of laser light by far exceeds all reasonable gap widths, and interference needs to be fully taken into account. LEDs have limited coherence length, and only a few resonances occur. White light would avoid resonances but is not what one usually deals with.

If two polarization-maintaining fibers are to be joined, there is the additional requirement that the orientation of the birefringent axes must match (see Sect. 4.6.2).

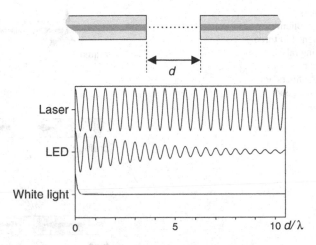

Fig. 8.5 Depending on the degree of coherence of the light, there can be more or less obvious Fabry–Perot resonances in the coupling efficiency as the gap width between fibers is varied. The coherence length of laser light always exceeds the gap width. In the case of luminescent diodes (LEDs), the coherence length is often just a couple of wavelengths; the resonances then quickly decay as the gap width is increased. For white light, e.g., from a tungsten filament light bulb, the coherence length is on the order of one central wavelength, and no oscillations of the coupling efficiency are observed. If the fibers are brought into physical contact (gap width zero), Fresnel loss vanishes altogether

There are dedicated versions of connectors which have a special locking pin so that they always lock at the desired angular orientation and cannot rotate.

8.3.2 Permanent Connections

Permanent connections are known as splices; the expression comes from sailor's language where it denotes a way to join two ropes by unravelling the strands, then twisting them together. Fiber splices can be made either by gluing or by fusing. Gluing is a low cost technique; fusion is more durable and has lower and more reproducible loss.

For gluing, both fibers are inserted in some tight guiding tube, which provides some centering of the fibers with respect to each other. One can manually move the fibers somewhat and can try to find the optimum position of lowest loss.

The tube is filled with a transparent fluid adhesive which cures under ultraviolet light. As soon as the desired position is found, one turns on an ultraviolet lamp and hopes that the positions are kept until the adhesive sets. Loss of 0.3 dB can be obtained with some routine, and with luck, even better than that.

The professional procedure is to fuse the fibers. This involves heating the glass until it softens. As heat sources various options have been tried, including

Fig. 8.6 Schematic
representation of splicing:
Fibers are positioned in three
axes. A premelting (also
called prefusing) cleans the
fiber tips, then the fibers are
fused. Afterward, a good
splice is nearly invisible

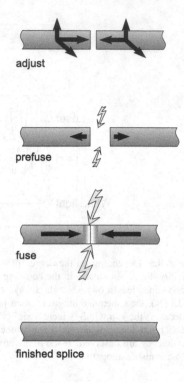

microscopic gas flames. However, it is now standard to use an electric arc; it has the
advantage of being easily controlled by a computer.

Figure 8.6 shows how the splicing procedure goes about. Both fibers are
positioned and moved closely together. Then during the so-called premelting a very
weak arc discharge, not hot enough to soften the glass, is applied, often with a slight
increase of the gap width. Premelting serves to remove possible dirt from the fiber
tips. Next is the fusing process proper: Microprocessors control the precise amount
of discharge current and arc duration to obtain the best possible result. While the arc
is on, the fibers are advanced toward each other, actually beyond the zero position
so that they are slightly pushed into each other.

The optical loss in a splice can be discussed in close analogy to that of a
connector [7] (see Fig. 8.4); of course, there is no air gap. Transverse offset is also
not a major problem because when the fiber tips are molten, surface tension moves
the fibers into that position where their outsides connect smoothly. As long as the
cores are centered well in the fiber, this automatically means a minimal transverse
offset. Fibers usually are well-centered these days.

When two fibers with the same mode profile, i.e., fibers of the same type, are
joined, one can obtain losses well below 0.1 dB and with the fanciest fusion splicers
down to 0.02 dB. As soon as dissimilar fibers are joined, the mode mismatch creates
an additional loss. For multimode fibers, the situation is more complicated because
the mode partition is modified; for detail see [9].

8.4 Elements for Spectral Manipulation

8.4.1 Fabry–Perot Filters

Selective filters can be produced in fiber technology [8]. Figure 8.7 shows an all-fiber Fabry-Perot interferometer which uses partially mirrored end faces, with a gap of width d in between. Transmission peaks occur when the round trip path length $2d$ (or $2nd$ if the gap is filled with a medium with $n \neq 1$) equals an integer multiple of the wavelength. This translates to resonance frequencies at all integer multiples of $\nu = c/(2nd)$: A Fabry-Perot filter is a comb filter. For example, at $d = 15\,\mu m$ and $n = 1$, resonances occur at all multiples of $10\,THz$. Tuning is accomplished

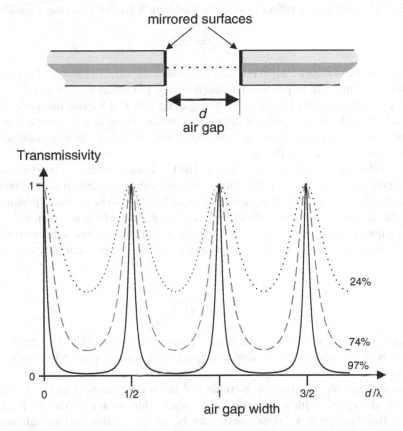

Fig. 8.7 Two fibers have their end faces coated with a partially reflecting layer to give a reflectivity R and are then combined into a Fabry–Perot interferometer. Its transmissivity is shown here for three selected values of R. The curves are valid for a very small gap; if the gap is wider than a couple of wavelengths, additional loss arises from the widening of the light exit cone and the beginning curvature of the wavefronts. Also, short coherence length light will wash out the fringes. Compare with Fig. 8.5 where $R \approx 0.035$ was assumed

through tiny adjustment of d by means of a piezoceramic transducer. The mirror reflectivity determines the sharpness of the resonances. For the extra loss due to beam divergence consult (Fig. 8.4c).

8.4.2 Fiber–Bragg Structures

A very different type of in-fiber filters is increasingly used: so-called fiber Bragg gratings. The underlying idea stems from the observation that a germanium-doped fiber core can suffer lasting changes of its refractive index after irradiation with ultraviolet light. This effect can be used to write a periodic variation of the refractive index into the fiber core. The modification is permanent; it is known as a Bragg grating. The grating will reflect light at the wavelength given by the Bragg condition

$$\lambda_{\text{Bragg}} = 2n\Lambda$$

where Λ is the grating constant (i.e. its pitch) and n the effective index of the glass.

Bragg gratings can be produced in several ways. For direct writing, one focusses UV light to the intended position of first grating line and expose for some time, then advances the writing position stepwise in increments of Λ and repeats, until all grating lines are written. This is a slow procedure, and it is extremely challenging in terms of mechanical accuracy.

More elegant is interferometric writing: The UV laser beam is split into two equal parts; both are then steered so that they cross each other at a certain small angle θ. In the overlap area a standing wave is generated. The fiber to be treated is positioned such that its core is exposed to this standing wave pattern which will then write the grating lines (see Fig. 8.8). The spatial period of the standing wave is determined by the UV wavelength λ_{UV} and the crossing angle; with a little geometry one can show that

$$\Lambda = \frac{\lambda_{\text{UV}}}{2\sin(\theta/2)} \quad .$$

It is therefore quite easy to tune the grating period to a desired value by varying the angle (most UV lasers have fixed wavelength). Still, some precision in positioning is required.

Alternately, one can irradiate a single UV beam perpendicular to the fiber, but place a glass plate with a periodic index modulation on top of the fiber. Light transmitted through this 'phase mask' can be diffracted into various diffraction orders, but one can design the phase mask structure such that the diffraction orders $+1$ and -1 dominate. The corresponding beams overlap in space and create a standing wave similar to the situation described above. For a given phase mask, Λ is not tunable. However, this sacrifice of flexibility implies a gain in reproducibility,

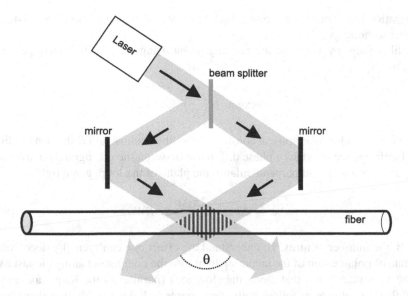

Fig. 8.8 Sketch of a setup to write fiber Bragg gratings interferometrically. Intersecting beams from a UV laser create a standing wave pattern in the fiber

so that the mask technique is easy to apply and suitable for mass production. Fiber Bragg gratings can be purchased off-the-shelf as commercial products.

Beyond controlling the center wavelength of the filter characteristics, also the resonance width can be tailored. Depending on the total length of the grating (which may range from millimeters to a few centimeters) one can obtain narrower or wider filter curves, with reflectivities at the center wavelength very close to 100 %. This is why such Bragg filters can be used as selective end mirrors in fiber lasers (see Sect. 9.7.2). A further development are 'chirped' gratings which have a sliding grating period; they find use as band filters.

8.5   Elements for Polarization Manipulation

8.5.1   Polarization Adjusters

It is well known that a given state of polarization can be translated into some other state of the same degree of polarization by inserting a suitable retardation plate (birefringent plate) into the beam. The most common plates are half-wave plates ($\lambda/2$ plates) with a retardation of one half wavelength which allow, e.g., to rotate the plane of polarization of a linearly polarized light beam by any angle, and quarter-wave plates ($\lambda/4$ plates) which can transform, e.g., linear polarization into circular polarization or the other way round. Such birefringent elements can also be built in all-fiber technology.

A fiber loop, by virtue of the bending, is birefringent, and its birefringence is given by [6]

$$\Delta n = b \left(\frac{r}{R}\right)^2 , \tag{8.1}$$

where $b = 0.133$ is an empirical constant, r the fiber radius, and R the bend radius. This birefringence provides a phase difference between the orthogonal polarization components parallel and perpendicular to the plane of the loop, given by

$$\Delta\varphi = \Delta k z = \frac{2\pi\Delta n}{\lambda} 2\pi R W \tag{8.2}$$

with W the number of turns. To understand the effect one can mentally decompose the state of polarization of the incoming light into the component along the fast axis (in the loop plane) and that along the slow axis (parallel to the loop's axis) (see Fig. 8.9). This allows to follow both components individually. Putting them back together after the loop then produces the resulting state of polarization. If one now

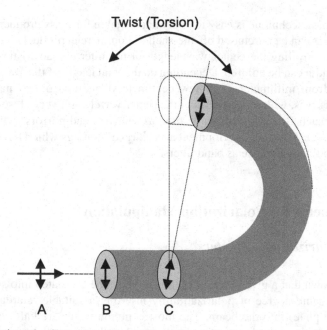

Fig. 8.9 Torsion of a fiber rotates the birefringent axes. If, e.g., the incoming light is linearly polarized as A, then it oscillates in the fast axis of the non-twisted fiber (B) and the state of polarization is maintained. If the fiber gets twisted, though, the polarization plane is at an angle with the resulting axis (C); then the state of polarization will evolve upon further propagation in the loop

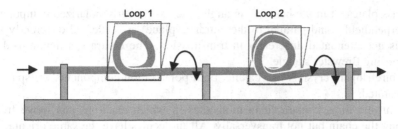

Fig. 8.10 Fiber loops as polarization controllers. The individual loops can be designed as half-wave or quarter-wave elements and are adjusted by rotation (*arrows*). For a realization see the picture on p. 99

rotates the loop around the axis along the incoming fiber, one changes the projection of the filed onto the slow and fast axes, whereby the state of polarization is modified. The effect is equivalent to the rotation of a conventional wave plate around the beam direction.

Specifically, this is what one finds: If a quarter wave of retardation is desired and if one chooses $W = 1$ (a single turn), then

$$\Delta\varphi = \frac{\pi}{2} \quad \Rightarrow \quad R = \frac{8\pi b r^2}{\lambda}. \tag{8.3}$$

At $\lambda = 1.5\,\mu\mathrm{m}$ and for a fiber with $2r = 125\,\mu\mathrm{m}$, a single loop of radius $R = 8.7\,\mathrm{mm}$ constitutes a quarter-wave plate. Equations (8.2) and (8.3) are only approximate because as the loops are rotated there is also some circular birefringence generated which counteracts the linear birefringence. To obtain a universally useful polarization controller one takes two or, more often, three loops with diameters on the order of a few centimeters, which may have a single, then two, and a single turn again to form a quarter-wave, a half-wave, and another quarter-wave plate. The loops are hinged so that they can be rotated easily (Fig. 8.10). Such a device acts as a polarization controller and is capable of transforming any incoming state into any outgoing state of polarization [6].

This construction is helpful in the laboratory but requires mechanically moving parts. It is therefore not very suitable for automatic polarization control. For the latter, a concept is preferred which generates birefringence by squeezing the fiber by mechanical force applied transversally. In a practical design, the fiber is squeezed at several positions in different directions by piezoceramic actuators [12].

8.5.2   Polarizers

A polarizer creates losses selectively for one of two possible orthogonal states of polarization. Three technical realizations are well known in optics.

- Glass plates sit in the beam at an angle; the two linearly polarized components (perpendicular and parallel to the entrance plane) are reflected differently and thus get attenuated differently in transmission. The contrast is maximized by choosing Brewster's angle.
- In birefringent crystals like calcite, both polarization components are spatially separated.
- Dichroitic films contain chain molecules in which electrons can move freely along the chain but not transversally. All molecules have the same orientation. That part of the light that is polarized parallel to the chains is absorbed so that only the orthogonal state of polarization is transmitted.

In order to make a fiber-optic polarizer, one can insert a slab of dichroitic material in a gap in the fiber. Alternatively, one can polish down a fiber from the side until its cross-section has the form of the letter D and the core is nearly exposed just beneath the flat surface. If the flat surface is then coated with metal, one obtains polarization-dependent losses.

8.6   Direction-Dependent Devices

8.6.1   Isolators

Isolators are well known in conventional bulk optics and play a role in laser technology. These are devices which let light pass through in one direction, but block it in the opposite direction. They are also called *optical diodes*.

Optical diodes rely on the Faraday effect, the rotation of the plane of polarization of linearly polarized light in a material subject to an external longitudinal magnetic field. The physical mechanism is based on the splitting of atomic energy levels into Zeeman substates due to the magnetic field; this yields a circular birefringence. The resulting angle of rotation of the plane of polarization ϵ is given, assuming a homogenous magnetic field, by

$$\epsilon = VHL \, , \tag{8.4}$$

where H is the magnetic field strength and L the length of the light path through the material. V is *Verdet's constant*.[2] This material constant has units of $\mathrm{rad/(m\,A/m)} = \mathrm{rad\,A^{-1}}$. Since in most instances nonmagnetic materials are considered, often Eq. (8.4) is written using B instead of H; then, units are $\mathrm{rad/(T\,m)}$. Verdet's constant depends on wavelength; according to classical theory it is given by

$$V = \frac{e}{2m_e c} \, \lambda \, \frac{dn}{d\lambda} \, , \tag{8.5}$$

---

[2]Marcel Emile Verdet 1824–1866.

Table 8.1 Selected values of Verdet's constant

Material	Wavelength (λ/nm)	Verdet's constant $\lvert V \rvert / \left(\dfrac{\text{rad}}{\text{T m}} \right)$
Water	632	3.8
Light flint glass	589	9
Heavy flint glass	589	20
Fused silica	589	4.8
	632	3.7
TGG	632	134
	1064	40
YIG	1310	200
	1550	1700

where e is the elementary charge and m_e the electron mass. The Table 8.1 shows selected typical values of V; for a measurement across the entire visible range for fused silica (Suprasil), see [13].

In the context of practical components, it is not only Verdet's constant that is relevant, but also the ratio of this constant and the optical loss at the operational wavelength. For visible light, it turns out that TGG (*terbium gallium garnet* $Tb_3Ga_5O_{12}$) is a useful material. In the near infrared, YIG (*yttrium iron garnet* $Y_3Fe_5O_{12}$) is important. Using YIG and a powerful permanent magnet (e.g., a samarium–cobalt type), path lengths of a few millimeters suffice to obtain a rotation of $\epsilon = 45°$. In the interest of long-term stability, one chooses the magnetic field strength high enough to drive the material into magnetic saturation. That typically happens at $B \approx 1\,\text{T}$, in the case of YIG at $0.178\,\text{T}$. Then the rotational angle becomes independent of fluctuations of the magnetic field strength.

If one places one polarizer each before and after the Faraday rotator and sets their angle at 45° relative to each other, light can pass with minimal loss (in principle, lossless; in practice, often under 1 dB). Light propagating in the opposite direction is projected onto the 45° direction at the rear polarizer, is rotated by another 45°, and arrives at the front polarizer with a total rotation of 90° so that it is perfectly blocked (Fig. 8.11). In practical devices the blocking is not perfect; one obtains attenuations around 30 dB, in stark contrast to the forward attenuation of $\approx$1 dB (Fig. 8.12).

Occasionally, the fiber itself has been used as a Faraday rotator, in order to make an all-fiber isolator [11]. Unfortunately, Verdet's constant for fused silica is quite small so that extremely powerful (bulky, power-hungry, expensive) magnets are required. Even with superconducting magnets, one still needs to use many meters of fiber to obtain a rotation angle of 45°. This is why in practical devices, almost always TGG or YIG is used.

There is a distinction between polarizing and polarization-independent isolators. The former are built as just described. Polarization-independent isolators first split the incoming light with birefringent polarizers into two polarization components.

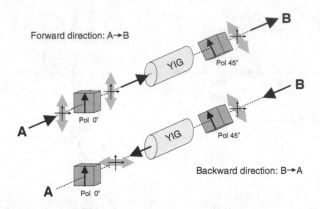

Fig. 8.11 Principle of an optical isolator based on the Faraday effect: As linearly polarized light in 0° orientation passes in forward direction (A → B), it can pass through both polarizers without attenuation because they have just the right position. Backtraveling light (B → A) may be partially blocked by the rear polarizer, but inasmuch as it passes, it is rotated further and hits the front polarizer at 90° polarization orientation so that it is blocked there

Fig. 8.12 An optical isolator. It comes with two fibers attached, known as "pigtails". The engraved arrow indicates the forward direction. A 1 Euro coin is shown for size comparison

These components are then sent through the isolator on parallel but separate paths; each is rotated. Finally both components are recombined. The result is an optical diode which is "transparent" to any forward light, but blocks any backtraveling light, in full independence of its state of polarization.

8.6.2   Circulators

Circulators are well-known devices in microwave engineering and have been introduced recently to fiber optics. These are multiport components (at least three "ports"). Each port can serve as an input or an output for signals. A signal launched into port 1 appears as an output at port 2, a signal launched into port 2 appears as an output at port 3, and so on—in the ideal case with cyclic permutation.

An optical circulator is based on an optical isolator (Fig. 8.13). The only modification is that the front polarizer is replaced with a version which acts as a

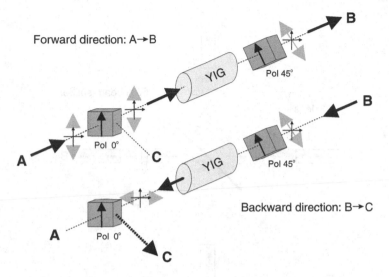

Fig. 8.13 An optical circulator conveys signals in the directions A → B and B → C

polarizing beam splitter. The backtraveling beam is then not absorbed (eliminated) but directed to an additional output, the third port where another fiber is attached. This way one obtains a three-port circulator with the functions A → B and B → C. This already suffices for a variety of useful applications. A typical example is the combination with a fiber-Bragg grating (see Sect. 9.7.2). Ports A and C are placed in the signal path; the grating is attached at port B. Fiber-Bragg gratings, which are band reject filters by their nature, are thus converted into band-pass filters which can be used to filter out a single wavelength from a wide spectrum.

8.7 Couplers

There would be no way to set up a network of fiber-optic links without having the possibility to branch between several fibers. Often it is required that a signal be split into two fibers, or two signals from two fibers are to be combined into one fiber. The same goes for larger numbers of fibers.

8.7.1 Power Splitting/Combining Couplers

The simplest case of coupling is shown schematically in Fig. 8.14. Such a coupler can be made of discrete bulk optical elements, but is neither practical, cost-effective, nor lossless.

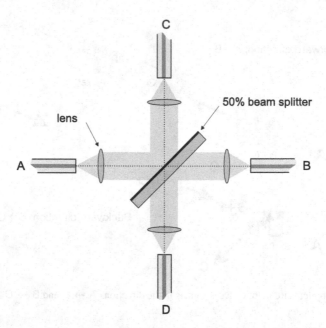

Fig. 8.14 A discrete fiber coupler connects four fibers with the help of four collimation lenses and a beam splitter of, e.g., 50 % reflection. However, such a setup requires delicate adjustments and has a rather large footprint; therefore, it is not of practical relevance. We show it solely to demonstrate the concept

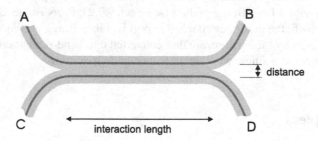

Fig. 8.15 Construction of a fused fiber coupler. Two fibers are fused together over a certain well-defined length such that both cores are at a well-defined mutual distance. The distance sets the coupling coefficient of the modes; in combination with the interaction length, the branching ratio is defined. The power of a signal which is launched at A is split between B and D according to the branching ratio, etc.

Fortunately one can obtain nearly the same functionality in an all-fiber concept. Two fibers are brought together side by side over a length of a few centimeters (Fig. 8.15). Then both are fused together by heating. The modes in each fiber penetrate into the cladding as we have seen; in the fused coupler, the mode of one fiber has a nonvanishing spatial overlap with the mode of the other fiber. This implies that they are coupled to a certain degree. When part of the energy guided in one fiber

Fig. 8.16 A typical fused fiber coupler. It comes with four pigtails

can make the transfer to the other fiber, in that second fiber one obtains a buildup of power—accompanied by a corresponding reduction of power in the first fiber, of course. Let us consider a symmetric coupler (two like fibers) in which the phases of the wave in both fibers evolve in the same way. Then, the powers as a function of common path length z evolve as

$$P_1 \propto \cos^2(\kappa z), \tag{8.6}$$

$$P_2 \propto \sin^2(\kappa z), \tag{8.7}$$

where the coupling coefficient κ is sensitively dependent on the spatial distance of both fiber cores. By judicious choice of coupling coefficient and interaction length, one can tune the branching ratio of the coupler to virtually any desired value; 0 % does not make much sense, but 100 % is possible; more useful are values in between. Very often a 50:50 branching ratio is required. In that case there is a 3-dB attenuation for each direction so that this case is called a 3-dB coupler. Also, 10:90 branching ratio couplers (10 dB couplers) and some other values are employed (Fig. 8.16).

In an alternate procedure, two fibers are polished down from the side until the (initially circular) cross-section acquires the shape of the letter D. Then, the flat sides are brought into contact and adjusted; this gives a fiber coupler with a tunable coupling ratio.

In either case couplers are four-port devices. If only three ports are used, the device acts as a splitter or as a combiner. One can add more devices in order to split/combine among more channels: E.g., three couplers allow to make a 1-to-4 splitter (*four-way splitter*); four couplers can be combined into a 4-to-4 coupler (*4-by-4 broadcast star*) (see Fig. 8.17).

In the context of photonic components, there is also a technology of optical components integrated on a microchip. When the application demands that light is coupled out of a fiber and into a photonic chip anyway, it may make sense to include the couplers on-chip.

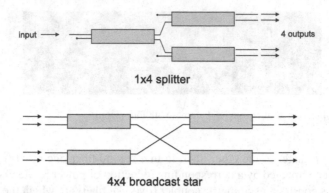

1x4 splitter

4x4 broadcast star

Fig. 8.17 In a four-way splitter made from 3-dB couplers, each output presents one fourth of the input power. If inputs and outputs switch their roles, one obtains a four-way combiner which presents the sum of four inputs. A 4-to-4 coupler made from 3-dB couplers presents one fourth of the sum of four inputs at each of its four outputs. This principle can be extended to practically any arbitrary number of inputs and outputs

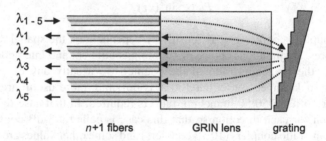

Fig. 8.18 Basic idea of a wavelength-dependent coupler (WDM coupler) in bulk optics using a diffraction grating. The acronym GRIN is for gradient index; GRIN lenses are offered commercially. In this example, five wavelengths from an input fiber are split to as many output fibers. Of course, the direction can be reversed, and one obtains a five-way combiner

8.7.2 Wavelength-Dependent Couplers

Quite often, it is desired to split or combine various signals in fibers not all in the same way but according to their wavelength. This is the prerequisite for wavelength division multiplexing (WDM, see Sect. 11.1.5) which in turn is the basis for utilizing the enormous bandwidth provided by the fiber (25 THz in the third window) to anything more than a ridiculously small fraction.

Such wavelength-dependent couplers (WDM couplers) can be made in principle with bulk optics. Figure 8.18 shows the idea for the case of a 5-to-1 WDM coupler using a diffraction grating and a GRIN lens (GRIN = gradient index). In practical devices all-fiber versions are desirable. There are constructions using the wavelength dependence of the branching ratio in fused fiber couplers; this may be augmented with grating structures. There are also constructions using interference filters.

8.8 Optical Amplifiers

Signal power is lost in long pieces of fiber; more is lost in couplers. Often it is required to make up for the losses by amplifying the optical signals. The conventional technology, used until a couple of years ago, relied on so-called "repeaters" in which the optical signal was converted into an electronic signal, then was amplified and possibly reshaped by electronic means, and finally was converted back to an optical format. This is not only quite involved; it also creates a bottleneck for the data rates that can be transmitted over an optical fiber. The theoretically available bandwidth of the fiber of tens of terahertz would be reduced to whatever can be handled by electronics, which is perhaps 10 GHz. One does not easily give up three orders of magnitude of opportunity!

Fortunately enough, there are also all-optical amplifiers. They are subject to the same constraints as any other amplifier: there is no amplification without noise. Any amplifier adds some extra noise to the signal, and part of this extra noise is unavoidable due to fundamental physical reasons. The origin of that contribution can be traced back to Heisenberg's uncertainty relation of quantum mechanics [4] as follows.

The uncertainty relation

$$\Delta E \, \Delta t > \hbar/2$$

can also be interpreted as

$$\Delta n \, \Delta \phi \geq \frac{1}{2} \, ,$$

where $n = E/(\hbar\omega)$ is the photon number and $\phi = \omega t$ is the phase of the light wave.

In a (linear) amplifier, the gain factor G represents the ratio of output signal power to input signal power. An ideal amplifier would just multiply the photon number such that each input photon would produce exactly G output photons. In this ideal case, there would be no change to the phase, except for a trivial overall shift ϕ_0 due to the transit time. Then, an input signal with photon number n_{in} would produce an output with $n_{\text{out}} = Gn_{\text{in}}$. ϕ_{in} would be converted to $\phi_{\text{out}} = \phi_{\text{in}} + \phi_0$.

Let us now continue our gedanken experiment and place an ideal detector at the output of the ideal amplifier. "Ideal" means that it can detect photons such that the equality is fulfilled in the uncertainty relation:

$$\Delta n_{\text{out}} \, \Delta \phi_{\text{out}} = \frac{1}{2} \, .$$

The detector will thus register $n_{\text{out}} \pm \Delta n_{\text{out}}$ photons and a phase of $\phi_{\text{out}} \pm \Delta \phi_{\text{out}}$.

There is no reason why it should be wrong to think of the combination of amplifier and detector as one unit, which would serve as a particularly sensitive detector. This internal-gain detector then measures a signal with

$$\Delta n_{\text{in}} \, \Delta \phi_{\text{in}} = \frac{1}{2G} \, ,$$

which violates the uncertainty relation whenever the amplifier deserves its name, i.e., whenever $G > 1$.

The contradiction is resolved when one accepts the following: Any amplifier adds as much noise to a signal with frequency ν as a hypothetical noise source at the amplifier input would when the amplifier were ideal, and the noise source had a spectral power density of

$$\frac{dP}{d\nu} = \left(1 - \frac{1}{G}\right) h\nu \, .$$

This immediately shows: the only possible noise-free amplifier has $G = 1$ in which case the word amplifier would be a misnomer.

What does this mean? One might naïvely think that an attenuation of some signal and subsequent amplification by the same factor would faithfully reconstitute the original signal. This is not so! There will be an additional noise contribution. This extra noise may be strong if the amplifier is of mediocre engineering, but even the best amplifiers will always add at least some noise. Fortunately, engineering of optical amplifiers has matured so far that the best commercially available types are extremely close to the theoretical limit.

For a practical realization of optical amplifiers, there are two quite different approaches or technologies: active fibers and semiconductor elements.

8.8.1 Amplifiers Involving Active Fibers

It seems that the interest is shifting toward amplifiers which consist simply of a piece of special fiber. Amplifying fiber is doped with suitable materials and receives power from an auxiliary light source. In the third transmission window, erbium is the most suitable dopant [2]; at several other wavelengths, useful dopants are also known, like neodymium at $1.06 \, \mu$m. Figure 8.19 schematically shows the relevant energy levels of these substances. When the transition at 980 or 1480 nm is pumped, an inversion of the $4I_{13/2}$ with respect to the $4I_{15/2}$ ground state is created; this implies an optical gain. With a few tens of milliwatts pump power, one can achieve 30 dB gain in about 10 m of erbium-doped fiber. The gain bandwidth extends from 1530 to 1570 nm (Fig. 8.20). The lifetime of the upper state is extremely long (10 ms), and therefore gain saturation and thus channel crosstalk among wavelength channels are practically absent. This is important because otherwise the huge bandwidth would not be useable. While the gain is not flat throughout the gain bandwidth, it can be equalized to a large extent with filters.

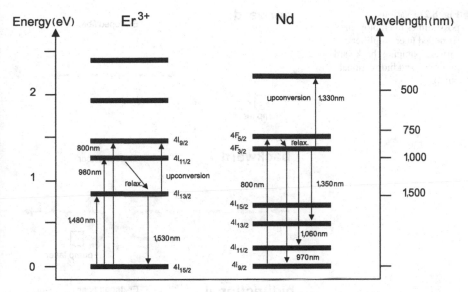

Fig. 8.19 Level scheme for optical fibers doped with Er ions or Nd ions. Shown are the energetic levels in electron volts and the transition wavelengths in nanometers, both referred to the ground state

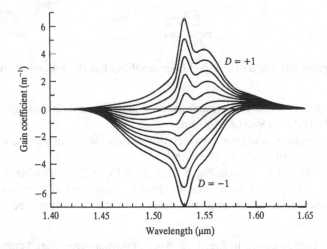

Fig. 8.20 Gain spectrum of an Er-doped fiber for various levels of inversion. Without any inversion (*bottom curve*) the fiber absorbs light. As the inversion increases, gain first appears at the long-wavelength side. At the highest inversion shown (*top curve*), the gain has spread across the entire band. From [3]

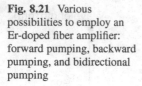

Fig. 8.21 Various possibilities to employ an Er-doped fiber amplifier: forward pumping, backward pumping, and bidirectional pumping

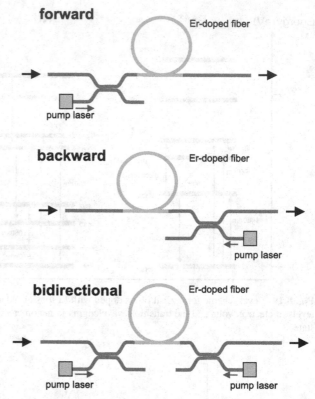

The complete setup of an Er-doped fiber amplifier has the following components (Fig. 8.21):

- a pump source, typically a continuous-wave laser diode with high power (on the order of 100 mW) at 980 or 1480 nm;
- a wavelength-dependent coupler which inserts the pump light into the signal path;
- a suitable length of Er-doped fiber; and
- optical isolators (not shown in Fig. 8.21, but see Fig. 8.22) which block backtraveling light and therefore make sure that both amplification of spontaneous emission in backward direction and stimulated Brillouin scattering (see Sect. 9.7.1) are suppressed.

A typical setup is shown in Fig. 8.22. Such amplifiers are offered commercially. They can be employed in a variety of ways:

As booster	for postamplification of a low-power light source at the beginning of a transmission line.
As intermediate amplifier	for compensation of loss inserted somewhere along the line.
as preamplifier	to increase the sensitivity of photodetectors for weak signals at the end of a transmission line.

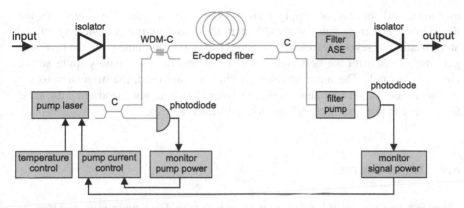

Fig. 8.22 Setup of a realistic Er-doped fiber amplifier in forward-pumping geometry. C: coupler (tap, e.g., 95:5 coupler), WDM-C: wavelength-dependent coupler to separate signal and pump wavelengths. Filter ASE: selective filter to suppress amplified spontaneous emission. Filter pump: selective filter to suppress pump light

As distribution amplifier   for compensation of losses where the signal power is split into several branches, to restore the original power level in all branches.

As an oscillator   by optical feedback, the amplifier is turned into a laser (typically in conjunction with wavelength-tuning elements). We note that a fiber amplifier with an optical resonator to provide feedback operates as a fiber laser; we will return to this aspect in Sect. 8.9.4.

In the second window neodymium or praseodymium is used to dope the fibers. They do not make quite as near-perfect amplifiers as erbium does, but there is a large volume of installed fiber-optic systems operating in the second window, and therefore there is considerable interest in making amplifiers for this wavelength regime.

8.8.2   Amplifiers Involving Semiconductor Devices

In suitable semiconductor materials with a p–n junction one can excite carriers from the valence band into the conduction band by running an electric current through the junction (see Sect. 8.9.1 below). The current thus produces an inversion, a nonequilibrium excess population in a higher-energy state. The excited carriers can then return to the valence band by emission of a photon. When this return is triggered ("stimulated") by a signal photon, the process constitutes an amplification. The mechanism is also central for the operation of semiconductor lasers (see below).

If a laser diode is operated without optical feedback, it never reaches the threshold for laser oscillation and functions as an amplifier: Stimulated emission amplifies light sent in. Advantages of this technology are that devices are readily available, and the energy supply could not be any simpler. Disadvantages are the relatively narrow gain bandwidth and the less-than-perfect linearity of the amplification. If several wavelength channels are used simultaneously, the inversion gets modulated with the beat frequency. Then one finds a possibly quite severe channel crosstalk. The more wavelength channels are used, the more serious the problems get. This is why semiconductor amplifiers have not found quite the same acceptance in practical applications as doped-fiber devices.

8.9   Light Sources

There is a vast variety of light sources known to man. Discounting sun and stars as well as flames, we name just a few:

- Tungsten filament light bulbs
- Bulk lasers
- Luminescent diodes (LEDs)
- Laser diodes
- Fiber lasers

For fiber technology, there are certain demands which a light source must meet:

- Must be possible to couple into fiber with good efficiency
- Must have low energy requirements
- Must be cost-effective
- Must have long lifetime
- Must be virtually maintenance-free
- Must provide means of modulation

The first five items are based on economic considerations because one has to expect that in a vast fiber network there may be huge numbers of light sources, many of which are located in far-flung and hard-to-reach places. Modulation is a requirement dictated immediately by the application: to transmit information.

Of course, light bulbs are ruled out. Their coupling efficiency is minimal, their lifetime is inadequate, and their capability for modulation exists only for frequencies up to a few hertz, certainly not gigahertz. Bulk lasers as found in many physics laboratories (think He-Ne lasers, Nd:YAG lasers, etc.) have good spatial coherence and thus good coupling efficiency. On the other hand, cost, energy, and maintenance requirements are definite disadvantages for any application outside a research lab and so is the lack of modulation capability at least in most types.

Therefore, from the above list, only luminescent diodes (LEDs), laser diodes, and fiber lasers are left as viable light sources. We will now discuss these choices in somewhat more detail.

8.9.1   Light from Semiconductors

The mechanism of light generation in semiconductors is the recombination of carriers at a p–n junction. An electric current provides the energy required to excite electrons to the conduction band; as they relax back to the valence band, the amount of energy corresponding to the band gap is released in form of a photon. For a semiconductor with band gap E_{gap}, one finds light with a frequency of $\nu \approx E_{gap}/h$.

8.9.2   Luminescent Diodes

The recombination radiation has, without extra steps, no preference for any particular spatial direction. In LEDs, one does not attempt to achieve much directionality other than getting the light out of the component on one side, typically in a wide cone. LEDs perfectly fulfill the above requirements of low cost, low power operation, and long lifetime without maintenance. They can be modulated up to perhaps 100 MHz, which is sufficient for many applications. However, for fundamental reasons the fiber-coupling efficiency is not impressive, and the power actually launched into a fiber is low, well under 1 mW. In this situation, LEDs find applications for short distances, like in LANs (*local area networks*) within premises where highest data rates are less important than lowest cost. With certain geometries, it has been attempted to optimize the coupling efficiency. Figure 8.23 shows the design of a "Burrus LED" where the fiber is butt-coupled to the light-emitting chip.

Fig. 8.23 Construction of a "Burrus LED." The fiber is butt-coupled to the light-emitting chip and is permanently held in place to avoid the need of later adjustment

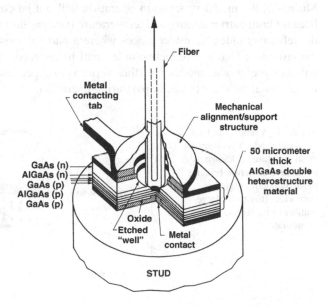

8.9.3   Laser Diodes

The same principle of generation of light can be refined into the concept of laser diodes. One shapes the semiconductor chip in such a way that optical feedback is obtained. Then a stimulated process takes over, and coherent emission of light results. Coherent light is tremendously much easier to focus and couple into a fiber than incoherent light.

The first laser diodes in the 1960s consisted of little more than a semiconductor chip with p-doped and n-doped material. They had smoothly cleaved end facets with a natural reflectivity (Fresnel reflection) on the order of 30 %, due to the high refractive index of semiconductors like GaAs of about $n = 3$ (Fig. 8.24). This reflectivity is fully sufficient for resonator mirrors. The side faces of the chip remain unpolished and rough and are therefore no good reflectors. The length of the chip on the order of 300 μm is basically defined by the required gain length. The thickness of the active layer is on the order of 0.5 μm. These dimensions have immediate consequences for the modal structure of the laser resonator.

The resonator is hundreds of wavelengths long. Then the frequencies of adjacent longitudinal modes differ by fractions of 1 %. As the gain bandwidth amounts to several percent of the central frequency, one can expect the simultaneous oscillation of several longitudinal modes. In some cases that may even be desirable: The process of mode locking can be used to generate short pulses of light. As for transverse modes, the active layer is thin; in the direction perpendicular to the active layer only a single mode can oscillate. On the other hand, in the lateral direction (perpendicular to the optical axis and parallel to the active layer), there is a 100 μm or so wide gain structure which gives rise to a multiplicity of lateral modes. Moreover, the modal structure in operation will not be constant during operation because both carrier density and temperature (heating during operation) will affect the refractive index. Consider places where a particular oscillating mode depletes the inversion: Here the refractive index will be reduced, and the gain mechanism will then prefer other modes. One therefore has to expect undesired sudden changes in the modal structure (mode hops) during operation.

Fig. 8.24 A schematic view of a laser diode. This simplest of all structures is known as *broad area structure* because the active (gain) region is very wide. It allows a large number of transverse modes to oscillate

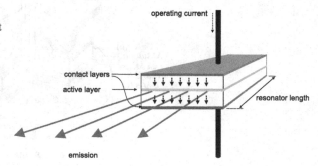

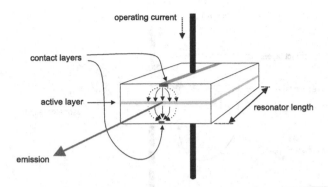

Fig. 8.25 Possible structure of a gain-guided laser diode. Oxide layers restrict the electrical contacts, and thus the current flow, to a narrow zone. Carriers are injected into the active zone only where sufficient current density exists. The resulting rise in refractive index guides the light and restricts the emission to basically the same narrow zone

8.9.3.1   Gain Guiding

In view of these problems it was a first improvement to modify the geometry of current flow through the chip. Narrow contact stripes, or the introduction of insulating zones, make it possible to restrict the current flow to a narrow region in the active layer which may be just a couple of micrometers wide. This is shown schematically in Fig. 8.25. Gain occurs only where there is sufficient current density which is now only a small part of the active layer; this is called a gain-guided geometry. The advantages are that (1) the current density is increased in the relevant position which lowers the laser threshold and (2) the transverse mode profile is strongly restricted. Nevertheless, mode hops are not entirely eliminated; this is easily seen in kinks of the output power vs. pump current characteristics of such lasers. When it comes to coupling light from such lasers into fibers, there are nasty consequences: Changes in the modal structure give rise to modified overlap with the fiber's modal profile and result in jumps of the incoupled power. Unpredictable severe fluctuations of power are certainly not desirable for any application.

8.9.3.2   Index Guiding

The next improvement was the introduction of the index-guided laser diode geometry. In a considerably more involved production process (which, however, has become routine now), there are lateral steps of the refractive index built into the active layer by use of differently doped material. This is shown in Fig. 8.26. Even weak index guiding with index steps on the order of 1 % make sure that the index modifications through the concentration of carriers are overwhelmed. Therefore these structures can run in lateral single-mode operation.

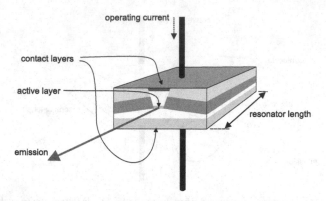

Fig. 8.26 Possible structure of an index-guided laser diode. The active zone is surrounded on all sides by material with larger band gap; this is also known as "buried heterostructure." A lateral index step of ≈ 0.2 provides strong guidance of the light. By way of the intricate structure, it is assured that also the current flows only through the relevant part of the active zone, resulting in a low laser threshold. Many different geometries of buried heterostructure lasers have been suggested and realized; the one shown here is called an "etched-mesa" structure

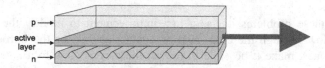

Fig. 8.27 Schematic representation of a distributed feedback laser (DFB laser). A periodic index modulation is introduced over the entire resonator length; it acts as a grating and selects a particular frequency

8.9.3.3 Distributed Feedback

Finally let us look at the longitudinal modes. As long as the resonator mirrors at the chip facets set the mode spacing, there is little one can do to achieve single-mode operation: One cannot make the chip shorter because a certain length is required to provide adequate gain. In this situation only frequency-selective means can help. When longitudinal single-mode operation is required, one uses laser diodes into which a grating has been incorporated for wavelength selectivity (see Fig. 8.27). The grating favors feedback at the frequencies defined by a Bragg condition for the grating. One can obtain both the selection of a single mode and an improved frequency stability of this particular mode. The grating may be extended over the entire resonator length; then this is known as distributed feedback or DFB laser. The grating may alternatively be formed only on short segments toward the resonator ends in a zone with little gain; this is then called distributed Bragg reflector or DBR laser. Both DFB and DBR lasers have become something of a standard for long-haul transmission because single-mode operation is favorable, and the frequency stability is good.

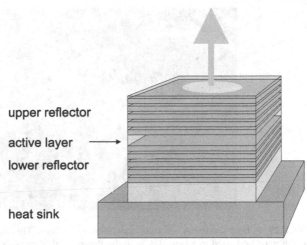

Fig. 8.28 Schematic representation of a VCSEL. In this figure, the vertical dimensions are exaggerated for the sake of clarity. Bragg reflectors above and below the active layer provide feedback; the thickness of the active layer is only one wavelength or so

upper reflector

active layer ⟶

lower reflector

heat sink

8.9.3.4 VCSELs

In a more recent development laser diodes are also made with an entirely different geometry: They are called vertical cavity surface emitting lasers or VCSELs (rhymes with pixels; see Fig. 8.28). In these lasers, the light does not travel the length but the width of the active layer. This implies that the optical axis (the direction of light propagation) is parallel to the direction of the pump current. Above and below the active layer, there are multiple layer reflectors acting as wavelength-selective mirrors similar to a DBR structure. In this concept, the resonator length is very short—hardly any longer than the wavelength. This enforces longitudinal single-mode operation, which is a definite advantage. The lateral beam profile can be optimized by suitable structuring. In fact, meanwhile VCSELs can produce better beam geometries than side-emitting lasers. A downside is that since their resonator and thus their gain length are so short, it has been difficult to generate high output powers from VCSELs. On the other hand, they can be modulated at high speed (well above 10 GHz instead of a few gigahertz). By this token, it appears likely that they will find applications in fiber optics.

Laser diodes provide the best combination of properties of lasers (spatial coherence assures good incoupling efficiency) with those of LEDs: long lifetime and low operational power requirement. They can be modulated into the gigahertz regime, which is often good enough; if not, they are operated continuous wave, and an external modulator serves to carve out pulses as required. Cost ranges from just a few euros for the simplest types up to a few thousand euros, for the fanciest DFB lasers and other specialty constructions; however, in relation to the complete system this may still be considered low cost. All told, laser diodes fulfill all the important requirements of fiber technology and are therefore the de facto standard. Figure 8.29 shows a typical laser diode. The device shown has a piece of fiber (the "pigtail") connected to the chip by the manufacturer. Coupling the light of laser diodes into

Fig. 8.29 A laser diode made for communication purposes. This device is enclosed in a housing which must be bolted to a heat sink with the flange visible on the right, in order to remove heat. The electrical connections are by pins on the bottom and are not visible in this picture. The optical output goes into a piece of fiber attached to the laser (the pigtail). The pigtail is protected with a plastic coating and a rubber bend relief where it leaves the case. A small coin is shown for size comparison

single-mode fibers requires precision; it rarely exceeds 50 % in spite of all serious attempts. Therefore it is a great simplification for the end user when this critical step has already been taken care of by the manufacturer.

8.9.4 Fiber Lasers

A fiber laser is created when a fiber amplifier is inserted in an optical resonator [3]. The optical feedback provided by the resonator allows a light field (initially starting from spontaneous emission) to build upon itself until saturation effects arrest further growth: This describes a laser. The energy supply (pump source) must be provided by optical means and is therefore more complex than for laser diodes. On the other hand, fiber lasers are perfectly suited to coupling their light into a fiber: All it takes is a splice. Transverse or lateral modes cannot arise, but the longitudinal mode spectrum has a particularly high number of modes. This is due to the long resonator length and the often very wide gain bandwidth, but is not a disadvantage when modelocked operation is required. A big disadvantage is that by virtue of the long lifetime of the upper laser state fiber lasers cannot be modulated. Any modulation imposed on the pump power would be low-pass filtered; in the case of Er-doped fiber lasers the time constant is $\tau = 10$ ms. Then, the highest possible modulation frequency is a ridiculous $v_{max} = 1/(2\pi\tau) \approx 16$ Hz! As mentioned above, external modulation is used for laser diodes only when the fastest data rates are required; for fiber lasers one always has to resort to an external modulator. This makes them unlikely candidates for undemanding, low cost applications with low data rates.

Both erbium and praseodymium fiber lasers are now investigated by many researchers, and there are also commercial products. One may expect that they will find many applications. Particularly successful are neodymium-doped fiber lasers which can produce enormous output powers but mostly work at a wavelength of 1.06 μm which is not very interesting in the context of fiber-optic data transmission. They are pumped at 800 nm; this pump wavelength can be produced at low cost by GaAs diode lasers. We mention in passing that by absorption of more pump photons one can have upconversion so that fiber lasers are also capable to generate light with shorter wavelength, including visible light.

A particular wavelength within the gain bandwidth can be preferred by selective means. In linear resonators, fiber-Bragg gratings are often used (compare Fig. 9.34); in ring resonators, a combination of a fiber-Bragg grating and a circulator (see Sect. 8.6.2) can be used.

8.10 Optical Receivers

As receivers for light one might consider the following options:

- Photomultipliers
- Photodiodes (pn and pin type)
- Photodiodes (avalanche type)

However, for data transmission applications, photomultiplier are ruled out because for the infrared wavelengths of the second and third windows there simply are no photocathode materials. Photodiodes are economic, small, fast, and reliable. They can also be integrated very well together with other circuitry like preamplifiers. There is a special version of photodiodes called avalanche diodes. Avalanche diodes have an internal amplification mechanism and are therefore more sensitive to weak light signals, but their inner structure is more complex, and they require more complex circuitry. Therefore they are only used when the added sensitivity is definitely a requirement. This applies in the latest generation of transatlantic fiber cables.

8.10.1 Principle of pn and pin Photodiodes

Any photodiode relies on a p–n junction into which light can be irradiated. At the junction, a photon can generate an electron–hole pair provided that $h\nu > E_{gap}$. h is Planck's constant, ν is the frequency of the quanta of light, and E_{gap} is the energetic band gap between valence and conduction bands of the detector material. In other words, the photon energy must exceed the band gap. The electric charges thus generated can then be measured as an external current, called the photocurrent. The current in a photodiode is given by

$$I = I_0 \left(e^{\frac{eU}{mkT}} - 1 \right) - I_p , \tag{8.8}$$

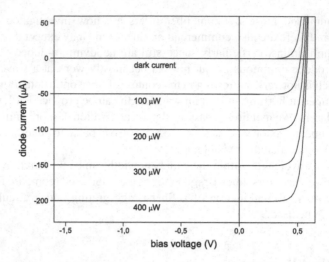

Fig. 8.30 Voltage-current characteristic of a photodiode, with the absorbed light power as a parameter. For this figure, it was assumed that $\mathcal{R} = 0.5\,\mathrm{A/W}$ and $I_0 = 100\,\mathrm{pA}$. At constant negative bias voltage, i.e., when one crosses the set of curves on a vertical path, the reverse current is practically identical to the photocurrent, which is proportional to the received power and almost independent of bias. For open circuit (zero current), one crosses the set of curves on a path given by the horizontal axis; then the diode generates a photovoltage which is proportional to the logarithm of the received power

where I is the current, U is the voltage, I_0 is a constant current given by the material and the temperature, e is the electron charge, $m \approx 1.5$ is known as the Shockley factor, k is Boltzmann's constant, and T is the temperature (in Kelvin). The equation differs from that of any ordinary diode only in the additional term for the photocurrent, I_p (Fig. 8.30).

We have to distinguish two very different modes of operation:

As a current source   This requires a constant bias voltage $U = $ const. which is applied in reverse direction or may be zero. Then the photocurrent is proportional to the received light power over many orders of magnitude and almost independent of the bias voltage. The main effect of the nonzero bias is to reduce the diode capacity and improve the temporal response.

As a voltage source   This is the operation with high impedance load so that basically $I = 0$. Then the diode generates a photovoltage which is proportional to the logarithm of the received light power.

While there can be uses for the voltage source mode occasionally, in our context the constant current mode is almost always preferred due to its linearity.

Unfortunately, not every photon actually generates a free charge which contributes to the photocurrent, but only a certain percentage. This percentage is

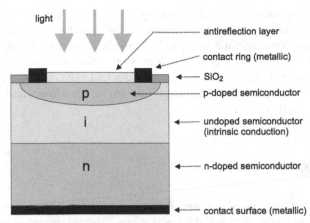

Fig. 8.31 Structure of a pin photodiode. Light enters through an antireflection layer which occupies the free aperture inside the contact ring. The remainder of the chip surface is passivated with SiO_2. Absorption takes place mostly in the i layer

called the quantum efficiency η and is an important characteristic of the detector. η is always smaller than unity because some photons fail to enter the detector (Fresnel reflection at the surface) or are not absorbed near the junction; some carriers may also recombine before they contribute to the external photocurrent.

In order to obtain the most efficient absorption of the impinging light, there is usually an extra layer inserted between the p-doped and the n-doped layers, which is undoped and thus has only the intrinsic conductivity of the base material. This intermediate layer is called the i layer, as in "intrinsic conductivity". The diode is then described as a pin diode for its three layers (Fig. 8.31). pin diodes are the most frequently used type of photodiodes and can reach quantum efficiencies up to about 90 %.

To characterize photodiodes, the sensitivity (sometimes called responsivity) $\mathcal{R}$ is an important quantity:

$$\mathcal{R} = \frac{I_p}{P_p} = \frac{n_e e}{n_p h\nu} = \frac{\eta e}{h\nu} . \tag{8.9}$$

Here n_e and n_p denote the number per second of electrons or photons, respectively. If $\eta \approx 1$ one can write the numerical relation

$$\mathcal{R} = \frac{e}{hc} \lambda = \frac{\lambda(\mu m)}{1,24} \left[\frac{A}{W}\right] . \tag{8.10}$$

We see that the sensitivity is of the order of 1 A/W in the near infrared.

8.10.2 Materials

Silicon photodiodes are a product for mass markets. They are found in TV remote controls, in supermarket scanners, in CD and DVD players, as individual pixels in electronic cameras, etc. The band gap of silicon corresponds to $\lambda = 1.1\,\mu m$.

This means that the visible light is covered well, but for fiber optics, silicon is useful only in the first window. At the longer wavelengths of the second and third windows silicon is just simply transparent and does not absorb at all. The band edge of germanium sets on at $1.7\,\mu m$ so that the entire wavelength regime of interest is covered. It turns out, though, that diodes from composite materials such as InGaAs have a similar band gap as germanium but have lower dark current and are therefore usually preferred.

8.10.3   Speed

The fundamental speed limitation for a photodetector is that photo-generated carriers must transit the junction area. The maximum speed at which carriers move through the lattice depends on the material and is defined by scattering processes at the lattice atoms. Near the junction they are accelerated by the electric field of an external bias voltage. Some of the carriers, however, are not generated near the junction but in the p-layer or n-layer before or after. There the electric field strength is much lower so that these carriers must diffuse away; they contribute a slow portion to the electric signal. Then there is the external time constant defined by the circuitry: Unavoidable capacitances both inside and outside the diode, in combination with resistance in the circuit, defines an RC low pass. Careful construction allows to produce photodiodes with a couple of picosecond response time; commercial products are available with bandwidths up to about 100 GHz.

8.10.4   Noise

Noise is a fundamental limitation to any measurement. When it comes to the detection of light with photodetectors, several noise mechanisms contribute:

Quantum noise of the light: This is a property of the light itself. Light consists of photons which arrive according to some statistics. This is reflected in the temporal distribution of photo-generated carriers and thus causes a noise contribution to the photocurrent.

Dark current noise: This is a property of the detector. The effect depends on material and temperature; it can be reduced by careful choice of material and by cooling.

Surface leak current noise: This again is a property of the detector. Improvements can be obtained by precautions in manufacturing and possibly to some degree by cooling.

Noise from resistors and amplifiers: This is a property of the external circuitry. The effect, also known as Johnson noise or Nyquist noise, can be minimized by optimizing the circuit.

Quantum noise turns out to be the only contribution which cannot be modified by engineering; it thus constitutes a fundamental limit. We will discuss it in more detail in Sect. 11.1.7.

8.10.5 Avalanche Diodes

Avalanche diodes are the solid-state equivalent to photomultipliers. In comparison to pn or pin photodiodes, avalanche diodes are operated at a considerable reverse bias voltage. The high voltage internally generates strong electric fields which accelerate carriers to the point that upon collision with lattice atoms, they can generate more carriers by impact ionization. The photocurrent then grows like an avalanche and is amplified by a gain factor M.

It is not worth it to push M too far. The internal amplification process has its own noise component which actually grows faster than M. For each combination of avalanche diode and associated external circuit, there is an optimum gain at which the inherent noise contribution of the first amplifier stage or other subsequent electronic circuitry just ceases to be the dominant contribution to overall noise.

References

1. G. P. Agrawal, *Fiber-Optic Communication Systems*, John Wiley & Sons, New York (1992)
2. P. C. Becker, N. A. Olsson, J. R. Simpson, *Erbium-Doped Fiber Amplifiers: Fundamentals and Technology*, Academic Press, London (1999)
3. E. Desurvire, *Erbium-Doped Fiber Amplifiers*, John Wiley & Sons, New York (1994)
4. H. Heffner, *The Fundamental Noise Limit of Linear Amplifiers*, Proceedings of the IRE 1604 (July 1962)
5. A. F. Judy, H. E. S. Neysmith, *Reflections from Polished Single Mode Fiber Ends*, Fiber and Integrated Optics **7**, 17 (1987)
6. H. C. Lefevre, *Single-Mode Fibre Fractional Wave Devices and Polarization Controllers*, Electronics Letters **16**, 778 (1980)
7. D. Marcuse, *Loss Analysis of Single-Mode Fiber Splices*, The Bell System Technical Journal **56**, 703 (1977)
8. D. Marcuse, J. Stone, *Fiber-Coupled Short Fabry-Perot Resonators*, Journal Lightwave Technology **7**, 869 (1989)
9. S. C. Mettler, C. M. Miller, *Optical Fiber Splicing*, in [10] p. 263.
10. S. E. Miller, A. G. Chinoweth (Eds.), *Optical Fiber Telecommunications II*, Academic Press, London (1988)
11. R. M. Shelby, M. D. Levenson, D. F. Walls, A. Aspect, *Generation of Squeezed States of Light with a Fiber-Optic Ring Interferometer*, Physical Review A **33**, 4008 (1986)
12. H. Shimizu, S, Yamasaki, T. Ono, K. Emura, *Highly Practical Fiber Squeezer Polarization Controller*, Journal of Lightwave Technology **9**, 1217 (1991)
13. C. Z. Tan, J. Arndt, *Wavelength Dependence of the Faraday Effect in Glassy SiO_2*, Journal Physics and Chemistry of Solids **60**, 1689 (1999)

Part IV
Nonlinear Phenomena in Fibers

Experiment to investigate stimulated Brillouin scattering in optical fiber with visible light (ca. 590 nm). Compare Figs. 9.28, 9.29 and 9.30.

Part IV
Nonlinear Phenomena in Fibers

Chapter 9
Basics of Nonlinear Processes

It is well known from acoustics that when it comes to oscillations, nonlinearity leads to the appearance of overtones. The same phenomenon also exists in optics. A first experimental demonstration succeeded in the early 1960s [14] when the generation of twice the irradiated frequency was shown in a nonlinear crystal. The mechanism relied on the anharmonicity of the oscillation of the medium's polarization as produced by an intense light wave. Shortly thereafter, the third harmonic was also demonstrated. Since then, nonlinear optics has evolved into a field of research in its own right. Processes under study are optical rectification, parametric amplification, self-focusing, and self-phase modulation, to name just a few. Optical nonlinearity is responsible when optical properties of some material show intensity-dependent modifications, when light waves with frequencies are generated that are not present in the irradiated light, or when—speaking in more general terms—power is redistributed between different Fourier components of a light field. As a rule, nonlinear effects get more pronounced as the light intensity is increased. The reverse is also true: When the light intensity is sufficiently weak, nonlinear processes may safely be neglected. All of classical optics is therefore linear optics.

9.1 Nonlinearity in Fibers vs. in Bulk

Nonlinear processes are also observed in optical fiber: actually, often in a more pronounced form than in bulk optics. This is due to two peculiarities of fibers: By virtue of the very small mode cross-section, there is high intensity even at moderate power. And the waveguiding allows very long interaction lengths.

These two peculiarities belong together: one could also have high intensity in bulk optics by suitably focusing down to a tiny spot. But then typically the length of the interaction zone goes down due to diffraction of light. It may be best to discuss

© Springer-Verlag Berlin Heidelberg 2016
F. Mitschke, *Fiber Optics*, DOI 10.1007/978-3-662-52764-1_9

this for Gaussian beams: Beams generated by lasers typically have a Gaussian intensity profile (see Chap. 16).

For a Gaussian beam, there is a characteristic length called Rayleigh range. Its significance is that near a focus the beam diameter stays nearly constant over this length. (Speaking more precisely, the beam radius widens by no more than a factor of $\sqrt{2}$ over that at the beam waist.) Then the axial intensity drops by only very little (no more than a factor of 2). The Rayleigh length is given by

$$z_R = \frac{\pi w_0^2}{\lambda};$$ (9.1)

on the other hand, the intensity at the beam waist is obtained from the total power P_0 by the expression

$$I_0 = \frac{2P_0}{\pi w_0^2}.$$ (9.2)

This is the maximum; at the end of the Rayleigh zone $I_R = I_0/2$.

Let us consider that class of nonlinear interactions which are proportional to intensity and cumulate with interaction length. Then the product of intensity and diffraction-limited interaction length is a metric for the strength of the nonlinear effects. We obtain

$$I_0 z_R = \frac{2P_0}{\pi w_0^2} \frac{\pi w_0^2}{\lambda} = \frac{2}{\lambda} P_0.$$ (9.3)

This expression shows that there is no way how one could increase the strength of the nonlinear process by geometrical means, like nifty focusing arrangements or whatever.

In stark contrast, the same limitation does not occur in optical fibers. In the fiber the wave is guided, and the interaction can build up over nearly arbitrary distances. Of course, losses reduce the intensity in the fiber after some distance. We take that into account by integrating along the fiber:

$$\int_0^L I(z)\, dz = \int_0^L I_0\, e^{-\alpha z}\, dz = \frac{I_0}{\alpha} \left(1 - e^{-\alpha L}\right) = I_0 L_{\text{eff}},$$ (9.4)

where we introduced the *effective interaction length*

$$L_{\text{eff}} = \frac{1}{\alpha} \left(1 - e^{-\alpha L}\right).$$ (9.5)

This effective interaction length is quite a useful concept in nonlinear fiber optics. It is always shorter than the actual fiber length. This is because the power goes down so that remote parts of the fiber contribute only little to the nonlinear effect.

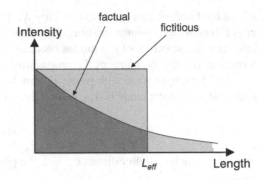

Fig. 9.1 Sketch to explain the concept of the effective interaction length

The definition amounts to replacing the actual decreasing power by a constant, i.e., its initial value, but limiting the interaction length to the effective value (see Fig. 9.1). In the limiting case that the actual fiber length tends to infinity, L_{eff} tends to $1/\alpha$. In practical numbers, assuming a loss of $0.2\,\text{dB/km}$, $L_{\text{eff,max}} \approx 22\,\text{km}$.

In order to convert the transmitted power to intensity, we need to be more specific about the mode's cross sectional area. We obtain it from integrating the field amplitude $E(x, y)$ across the entire cross-section with suitable normalization:

$$A_{\text{eff}} = \frac{\left(\int_{-\infty}^{\infty} \int_{-\infty}^{\infty} |E(x, y)|^2 \, dxdy\right)^2}{\int_{-\infty}^{\infty} \int_{-\infty}^{\infty} |E(x, y)|^4 \, dxdy}. \tag{9.6}$$

A_{eff} is the effective mode area. If one uses the Gaussian approximation for the fiber mode, the effective area simply becomes $A_{\text{eff}} = \pi w^2$.

Let us wrap up: Nonlinear interaction is much stronger in fiber than in bulk. Compared with a figure of merit $\frac{2}{\lambda} P_0$ in bulk, for fiber one finds $P_0 L_{\text{eff}}/A_{\text{eff}}$. The ratio of the two,

$$\frac{\lambda L_{\text{eff}}}{2A_{\text{eff}}},$$

is quite large. Take as typical values $\lambda = 1\,\mu\text{m}$, $A_{\text{eff}} = 50\,\mu\text{m}^2$, and $L_{\text{eff}} = 20\,\text{km}$; then it is 2×10^8. We conclude that even mild nonlinearity can have very noticeable consequences in fibers.

9.2 Kerr Nonlinearity

It was shown in Chap. 4 that the refractive index can have an intensity dependence

$$n = n_0 + n_2 I,$$

where for fused silica $n_2 \approx 3 \times 10^{-20}$ m^2/W. This is also known as the optical Kerr effect. It provides a minute modification of the index by the n_2 term, which does not influence the structure of the modes because $n_2 I \ll (n_K - n_M)$. However, the $n_2 I$ term does modify the *phase* of the propagating light.

Consider a light wave with power P launched into the fiber. The effective cross-sectional area of the mode is A_{eff}. Generally, a wave of the type

$$\cos(\omega t - kz)$$

propagates such that after distance $z = L$ the phase has the value kL. Therefore,

$$\phi = kL = k_0 n L = k_0 \left(n_0 + n_2 I\right) L = \frac{2\pi}{\lambda} \left(n_0 + \frac{n_2}{A_{\text{eff}}} P\right) L. \tag{9.7}$$

This phase can be split into a linear and a nonlinear contribution:

$$\phi_{\text{lin}} = \frac{2\pi}{\lambda} n_0 L$$

and

$$\phi_{\text{nl}} = \frac{2\pi}{\lambda} \frac{n_2}{A_{\text{eff}}} PL = \frac{\omega_0 n_2}{c A_{\text{eff}}} PL = \gamma PL \quad \text{with} \quad \gamma = \frac{\omega_0 n_2}{c A_{\text{eff}}}. \tag{9.8}$$

We will extensively use the coefficient of nonlinearity γ below.

With the help of a reference wave providing a phase reference, one could easily measure the nonlinear phase shift. Let, e.g., $\lambda = 1.5 \, \mu$m, $n_2 = 3 \times 10^{-20}$ m^2/W, $A_{\text{eff}} = 40 \, \mu$m^2, $P = 1$ W, and $L = 1$ km. Then we obtain $\gamma = 3.14 \times 10^{-3}/$(Wm) and $\phi_{\text{nl}} = 3.14$ rad which corresponds to one half of a wavelength—certainly easily measured interferometrically.

9.3 Nonlinear Wave Equation

Now we will set up a wave equation which takes all relevant effects into account. As it will turn out, it fully suffices to write an equation for the *field envelope*. This implies that the variable in the equation is not the field strength itself, but only its amplitude; a term oscillating at the optical frequency is removed. This is justified whenever the envelope changes much more slowly than the field, in other words, when within the duration of the shortest pulses of light there are many oscillation periods of the field. Some current research at the forefront of laser physics now pushes this limit, but there are no direct consequences for fiber-optic applications.

9.3.1 Envelope Equation Without Dispersion

At first we want to see how an envelope equation is set up in a simple case. We start from the linear wave equation derived in Chap. 3 [compare Eq. (3.23)]

$$\nabla^2 E - \frac{n^2}{c^2} \frac{\partial^2 E}{\partial t^2} = 0 \tag{9.9}$$

and use the following ansatz for E:

$$E(x, y, z, t) = A(x, y, z, t)\, e^{i(\omega_0 t - \beta_0 z)}. \tag{9.10}$$

This describes a wave traveling in positive z direction, with carrier frequency ω_0 and propagation constant $\beta_0 = \omega_0 n/c$. All other space and time dependence is lumped into the *envelope* $A(x, y, z, t)$. We assume that these dependencies are variable only at a much slower scale, so that any spatial or temporal derivative of A, in comparison to the same derivative of the exponential term, is of order $\epsilon \ll 1$. In physical terms this means that the envelope changes only very little during one oscillation period or over one wavelength. This approximation is justified as long as even the shortest light pulses contain several oscillations of the field.

In order to insert Eq. (9.10) in Eq. (9.9), we first take the derivatives (and write A for convenience, without reiterating its arguments):

$$\frac{\partial E}{\partial x} = \frac{\partial A}{\partial x} e^{i(\omega_0 t - \beta_0 z)}$$

$$\frac{\partial^2 E}{\partial x^2} = \frac{\partial^2 A}{\partial x^2} e^{i(\omega_0 t - \beta_0 z)}$$

$$\frac{\partial E}{\partial y} = \frac{\partial A}{\partial y} e^{i(\omega_0 t - \beta_0 z)}$$

$$\frac{\partial^2 E}{\partial y^2} = \frac{\partial^2 A}{\partial y^2} e^{i(\omega_0 t - \beta_0 z)}$$

$$\frac{\partial E}{\partial z} = \frac{\partial A}{\partial z} e^{i(\omega_0 t - \beta_0 z)} - i\beta_0 A\, e^{i(\omega_0 t - \beta_0 z)}$$

$$\frac{\partial^2 E}{\partial z^2} = \left(\frac{\partial^2 A}{\partial z^2} - 2i\beta_0 \frac{\partial A}{\partial z} - \beta_0^2 A \right) e^{i(\omega_0 t - \beta_0 z)}$$

$$\frac{\partial E}{\partial t} = \frac{\partial A}{\partial t} e^{i(\omega_0 t - \beta_0 z)} + i\omega_0 A\, e^{i(\omega_0 t - \beta_0 z)}$$

$$\frac{\partial^2 E}{\partial t^2} = \left(\frac{\partial^2 A}{\partial t^2} + 2i\omega_0 \frac{\partial A}{\partial t} - \omega_0^2 A \right) e^{i(\omega_0 t - \beta_0 z)}.$$

Now we insert

$$-2i\beta_0\frac{\partial A}{\partial z} + \frac{\partial^2 A}{\partial x^2} + \frac{\partial^2 A}{\partial y^2} + \frac{\partial^2 A}{\partial z^2} - \beta_0^2 A - \frac{n^2}{c^2}\frac{\partial^2 A}{\partial t^2} - 2i\omega_0\frac{n^2}{c^2}\frac{\partial A}{\partial t} + \omega_0^2\frac{n^2}{c^2}A = 0.$$

Obviously the exponential factor is cancelled out, and we succeeded in finding an equation of motion for the envelope! The two terms proportional to A mutually cancel because $\beta_0^2 = \omega_0^2 n^2/c^2$. The derivatives in x and y directions can be combined using the transverse nabla operator ∇_{xy}. Then this is what remains:

$$- 2i\beta_0\frac{\partial A}{\partial z} + \nabla_{xy}^2 A + \frac{\partial^2 A}{\partial z^2} - \frac{n^2}{c^2}\frac{\partial^2 A}{\partial t^2} - 2i\omega_0\frac{n^2}{c^2}\frac{\partial A}{\partial t} = 0. \qquad (9.11)$$

The kth derivative is of order $\epsilon^k \ll 1$; therefore, in leading order we only retain

$$2i\beta_0\frac{\partial A}{\partial z} + 2i\omega_0\frac{n^2}{c^2}\frac{\partial A}{\partial t} = 0$$

or

$$\frac{\partial A}{\partial z} + \frac{n}{c}\frac{\partial A}{\partial t} = 0. \qquad (9.12)$$

This describes an envelope which propagates with constant shape and with velocity $v = c/n$, the phase velocity. This is no wonder—we have neglected both dispersion and nonlinearity so far!

Indeed the equation is better than valid only in first order. In Eq. (9.12), we note that a z derivative of A (not of E!) is the same as a t derivative up to a factor of $-n/c$. One can do the same trick twice:

$$\frac{\partial^2 A}{\partial z^2} = \left(-\frac{n}{c}\right)^2\frac{\partial^2 A}{\partial t^2}.$$

Now one sees that the third and fourth term in Eq. (9.11) cancel out. At next higher order this is what remains:

$$\frac{\partial A}{\partial z} + \frac{n}{c}\frac{\partial A}{\partial t} + \frac{i}{2\beta_0}\nabla_{xy}^2 A = 0. \qquad (9.13)$$

This is different only in the term for transverse change. In a fiber, diffraction is compensated by the waveguiding mechanism so that derivatives with respect to x and y are zero.

The term containing the first temporal derivative can be scaled out. To do so we introduce a comoving frame of reference

$$\tau = t - \frac{n}{c}z$$

and obtain

$$\frac{\partial A}{\partial z} + \frac{i}{2\beta_0} \nabla_{xy}^2 A = 0. \tag{9.14}$$

This equation describes the transverse diffraction of a wave packet. Without transverse change the pulse shape is constant.

In a linear fiber, free from dispersion and loss, a wave packet propagates without change of shape with a velocity equal to the phase velocity.

Now we must incorporate the effects of dispersion, loss, and nonlinearity. This implies that n will now become a function of frequency (or wavelength) and power (or amplitude), and the amplitude a function of position (or distance).

9.3.2 Introducing Dispersion by a Fourier Technique

A Fourier transform is used to convert from a function in the temporal domain to the corresponding function in the frequency domain (or vice versa), or from spatial position to spatial frequency, etc. Let us begin with a time-frequency transformation of some function $F(t, z)$.

We use the abbreviation

$$\tilde{F}(\omega) = \mathsf{FT}\big(F(t)\big)$$

to denote the Fourier transform $\mathsf{FT}(\dots)$, spelled out as

$$\tilde{F}(\omega) = \int_{-\infty}^{+\infty} F(t)\,e^{i\omega t}\,dt.$$

Then it holds that

$$\mathsf{FT}\left(i\frac{\partial}{\partial t}F(t)\right) = \int_{-\infty}^{+\infty} i\frac{\partial F(t)}{\partial t}\,e^{i\omega t}\,dt.$$

By partial integration one finds:

$$\mathsf{FT}\left(i\frac{\partial}{\partial t}F(t)\right) = ie^{i\omega t}\,F(t)\Big|_{-\infty}^{+\infty} - \int_{-\infty}^{+\infty} iF(t)\,\frac{\partial}{\partial t}e^{i\omega t}\,dt$$

$$= 0 + \omega \int_{-\infty}^{+\infty} F(t)\,e^{i\omega t}\,dt$$

$$= \omega\,\mathsf{FT}\big(F(t)\big)$$

$$= \omega\,\tilde{F}(\omega).$$

In the second line, we assumed that $F(t = \pm\infty) = 0$, which means that for distant times $F(t)$ decays. Then we conclude that a Fourier transform acts simply to replace

$$\tilde{F}(\omega) \leftrightarrow F(t),$$

$$\omega\tilde{F}(\omega) \leftrightarrow i\frac{\partial}{\partial t} F(t),$$

$$\omega^k\tilde{F}(\omega) \leftrightarrow \left(i\frac{\partial}{\partial t}\right)^k F(t).$$

A factor ω in the frequency domain corresponds to an operator $i(\partial/\partial t)$ in the time domain. The same logic applies also if instead of ω we use $\Delta\omega = \omega - \omega_0$ throughout, with fixed ω_0.

Quite similarly we can do a position–spatial frequency transformation, i.e., a transformation between coordinate z and wave number β.

$$\tilde{F}(\beta) = \mathsf{FT}(F(z)) = \int_{-\infty}^{+\infty} F(z)\,e^{-i\beta z}\,dz. \tag{9.15}$$

The sign in the exponent stands for a wave traveling "to the right" or toward positive z. With an analogous calculation one finds the following correspondence:

$$\tilde{F}(\beta) \leftrightarrow F(z),$$

$$\beta\tilde{F}(\beta) \leftrightarrow -i\frac{\partial}{\partial z} F(z),$$

$$\beta^k\tilde{F}(\beta) \leftrightarrow \left(-i\frac{\partial}{\partial z}\right)^k F(z).$$

Again, this is also valid if $\Delta\beta = \beta - \beta_0$ instead of β. We will now apply this insight to the series expansion of the wave number as a function of frequency, which is

$$\beta = \beta_0 + \Delta\omega\beta_1 + \Delta\omega^2\frac{\beta_2}{2} + \Delta\omega^3\frac{\beta_3}{6} + \cdots. \tag{9.16}$$

Now we make the transition to the time domain by inserting operators and apply them to $A(z, t)$:

$$\Delta\beta = \beta_1\Delta\omega + \frac{\beta_2}{2}\Delta\omega^2 + \frac{\beta_3}{6}\Delta\omega^3 + \cdots \tag{9.17}$$

$$-i\frac{\partial}{\partial z}A = i\beta_1\frac{\partial}{\partial t}A - \frac{\beta_2}{2}\frac{\partial^2}{\partial t^2}A - i\frac{\beta_3}{6}\frac{\partial^3}{\partial t^3}A + \cdots. \tag{9.18}$$

Obviously $1/\beta_1$ is the group velocity. It is hardly a surprise (but it is nice!) that now we have an equation containing group velocity, not phase velocity.

It often suffices to truncate the series after the β_2 term. Then we are left with

$$i\frac{\partial}{\partial z}A + i\beta_1\frac{\partial}{\partial t}A - \frac{\beta_2}{2}\frac{\partial^2}{\partial t^2}A = 0. \tag{9.19}$$

If we now use a frame of reference comoving with group velocity $1/\beta_1$ by introducing $T = t - \beta_1 z$, we are left with

$$i\frac{\partial}{\partial z}A - \frac{\beta_2}{2}\frac{\partial^2}{\partial T^2}A = 0 \tag{9.20}$$

where A is shorthand for $A(z, T)$.

One might include further terms from the series (3.18); then, additional terms will appear in the wave equation (9.20). For third-order dispersion, the term would be $-(\beta_3/6)(\partial^3 A/\partial T^3)$ on the LHS. One can also introduce a modification of the wave number due to nonlinearity by adding to $\Delta\beta$ a term $\Delta\beta_{\mathrm{NL}} = n_2 I \beta_0$. Moreover, by using a complex wave number, one can describe power loss, e.g., with $\Delta\beta_{\mathrm{loss}} = i\alpha/2$ where α is Beer's absorption coefficient. By including all these, the equation becomes

$$\Delta\beta = \beta_1\Delta\omega + \frac{\beta_2}{2}\Delta\omega^2 + \frac{\beta_3}{6}\Delta\omega^3 + \beta_0 n_2 I + i\frac{\alpha}{2}, \tag{9.21}$$

$$-i\frac{\partial}{\partial z}A = i\beta_1\frac{\partial}{\partial T}A - \frac{\beta_2}{2}\frac{\partial^2}{\partial T^2}A - i\frac{\beta_3}{6}\frac{\partial^3}{\partial T^3}A + \beta_0 n_2 I A + i\frac{\alpha}{2}A. \tag{9.22}$$

The prefactor $\beta_0 n_2 I$ can also be written as $(\omega_0/c)\,n_2(|A|^2/A_{\mathrm{eff}})$. This is done by taking the amplitude A as the square root of power, a choice not in accord with SI conventions but quite common in the nonlinear optics literature. A_{eff} is the effective mode area in the fiber over which the power $|A|^2$ is distributed to give the intensity I. Then this is left:

$$i\frac{\partial}{\partial z}A - \frac{\beta_2}{2}\frac{\partial^2}{\partial T^2}A - i\frac{\beta_3}{6}\frac{\partial^3}{\partial T^3}A + \gamma|A|^2A + i\frac{\alpha}{2}A = 0. \tag{9.23}$$

9.3.3 The Canonical Wave Equation: NLSE

We now consider an important special case: By neglecting third-order dispersion and loss, one retains the *nonlinear Schrödinger equation* (NLSE):

$$i\frac{\partial}{\partial z}A - \frac{\beta_2}{2}\frac{\partial^2}{\partial T^2}A + \gamma|A|^2A = 0. \tag{9.24}$$

It derives its name from Erwin Schrödinger, of quantum mechanics fame, because it has a close similarity with the quantum mechanical Schrödinger equation

$$i\frac{\partial}{\partial t}\psi - \frac{\partial^2}{\partial z^2}\psi + V\psi = 0. \tag{9.25}$$

Coefficients β_2 and γ in Eq. (9.24) can be scaled out by using suitable units for time and amplitude, as we shall see in Sect. 9.3.5. The essential part of the comparison is this:

- $|A|^2$ stands in for the potential V; here it has a specific shape.

> The field itself generates the potential which then acts back on the field distribution.

- Space and time coordinates have switched roles.

> The quantum mechanic Schrödinger equation describes how a spatially localized wave packet spreads out spatially as time goes by.
>
> The nonlinear Schrödinger equation of fiber optics describes how a temporally short light pulse spreads out in its duration as it propagates over some distance.

To wrap up: n becomes a function of intensity, but since $n_2 I/n_0 \ll 1$ there is no influence on mode geometry, field distribution, etc. in leading order. We may safely neglect transverse changes in the waveguide. On the other hand, the wave equation is no more linear, and the superposition principle does not hold. With a Fourier technique we have introduced frequency dependence.

Below we will use the following form of the wave equation as the reference version:

$$i\frac{\partial A}{\partial z} = +\frac{\beta_2}{2}\frac{\partial^2 A}{\partial T^2} - \frac{i}{2}\alpha A - \gamma|A|^2 A. \tag{9.26}$$

A modification of the pulse shape (LHS) can occur through terms for dispersion, loss, and nonlinearity (RHS). For dispersion we only use the leading order. For the nonlinearity only the Kerr effect is considered. Of course one can go further; there are plenty of research papers dealing with higher-order dispersion, temporal effects in the nonlinearity, or polarization effects. But for now we will use Eq. (9.26): Already this simplified version presents us with a few surprises. To get familiar with this matter, let us first distinguish a few limiting cases.

9.3.4 Discussion of Contributions to the Wave Equation

9.3.4.1 Absorption Alone

For $\beta_2 = \gamma = 0$, Eq. (9.26) is reduced to

$$\frac{\partial A}{\partial z} = -\frac{\alpha}{2}A, \tag{9.27}$$

which is solved by

$$A = A_0 e^{-\frac{\alpha}{2}z}. \tag{9.28}$$

After a characteristic length $L_\alpha = 1/\alpha$, the amplitude decays to $e^{-1/2}$ of its initial value and thus the power to $1/e$. This makes it clear that α is Beer's absorption coefficient.

9.3.4.2 Dispersion Alone

For $\alpha = \gamma = 0$, Eq. (9.26) is reduced to

$$i\frac{\partial A}{\partial z} = \frac{1}{2}\beta_2\frac{\partial^2 A}{\partial T^2}. \tag{9.29}$$

This is formally similar to a paraxial wave equation for diffraction in only one spatial direction. A formal solution can be found with Fourier techniques; we have already seen some results in Chap. 4. Here it may suffice to convince ourselves: If one sets

$$A = A_0 e^{i(\Omega T + kz)},$$

then it follows that

$$k = \frac{\beta_2}{2}\Omega^2,$$

and the wave vector becomes frequency-dependent; this corresponds to the pulse broadening discussed before. Again we can define a characteristic length scale; we choose $L_D = T_0^2/|\beta_2|$. As already discussed, a Gaussian pulse of width T_0 will widen after this distance by a factor of $\sqrt{2}$.

9.3.4.3 Nonlinearity Alone

Finally we can also define a characteristic length for the nonlinearity. For $\beta_2 = \alpha = 0$, Eq. (9.26) is reduced to

$$i\frac{\partial A}{\partial z} = -\gamma |A|^2 A. \tag{9.30}$$

As before, $A = A(z, T)$. For convenience we write $|A(0, T)|^2 = P_0(T)$ for the initial power profile of the light pulse. Then the equation is solved by

$$A = \sqrt{P_0(T)}\; e^{i\gamma P_0(T)z}. \tag{9.31}$$

Several aspects are remarkable in this result. First, the power profile remains unchanged; $P_0(T)$ does not contain z. Second, there is a characteristic length $L_{NL} = 1/\gamma P_0(T)$; this does contain power and is therefore power-dependent. At pulse center (maximum of $P_0(T)$), the value is different from that in the slope. After propagation over characteristic length, the pulse peak has acquired a phase factor of e^i, corresponding to a phase shift of 1 rad.

Looking at it from another side, after some given distance the phase factor is different for different positions within the pulse profile. It is largest at pulse center and tapers off in the wings. In other words, the pulse acquires a phase modulation known as *self-phase modulation* (Fig. 9.2). This is a crucial insight for much of the remainder of this discussion, and it will be discussed in more detail below.

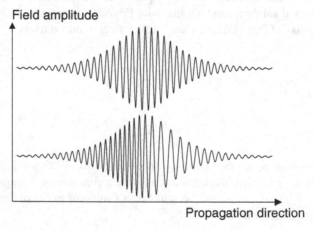

Fig. 9.2 Impact of self-phase modulation. *Top*: Within a light pulse, the optical field oscillates at a certain frequency; the envelope defines the pulse duration. Let such a pulse be launched into a medium where there is self-phase modulation. *Bottom*: After passing through that medium, phases have shifted. Propagation is maximally slowed down at pulse center; therefore, waves appear pulled apart in the rising slope (*right*) whereas they are compressed in the trailing slope (*left*)

Self-phase modulation may also be interpreted as self-frequency modulation since phase and frequency modulation are closely related. Using the nonlinear phase $\phi_{nl} = \gamma PL$, at some position L, the frequency deviation due to nonlinearity is

$$\Delta\omega = \frac{d\phi_{nl}}{dT} = \gamma L \frac{dP}{dT},$$

as shown in Fig. 9.3. A frequency modulation across the pulse is frequently called *chirp*; a self-frequency-modulated pulse is said to be chirped.

Fig. 9.3 Sketch to explain the connection between self-phase modulation and self-frequency modulation. *Top*: Let the power profile be bell-shaped as shown, e.g., $\text{sech}^2(T)$. *Center*: The nonlinear phase follows the power profile. *Bottom*: The instantaneous frequency is given by the temporal derivative of the phase and thus follows the temporal derivative of power

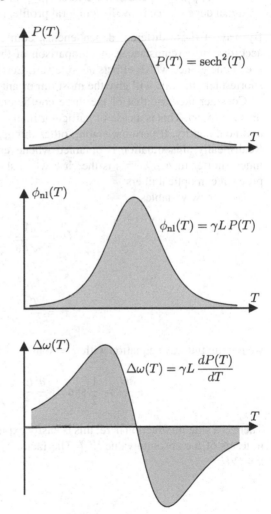

9.3.5 Dimensionless NLSE

Let us compare the characteristic lengths (we use $P_0(0) = P_0$):

- $L_\alpha = 1/\alpha$ depends solely on fiber properties.
- $L_{NL} = 1/\gamma P_0 = (cA_{eff})/(n_2\omega_0 P_0)$ depends on the fiber (on A_{eff} and n_2) and on the signal (ω_0 and P_0). The signal dependence is only by the instantaneous value, not the temporal profile.
- $L_D = T_0^2/|\beta_2|$ depends on the fiber ($|\beta_2|$) and on the signal (T_0). This time the signal dependence is by the temporal profile, not the absolute values.

By way of these different dependencies, combinations of all kinds are possible according to circumstances. A comparison of the characteristic lengths allows to see at one glance which effects are relevant. Certainly the effect represented by the shortest length scale will give the most important contribution.

Consider the case that of the three coefficients in Eq. (9.26), only $\alpha = 0$ and thus $L_\alpha \to \infty$. This is a case of particular interest, an interplay between dispersion and nonlinearity. It permits, among other things, pulse compression and solitons. Realistically, this situation is obtained whenever L_D and L_{NL} are comparable but much shorter than L_α. This is the case when short (e.g., picosecond) light pulses propagate in optical fibers.

Using new variables

$$U = \frac{A}{\sqrt{P_0}}$$
$$\zeta = \frac{z}{L_D} \tag{9.32}$$
$$\tau = \frac{T}{T_0}$$

we rewrite the wave equation with $\alpha = 0$ as

$$i \frac{\partial U}{\partial \zeta} = \frac{1}{2} \operatorname{sgn} \beta_2 \frac{\partial^2 U}{\partial \tau^2} - \frac{L_D}{L_{NL}} |U|^2 U. \tag{9.33}$$

There is a signum function here; this is easily explained by the fact that L_D is defined in terms of the absolute value $|\beta_2|$. The factor L_D/L_{NL} can be scaled out by using $u = NU$:

$$N = \sqrt{\frac{L_D}{L_{NL}}} = \sqrt{P_0 T_0^2 \frac{\gamma}{|\beta_2|}}. \tag{9.34}$$

Now the equation takes the form

$$i\frac{\partial u}{\partial \zeta} \pm \frac{1}{2}\frac{\partial^2 u}{\partial \tau^2} + |u|^2 u = 0.$$

(9.35)

This is the celebrated *nonlinear Schrödinger equation* in its dimensionless form, as it is most often found in literature. The sign corresponds to $-\mathrm{sgn}\,\beta_2$ and stands for anomalous $(+)$ or normal $(-)$ group velocity dispersion, respectively.

There are two possibilities for this sign. Therefore the reader may well guess at this point that there will be two distinct types, or classes, of solutions. Within each class the numerical values of parameters merely act as scale factors for the solution but do not affect its functional type. But, of course, there is also the very special case of $\beta_2 = 0$ which requires a careful analysis of its own. Just because the β_2 term in the series expansion, Eq. (4.19), is zero does not at all imply that all the higher-order terms vanish as well. In that case, higher-order dispersion must be taken into account.

9.4 Solutions of the NLSE

In this section, we study solutions of the nonlinear Schrödinger equation (9.35). First of all, there is the *trivial solution*

$$u \equiv 0.$$

(9.36)

There is no need to waste time on this case.

9.4.1 Modulational Instability

More interesting is the *continuous wave solution*

$$u = u_0\, e^{i|u_0|^2 \zeta};$$

(9.37)

in the dimensional units of Eq. (9.24) this reads

$$A = \sqrt{P_0}\, e^{i\gamma P_0 z}.$$

(9.38)

For this solution it is important to check the stability. This can be done by inserting a small perturbation away from the solution in a procedure called *linear stability analysis*. The solution is stable when the perturbation produces an opposite restoring action so that it decays with time. In the opposite case—when the perturbation keeps growing—the solution is unstable.

The continuous wave solution can be either stable or unstable depending on the sign of dispersion (this is shown, e.g., in Chap. 5 of [1]). For anomalous dispersion perturbations grow. The term *modulational instability* indicates that perturbations at certain frequencies grow faster than others.[1] The growth rate, or instability gain, has a frequency dependence [1]

$$g_\omega = |\beta_2| \omega \sqrt{\Omega^2 - \omega^2} \qquad (9.39)$$

with

$$\Omega^2 = \frac{4}{|\beta_2| L_{NL}} = \frac{4\gamma \hat{P}}{|\beta_2|},$$

where $\hat{P}$ denotes the peak value of the power.

The gain maximum is

$$g_{max} = 2\gamma \hat{P} \qquad (9.40)$$

and occurs at the frequency

$$\omega_{max} = \frac{1}{2}\sqrt{2}\,\Omega = \sqrt{\frac{2\gamma \hat{P}}{|\beta_2|}}. \qquad (9.41)$$

Figure 9.4 shows the spectral profile of this gain.

9.4.2 The Akhmediev Breather

In the presence of noise (and one never has *absence* of noise), perturbations will grow. Noise covers all frequencies alike, so it is most likely that the Fourier component at ω_{max} grows fastest, and soon dominates. The result is a periodic modulation with frequency ω_{max} on top of the continuous background wave, and the power of that modulation grows exponentially with g_{max}.

This exponential growth can not go on forever, of course. The 'small perturbation' approach breaks down once the modulation amplitude is no longer much smaller than the continuous wave background. Linear stability analysis tells us nothing about the further, ultimate fate of the modulation.

---

[1]In hydrodynamics there is the analogous phenomenon under the name of Benjamin–Feir instability [25].

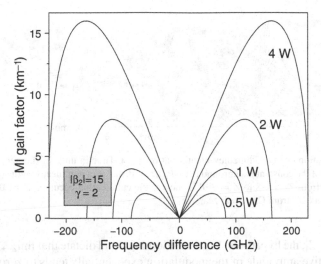

Fig. 9.4 Gain factor of modulational instability. It is assumed that $\beta_2 = 15\,\text{ps}^2/\text{km}$ and $\gamma = 2/(\text{Wkm})$. $\hat{P}$ is given as a parameter

It turns out that a full mathematical description of the continued process had been given as early as 1986 in a formal mathematical publication [4]; however, it took decades until this result was translated into an experimental verification. There is a formal solution of the NLSE which describes the complete development from infinitesimally perturbed continuous wave through exponential growth of a weak modulation to full-size modulation, and beyond. After the lead author of the original publication, this solution is now called an Akhmediev breather.

It takes the mathematical form

$$A(Z, T) = \sqrt{P_0} \left[1 + \frac{2(1 - 2a) \cosh(bZ) + ib \, \sinh(bZ)}{\sqrt{2a} \, \cos(\omega T) - \cosh(bZ)} \right] \exp(iZ) \qquad (9.42)$$

where distance is measured in nonlinear lengths, $Z = z/L_{\text{NL}}$, and

$$0 \leq a \leq 1/2 \qquad \qquad b = \sqrt{8a - 16a^2}$$

$$\omega = \omega_c \sqrt{1 - 2a} \qquad \omega_c = \sqrt{\frac{4\gamma P_0}{|\beta_2|}}$$

This expression describes a continuous wave of amplitude $\sqrt{P_0}$ evolving with a phase $\exp(iZ)$; inside the parenthesis, the fraction provides the modulation. As the $\cos(\omega t)$ term shows, this modulation is oscillatory in time. The real part of the remaining terms consists of $\cosh(Z)$ functions and is thus symmetric with respect to Z; the imaginary part contains $\sinh(Z)$ a is therefore antisymmetric. This indicates that the phase of the modulation passes through zero at $Z = 0$.

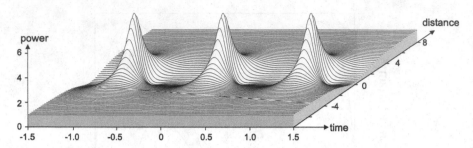

Fig. 9.5 Evolution of the Akhmediev breather at $a = 1/4$. Time in units of the period $(T\omega/2\pi)$, power in units of the constant background $(|A^2/P_0)$, and distance in nonlinear lengths $(Z = z/L_{NL})$. The evolution from $Z = -8$ to $Z = +8$ is shown; the culmination occurs at $Z = 0$. This figure was drawn with help from Christoph Mahnke

For large $|Z|$, the hyperbolic functions dominate and dictate that $\lim_{Z\to\pm\infty} |A|^2 = P_0$. The relative amplitude of the modulation exponentially tends to zero as the cos term is of order unity while the hyperbolic functions scale as $\exp(bZ)$. In the other extreme, at $Z = 0$ we find the largest modulation amplitude from inserting in Eq. (9.42) and rewriting as

$$A(0, T) = \sqrt{P_0} \, \frac{\frac{1-4a}{\sqrt{2a}} + \cos\omega T}{\cos\omega T - \frac{1}{\sqrt{2a}}} \, \exp(iZ). \tag{9.43}$$

In other words, the Akhmediev breather is a structure that emerges from a constant power background. As the light propagates down the fiber, a modulation grows exponentially, reaches its maximum amplitude at $Z = 0$, then decays again. The modulation is periodic in time; at the culmination point it takes the form of a sequence of pulses. For an illustration we concentrate on the case of maximum gain, $a = 1/4, b = 1$; see Figs. 9.5 and 9.6. Here Eq. (9.43) is reduced to

$$A(0, T)\Big|_{1/4} = \sqrt{P_0} \, \frac{\cos\omega T}{\cos\omega T - \sqrt{2}} \, \exp(iZ) \tag{9.44}$$

This is a periodic sequence of peaks, with their maximum amplitude at a value of $A_{max} = -(1 + \sqrt{2})\sqrt{P_0}$, so that $|A|^2_{max} \approx 5.828P_0$. In between the main peaks there are secondary peaks of amplitude $A_{sec} = (\sqrt{2} - 1)\sqrt{P_0}$ so that $|A_{sec}|^2 \approx 0.172P_0$. The zeroes of the amplitude appear as dimples in the power profile.

9.4.3 The Fundamental Soliton

The next solution of the nonlinear Schrödinger equation exists in the case of anomalous dispersion (for $\beta_2 < 0$, i.e., on the long wavelength side of the zero-

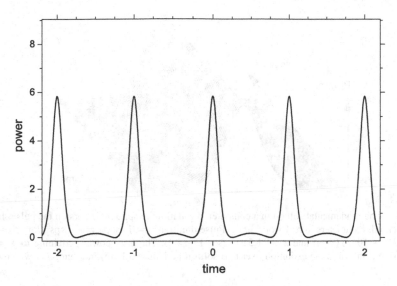

Fig. 9.6 Cross section of the Akhmediev breather in Fig. 9.5 at its culmination point at $Z = 0$. Units as in Fig. 9.5

dispersion wavelength) and takes the form

$$u = \text{sech}(\tau)\, e^{i\zeta/?} \tag{9.45}$$

or, in dimensional units,

$$A = \sqrt{P_1}\, \text{sech}\left(\frac{T}{T_0}\right) e^{i\gamma P_1 z/2}. \tag{9.46}$$

The time-dependent part is a hyperbolic secant (or "sech") function; it therefore describes a bell-shaped pulse. (Some information on sech is gathered in Chap. 17.) The position-dependent part is an exponential function acting as a phase factor; it rotates through 2π over the distance $\zeta = 4\pi$. The pulse shape (power profile) is constant since the only dependence on position is in the exponential, see Figs. 9.7, 9.8, and for comparison Fig. 9.9. The pulse shape is also stable in the sense that a certain perturbation away from the precise shape can heal out: a remarkable property which we are going to discuss some more!

This solution is a "solitary" solution in the sense that—in marked contrast to solutions of linear differential equations—the peak amplitude is fixed; if the solution is multiplied by any constant real factor other than unity, the result is not a solution. From this property derives the name of this solution: it is called a *soliton*. Indeed this is just one particular representative of a wider class of solitons, and it is more precisely called the *fundamental soliton* or $N = 1$ soliton for reasons which we will see shortly.

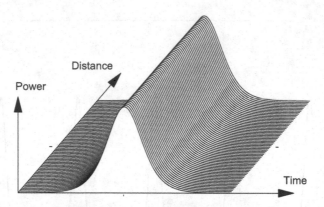

Fig. 9.7 The fundamental soliton in a computer simulation: in spite of dispersion the pulse shape is preserved. Parameters in real-world units: Pulse duration $\tau = 1$ ps, $\beta_2 = -18\,\text{ps}^2/\text{km}$, $\beta_3 = 0$, $\gamma = 2.5 \times 10^{-3}/(\text{Wm})$ and $\lambda = 1.5\,\mu\text{m}$ (see Fig. 4.5). The peak power pertaining to $N = 1$ is 22.37 W. Shown is the evolution over two soliton periods (56.16 m) in a temporal window of ± 5 ps

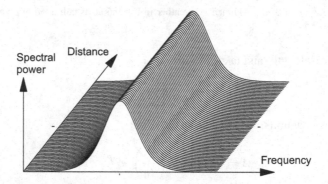

Fig. 9.8 The spectrum of the fundamental soliton in a computer simulation, using the same parameters as in Fig. 9.7. In spite of nonlinearity the spectral shape is preserved

Fiber solitons are light pulses which do not change their shape during propagation even though at the same time both dispersion and self-phase modulation act on it. It should be more than obvious that pulses with this property must be highly interesting for applications!

When we convert the dimensionless solution back to real-world units, we note: The condition for a fundamental soliton is

$$L_D = L_{NL} \quad \Leftrightarrow \quad N = 1. \tag{9.47}$$

By virtue of Eq. (9.34), the peak power of the $N = 1$ soliton is $\hat{P}_1 = |\beta_2|/\gamma T_0^2$. In other words: All $N = 1$ soliton share the property that the product of peak power and

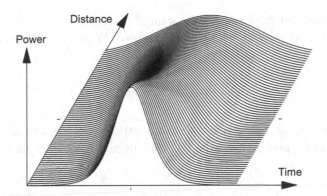

Fig. 9.9 For comparison, here we show the dispersive broadening of a sech^2 pulse. Shown is a pulse with an initial duration (FWHM) of 1 ps as it broadens over a propagation distance of 20 m. Of course, the spectrum is preserved in the process—exactly as in Fig. 9.8. The dispersion is $\beta_2 = -18\,\text{ps}^2/\text{km}$ and $\beta_3 = 0$

the square of the pulse duration is a constant, determined solely by fiber parameters:

$$\hat{P}_1 T_0^2 = \frac{|\beta_2|}{\gamma} \tag{9.48}$$

To illustrate the importance of this product, we note that the energy of the soliton is the time integral of power:

$$E_1 = \int_{-\infty}^{+\infty} P(t)\, dt = 2\hat{P}_1 T_0. \tag{9.49}$$

Then, $\hat{P}_1 T_0^2$ is something like the time integral of energy, a quantity which in classical mechanics is referred to as *action*. In quantum mechanics, the time integral of the amplitude envelope is called the pulse area [5]; here, action is thus the square of pulse area.

> In a given fiber a soliton can have just about any duration, peak power, or energy. However, these quantities always combine such that the action has the same value as given by Eq. (9.48).

If we also insert the relations between T_0 and τ ($\tau = 2\mathcal{Z}T_0$), between β_2 and D ($D = -(\omega/\lambda)\beta_2$), and between n_2 and γ ($\gamma = n_2(\omega_0/cA_{\text{eff}})$), then we obtain

$$\hat{P}_1 = \mathcal{Z}^2 \frac{\lambda^3}{\pi^2 c} \frac{|D|A_{\text{eff}}}{n_2} \frac{1}{\tau^2}. \tag{9.50}$$

Here we introduced the numerical constant $\mathcal{Z} = \cosh^{-1}\sqrt{2} = 0.8813\ldots$ for convenience. If we wish to find the average power $\bar{P}_1$ rather than the peak power $\hat{P}_1$, we write

$$\bar{P}_1 = \hat{P}_1 \cdot \frac{\tau}{T_{\text{rep}}} \frac{1}{\mathcal{Z}}, \tag{9.51}$$

where T_{rep} denotes the repetition rate of the experiment. If we finally insert the expression for $\hat{P}_1$, then we can write $\bar{P}_1$ explicitly as a function of easily measured quantities:

$$\bar{P}_1 = \mathcal{Z} \frac{\lambda^3 |D| A_{\text{eff}}}{\pi^2 c n_2 T_{\text{rep}}} \frac{1}{\tau}. \tag{9.52}$$

The soliton energy is then

$$E_1 = \frac{1}{\mathcal{Z}} \hat{P}_1 \tau. \tag{9.53}$$

Rather than expressing the energy in Joules, it can be interesting to write it as photon number which is found via $n_{\text{phot}} = E_1/(h\nu)$ (see Table 9.1).

From these equations we can draw several conclusions: Obviously a soliton can exist for any τ; it is straightforward to calculate its power. The shorter the duration, the higher the power required to form the soliton. Table 9.1 shows typical orders of magnitude. The table also mentions a characteristic length z_0 which is commonly called the soliton period and is given by $z_0 = (\pi/2)L_{\text{D}}$. Again we convert to real-world units:

$$z_0 = \frac{1}{(2\mathcal{Z})^2} \frac{\pi^2 c \tau^2}{|D| \lambda^2}. \tag{9.54}$$

Table 9.1 Typical orders of magnitude of characteristic soliton parameters. The table gives the peak power $\hat{P}$, the soliton period z_0, its energy, and the photon number, always rounded to three significant digits

τ	$\hat{P}$	z_0	E_1	n_{phot}
1 ns	22.4 μW	28,100 km	25.4 fJ	1.92×10^5
100 ps	2.24 mW	281 km	254 fJ	1.92×10^6
10 ps	224 mW	2810 m	2.54 pJ	1.92×10^7
1 ps	22.4 W	28.1 m	25.4 pJ	1.92×10^8
100 fs	2.24 kW	281 mm	254 pJ	1.92×10^9

Assumed values are a wavelength of 1.5 μm, a fiber dispersion of $\beta_2 = -18\,\text{ps}^2/\text{km}$ corresponding to about $D = 15\,\text{ps}/(\text{nm km})$, and a nonlinearity coefficient $\gamma = 2.5 \times 10^{-3}/(\text{Wm})$ corresponding to $n_2 = 3 \times 10^{-20}\,\text{m}^2/\text{W}$, and $A_{\text{eff}} \approx 50\,\mu\text{m}^2$. In all cases the action $W = |\beta_2|/\gamma = 7.2 \times 10^{-24}\,\text{W s}^2$

To interpret this quantity, we reinsert the transformation Eq. (9.32) into Eq. (9.45) and find that the phase of the fundamental soliton rotates full circle after a distance

$$\zeta = \frac{z}{L_D} \overset{!}{=} 4\pi$$

$$z = 4\pi L_D \quad \Rightarrow \quad z = 8z_0 \qquad (9.55)$$

> The phase of the soliton rotates with respect to the comoving frame of reference such that it repeats itself after $8z_0$.

The soliton period z_0 plays a central role in the propagation of higher-order solitons, described in Sect. 9.4.6 below. It will therefore turn out to be useful to write the spatial period of the phase, $z = 8z_0$, in terms of physical quantities. The soliton condition Eq. (9.47) gives us two variants:

$$z = 4\pi L_D = \frac{4\pi T_0^2}{|\beta_2|}$$

$$= 4\pi L_{NL} = \frac{4\pi}{\gamma \hat{P}_1}. \qquad (9.56)$$

In a fundamental soliton, there is a compensation of the linear chirp by dispersion and of the nonlinear chirp by self-phase modulation. This is why fundamental solitons propagate with no change of shape. This makes the fundamental soliton the natural bit of optical data transmission.

In a real fiber, there are some practical complications which are not taken into account in the nonlinear Schrödinger equation. In particular, in the presence of a gradual mild energy loss, the shape will not stay constant, but will readjust according to Eq. (9.48). That is, when a soliton loses some of its energy, it will acquire a somewhat wider pulse shape.

The key here is that the loss occurs gradually (adiabatically): If the power is abruptly reduced (at a splice, say) so that instantly $N < 1/2$, the soliton is destroyed, and only linear waves carry the remaining energy away. Adiabaticity implies that the loss is negligible over a distance on the order of z_0. If in a very long fiber energy is continually drained away from the soliton according to a factor $E(z) = E(0)\,e^{-\alpha z}$, the pulse duration T_0 will initially increase to accommodate the adiabatic energy loss. This, however, also increases $L_D = T_0^2/|\beta_2|$. Then, the loss per L_D (in contrast to the loss per unit of z) keeps growing until, inevitably, at some point the rate of loss exceeds the adiabatic limit. Beyond that point, the soliton is soon destroyed. This process is described in detail in [8]; however, practical systems will always be laid out such that it does not come to this.

9.4.4 How to Excite the Fundamental Soliton

What happens when a pulse is launched into a fiber which corresponds exactly to a soliton, except that the peak power is raised or lowered with respect to the soliton peak power? The stable solution of the wave equation is the fundamental soliton, and therefore a soliton will emerge. To accommodate the deviation in power, however, it will acquire a duration and peak power which is different from the start values. If the power is reduced, a somewhat longer soliton will be generated (see Fig. 9.10); if it is raised, a shorter soliton (Fig. 9.11). This is the "self-healing property" alluded to above which makes solitons particularly robust entities.

To find the final shape (duration and peak power) of the soliton quantitatively, we consider this: The coefficients of dispersion β_2 and nonlinearity γ define that particular value of the action $W = |\beta_2|/\gamma$ which any soliton in the fiber must have. Even when the launch pulse does not fulfill this condition, the soliton must.

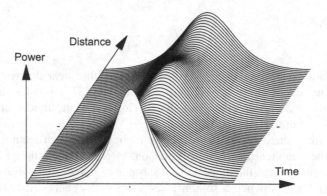

Fig. 9.10 Soliton formation when a pulse with $N = 0.8$ is launched in a computer simulation. Parameters as in Fig. 9.7 but with $N = 0.8$ and thus a peak power of 17.9 W

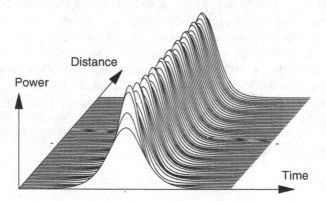

Fig. 9.11 Soliton formation when a pulse with $N = 1.2$ is launched in a computer simulation. Parameters as in Fig. 9.7 but with $N = 1.2$ and thus a peak power of 40.27 W

Therefore a rearrangement of simultaneously peak power, duration, and energy occurs. Of course there are many ways to vary three parameters at the same time, but they are not independent: $\hat{P}\tau = E$ and $E\tau = W$. Therefore, only a single additional constraint is required to make the rearrangement unique. This constraint comes from energy considerations:

By conservation of energy, certainly the soliton can not have more energy than the launch pulse. But it may have less: then the pulse sheds energy, and the energy of the soliton equals the launch energy minus the energy radiated off.

If the launched pulse happens to precisely match a fundamental soliton ($N = 1$), the radiated energy E_{rad} is zero. It is also zero for other integer values of N; larger integers describe higher-order solitons (see Sect. 9.4.6 below). Here, however, we are looking at noninteger N launch conditions. Let us specify $\hat{P}_{start} = (1 + \epsilon)^2 \hat{P}_1$ equivalent to $N = 1 + \epsilon$, but we keep $|\epsilon| < 1/2$.

The radiated energy is the initial energy minus the energy of the soliton. Using the energy of the $N = 1$ soliton as a convenient energy unit, this can be written as

$$E_{rad} = N^2 - (2N - 1).$$

Then

$$E_{rad} = \begin{cases} 0.25 & : & N = 0.5 \\ 0 & : & N = 1 \\ 0.25 & : & N = 1.5 \end{cases} ;$$

for intermediate values of N one finds values between 0 and 0.25. Clearly, one can specify the energy loss directly. A graphical representation is given in Fig. 9.12. We note in passing that the wiggle in the pulse shape as seen in Figs. 9.10 and 9.11 can be understood as a beating between the soliton and the radiation. As the radiation gradually disperses away, the wiggles eventually decay. By a technique explained in [6, 7] the beat note can be evaluated to obtain precise information about the soliton.

Since both action W and energy E are fixed, the remaining parameters are fixed, too:

$$W = |\beta_2|/\gamma, \tag{9.57}$$

$$E = E_{start} - E_{rad}, \tag{9.58}$$

$$\Rightarrow \tau = W/E, \tag{9.59}$$

$$\Rightarrow \hat{P} = E/\tau. \tag{9.60}$$

This is shown graphically in Fig. 9.13: At constant action, $\hat{P} = (1/\tau)^2$ and thus $\log \hat{P} = 2 \log(1/\tau)$. This curve, plotted in a $\log \hat{P}$–$\log(1/\tau)$ diagram, has the slope $(d \log \hat{P})/(d \log(1/\tau)) = 2$. At constant energy, on the other hand, the slope is unity; shown is a selection from this family of curves. One first identifies the curve of fixed energy pertaining to the launch condition. A second, lower curve is the one where

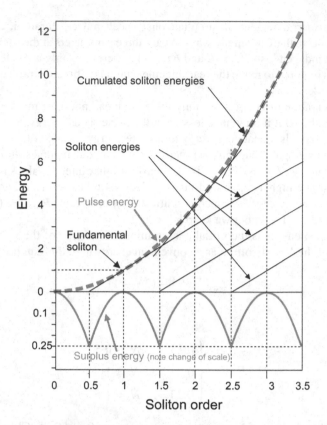

Fig. 9.12 Sketch to explain pulse energy and soliton content. If one takes a pulse with fixed duration and increases its energy from zero, beginning at $N = 0.5$ a soliton is formed. Its energy increases linearly as the pulse energy grows. Beginning at $N = 1.5$ a second soliton is generated, beginning at $N = 2.5$ a third, etc. The sum of all soliton energies is a piecewise linear function which runs close to the parabola $E \propto N^2$ (*dashed line*) and touches it wherever N is integer. These tangent points are the positions where all energy is invested in solitons. At all noninteger N some part of the energy is not invested into solitons but is radiated off. This part is given by the difference between the sum of solitonic energies and the parabola; for the sake of clarity, the lower part of the picture shows this difference on an expanded scale

the energy is reduced by just the amount which is radiated off. The final soliton must be on this curve. It must also be on the line designating the soliton action. Therefore, at the intersection of both, we have the final soliton. The coordinates of the intersection point indicate its pulse duration τ and peak power $\hat{P}$.

There is an interesting consequence from all this: If one launches a light pulse with $N < 1/2$ into the fiber, then no soliton will be generated because all the energy is converted to radiation. In a thought experiment, one may consider a fiber with a localized loss (at a splice, say). The condition right after the loss is equivalent to launching a weaker pulse in the fiber. If suddenly N is reduced to values below $1/2$—i.e., when a soliton is suddenly attenuated by at least a factor of 4, it is

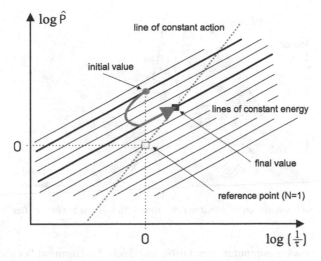

Fig. 9.13 Graphical construction of energy and duration of a soliton when the launch condition does not quite fit. In the $\log(\hat{P})$-$\log(1/\tau)$ diagram, there is a family of curves of constant energy. Also, a particular soliton (fixed τ and $\hat{P}$) is highlighted by a *white square*. All solitons in this fiber must lie on the curve of constant action (*single steeper line*). Let the launch condition (*gray circle*) have higher energy as required for the $N = 1$ soliton (the *upper of the bold lines*). Then, one first calculates the energy loss from Fig. 9.12; then one finds the curve pertaining to the remaining energy (the *lower bold line*). The intersection of the final energy and the constant action curves gives the soliton's $\hat{P}$ and τ

destroyed: 6 dB localized loss kills a soliton. In stark contrast, a gradual energy loss does not do much harm. This can be seen from the following consideration: If one attenuates first by less than a factor of 4, then allows for unperturbed propagation, a new soliton with lower energy and therefore longer duration will form; some energy will be radiated off and eventually go away by dispersion so that after settling of transients one clearly sees the new, lower-energy soliton which also has its own $N = 1$. Then one may attenuate again by less than a factor of 4, and the process repeats. A soliton survives if it is attenuated by a factor of 3 twice, but it dies when it is suddenly attenuated by a factor of 9. For continuously distributed loss, the soliton can survive for a long distance; it will continuously rearrange its width to accommodate its energy level. The long-term decay of a soliton in a fiber has been treated in [8].

9.4.5 Collisions of Solitons

The speed of propagation of a soliton in the fiber depends on the optical center frequency due to dispersion. If one launches two solitons with slightly different center frequency one shortly after the other, it can happen that both collide.

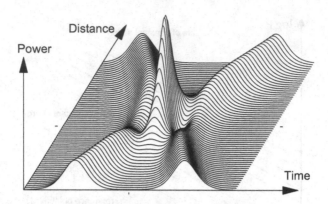

Fig. 9.14 Computer simulation of a soliton collision. Both solitons are intact after the collision

Figure 9.14 shows a computer simulation in which the common "center of mass" of both solitons rests in the reference frame. During the collision there are some pronounced interference spikes, but afterward both solitons continue their paths unharmed. Both shape and energy of the solitons is maintained; only a phase shift— not visible in the figure—remains. This behavior to remain intact in a collision is reminiscent of that of particles; the name "soliton" is meant to evoke that analogy (think proton, neutron, etc.).

In optical data transmission, in particular when several wavelength channels are transmitted at the same time (Chap. 11), such collisions can and will happen. The phase shift can have a mild influence there. Other than that it plays an important role in the context of so-called quantum nondemolition measurements in quantum optics (see, e.g., [10]).

9.4.6 Higher-Order Solitons

If one keeps increasing the power of the launched pulse beyond the $N = 1$ point, the soliton gets narrower. But then something remarkable occurs at four times the fundamental soliton power: The pulse goes through different shapes as it propagates; this is shown in Fig. 9.15. However, it does so in a periodic fashion, and at certain points along the fiber one finds the sech shape again. (An analytic expression for the complicated breathing shape was found in [29]). This behavior is also reflected in the shape of the power spectrum (see Fig. 9.16). Here we encounter the $N = 2$ soliton. We ask for its spatial period.

According to Fig. 9.12 at $N = 2$, there is a superposition of two solitons: One of them has the same energy as the fundamental soliton at $N = 1$, the other, three times as much. However, both are fundamental solitons: the one with higher power has correspondingly shorter duration. At three times the energy, the pulse width is one third and the peak power nine times that of the lower power soliton.

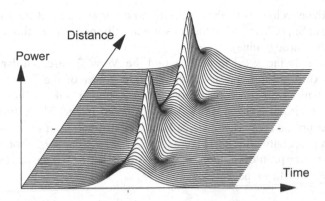

Fig. 9.15 An $N = 2$ soliton in a computer simulation. Propagation from 0 to $2z_0$ is shown, i.e., over two oscillation periods

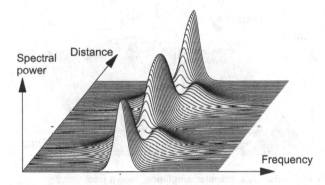

Fig. 9.16 Spectrum of an $N = 2$ soliton in a computer simulation, again from 0 to $2z_0$. Note that the spectrum is broadest where the pulse has the shortest duration (Fig. 9.15)

As both propagate together, their phases evolve at different rates due to their different power: Eq. (9.56) showed that the phase of the fundamental soliton has completed a full rotation after a distance $z = 4\pi L_{\mathrm{NL}} = (4\pi)/(\gamma \hat{P}_1)$. Then, for the higher-power soliton, the phase rotates nine times as fast.

As both phases rotate at different rates, a beat note is created. The beat pattern will repeat when the difference between both phases has gone through 2π:

$$\phi_2 - \phi_1 = \frac{9}{2}\zeta - \frac{1}{2}\zeta = 4\zeta \overset{!}{=} 2\pi,$$

$$\zeta = \frac{\pi}{2} \Rightarrow z = \frac{\pi}{2}L_D = z_0.$$

As a result we see the true significance of z_0: This is the spatial period of the beat note between the constituent fundamental solitons in a higher-order soliton, and hence the distance after which the power profile repeats itself. However, the phase underneath the envelope is repeated only up to a phase factor; it truly repeats for the first time at $8z_0$ (where one of the constituent solitons has gone through one full cycle, the other through nine).

If the power is increased further beyond the $N = 2$ case, similar logic can be applied. An N-soliton appears at the N^2-fold power of the $N = 1$ solitons. All solitons with $N > 1.5$, collectively known as higher-order solitons, have the property that their pulse shape varies periodically. At integer N the spatial period for the power profile is z_0. This disregards phase information; if phase is included, the pattern repeats only after $8z_0$. Figures 9.17, 9.18, 9.19 and 9.20 show temporal and spectral power profiles of the $N = 3$ and $N = 4$ case. The shapes can become quite complex, but they repeat after z_0.

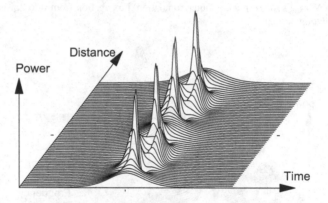

Fig. 9.17 An $N = 3$ soliton in a computer simulation, shown from 0 to $2z_0$

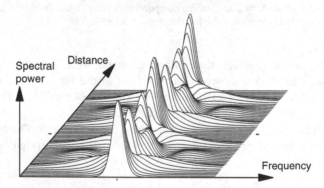

Fig. 9.18 Spectrum of an $N = 3$ soliton in a computer simulation, in correspondence with Fig. 9.17

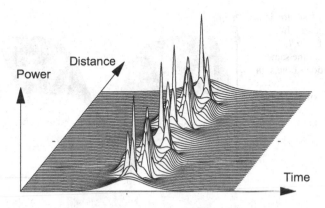

Fig. 9.19 An $N = 4$ soliton in a computer simulation, shown from 0 to $2z_0$

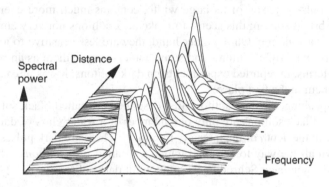

Fig. 9.20 Spectrum of an $N = 4$ soliton in a computer simulation, in correspondence with Fig. 9.19

9.4.7   Dark Solitons

For the case of normal dispersion, a solution of the nonlinear Schrödinger equation is given by

$$u = \tanh \tau \; e^{i\zeta}. \tag{9.61}$$

When the amplitude profile is described by a $\tanh(\tau)$ function, the power or intensity profile must follow $\tanh^2(\tau) = 1 - \mathrm{sech}^2(\tau)$. This implies that there is zero intensity at $\tau = 0$ but full intensity far away from the pulse center: a dip in a bright background. Dark solitons are notches in a constantly bright background (Fig. 9.21). From the tanh function, they inherit the special property that a phase jump of π occurs at center.

Both in experiment and numerical simulation, one does not have the chance to work with an infinitely wide bright background. A good approximation can be

Fig. 9.21 A dark soliton in a computer simulation. In comparison to Fig. 9.7, here all parameters are the same with the exception of the sign of dispersion; $\beta_2 = -18\,\text{ps}^2/\text{km}$ has been changed to $\beta_2 = +18\,\text{ps}^2/\text{km}$

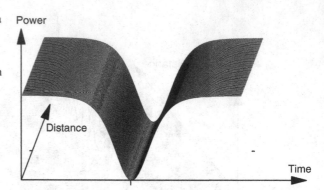

found by using a background pulse of considerably longer duration. Of course, this background pulse, by way of its large width, contains much more energy than a comparable bright soliton; this seems to make dark solitons not very attractive for optical data transmission. On the other hand, they are less sensitive to a variety of perturbations than bright solitons, and some authors pursue them as an alternative. In practical terms, in reported experiments on dark solitons, it was already difficult to produce them in the first place.

Strictly speaking, the dark solitons just described are called black solitons. The reason is that black solitons are only one member of a wider class of dark solitons which differ in the depth of the intensity minimum. There are dark pulses which do not dip down all the way to zero and which are called *gray solitons*. The general solution of the nonlinear Schrödinger equation for dark solitons is

$$u(\xi, \tau) = A_0 \sqrt{\frac{1}{B^2} - \text{sech}(A_0 \tau)}\; e^{i\left(\varphi(\tau') + \left(\frac{A_0}{B}\right)^2 \xi\right)} \tag{9.62}$$

with the abbreviations

$$\tau' = A_0 \tau + \frac{A_0^2}{B}\sqrt{1 - B^2}\,\xi$$

and

$$\varphi(t) = \arcsin \frac{B \tanh(t)}{\sqrt{1 - B^2 \text{sech}^2(t)}}.$$

The amplitude factor A_0 fixes the brightness of the background and B defines the "grayness". In the limit $\lim_{B \to 1}$ gray turns black, in a manner of speaking, and Eq. (9.62) reproduces the solution $A_0 \tanh(A_0 \tau)$. In this case there is the abrupt phase jump of π at soliton center; for gray solitons, the phase transits in a continuous way, not stepwise.

9.5 Digression: Solitons in Other Fields of Physics

Solitons, that is, nonlinear waves with special properties, certainly do not exist solely in optical fibers. Indeed, the term was coined following observations in other branches of science. The first reported conscious observation of a soliton phenomenon was written by a Scotsman, the civil engineer John Scott Russell. In 1838 he noticed a remarkable water wave in the Union Canal near Edinburgh and wrote this report [28]:

> I was observing the motion of a boat which was rapidly drawn along a narrow channel by a pair of horses, when the boat suddenly stopped—not so the mass of water in the channel which it had put in motion; it accumulated round the prow of the vessel in a state of violent agitation, then suddenly leaving it behind rolled forward with great velocity, assuming the form of a large solitary elevation, a rounded, smooth and well-defined heap of water, which continued its course along the channel apparently without change of form or diminution of speed. I followed it on horseback, and overtook it still rolling on at a rate of some eight or nine miles an hour, preserving its original figure some thirty feet long and a foot and a half in height. Its height gradually diminished, and after a chase of one or two miles I lost it in the windings of the channel.

A first attempt at a mathematical explanation was not published until 1895 when Diederik Korteweg and Gustav de Vries formulated a hydrodynamic wave equation. It was only after their work that in the interplay of dispersion and nonlinearity solitary waves became an expected feature. Their hydrodynamic wave equation is now called the Korteweg-de-Vries equation (KdV equation). It differs from the nonlinear Schrödinger equation of fiber solitons, which concerns us here, and therefore its solutions are somewhat different, too. One important difference is that unlike in the nonlinear Schrödinger equation, in the KdV equation the speed of propagation becomes amplitude-dependent. This is why water waves move faster in deep water than in shallow water, a fact which has considerable, indeed dramatic consequences in the case of a *tsunami* (Japanese for harbor wave). Triggered by undersea earthquakes, water surface waves with enormous energy propagate at rapid speed across the ocean, but they do so with extremely long wavelength ($\approx 100 \, \mathrm{km}$ and low amplitude (typically < 1 m. (This qualifies them as 'shallow water waves' in the sense that the wavelength is much longer than the ocean's depth). They therefore easily go unnoticed, but they can cross all of the Pacific Ocean within 2 days. As they approach a shore where the water depth is reduced, the energy transport remains conserved, but since the speed is reduced, the amplitude must increase and may generate crests of 20 m elevation. If such a wave hits a shore it will destroy whatever gets in its path. In December 2004, this happened in a particularly tragic form when

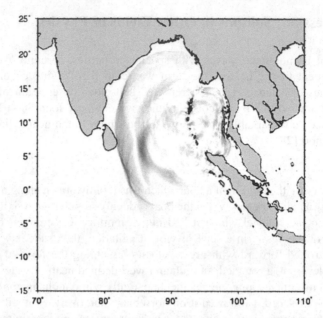

Fig. 9.22 A model calculation for the Indian Ocean tsunami on December 26, 2004. This is one frame of an animation, taken from [17] with kind permission

a quake off Sumatra triggered an Indian Ocean tsunami which wreaked havoc in Indonesia, Thailand, Sri Lanka and on to the East African shores and took about a quarter million human lives (Fig. 9.22).

In 1965, Norman Zabusky and Martin Kruskal studied interactions of solitary waves [35] and noticed their particle-like properties. Solitons can be reflected off each other without any harm to their structure. The moniker "soliton" reminds us of elementary particles.

Application of the soliton concept to fiber optics started in 1971 when Vladimir Evgen'evich Zakharov and Alexey B. Shabat [36] formulated a wave equation for pulse propagation in fibers and found solitonic solutions. The nonlinear Schrödinger equation turns out to be a good model for both deep water waves on the ocean and light pulses (strictly, their envelopes) in optical fibers. In 1973, Akira Hasegawa and Frederick Tappert predicted [16] that such optical solitons should be observable experimentally and that they hold promise for optical data transmission. Linn F. Mollenauer and collaborators succeeded in 1980 to experimentally demonstrate the existence of fiber solitons [24]. Then, various aspects of optical solitons were subject to closer investigation. F. Mitschke and Mollenauer showed the particle properties in 1986 by demonstrating for the first time the interaction forces between fiber-optic solitons [23].

A related concept, also in optics, is that of spatial or *transverse* solitons in various media. These are beams of light which stabilize their cross-sectional shape in the presence of nonlinearity and diffraction, in very close analogy (if one accepts to switch the roles of time and space) to the solitons described here. However, spatial solitons do not occur in single-mode fibers and are therefore beyond the scope of this book. The reader interested in a comparison is referred to [2, 20, 27, 37].

It is appropriate, however, to point out another similarity between light pulses in optical fibers and water surface waves because there has been much discussion about this recently. For centuries, sailors have reported of giant waves on the ocean that appeared suddenly, like out of nowhere, and disappeared as quickly as they came. Due to their enormous height they wreaked havoc on ships, and lucky were those who lived to report about it. Usually such reports were taken as tall stories: sailor's yarn, spun to impress those credulous enough to buy it. However, in recent years some hard evidence was found that such freak waves, or rogue waves, exist after all. Still, precious few photographs provide evidence, but sensors on seaborne structures like oil-drilling rigs have recorded such events, and surveys of the ocean surface from satellites confirm their existence. Monster waves may be rare at any given location, but it turns out not to be all that infrequent that such an event occurs somewhere in the vastness of the oceans.

Recently there was a report about the formation of rare but unusually high power peaks in optical fibers [30]; this occurred in the process of supercontinuum generation which is explained in Sect. 10.2.8. The analogy to rogue waves of the ocean was pointed out. It triggered a lively discussion in the optics community about *optical* rogue waves. While the term is used in a loose sense, there is some consensus that its characteristics are (1) far-above average amplitudes, (2) very sudden appearance and disappearance, and (3) that the peak height obeys statistics which develops a 'fat tail' towards large amplitudes, much exceeding anything one would expect from ordinary statistics (normal distribution etc.).

Prompted by Solli et al. [30], many researchers developed further hypotheses about optical rogue waves. One such idea [3] invokes a family of solutions of the nonlinear Schrödinger equation with the shared property that they are modulations of a constant background power. We already discussed one member of this family, the Akhmediev breather, in Sect. 9.4.2. The Akhmediev breather is governed by a single parameter, $0 \leq a \leq 1/2$. However, Eq. (9.42) also admits solutions for $a > 1/2$. While the Akhmediev breather is a temporally periodic structure that appears once and disappears again, for $a > 1/2$ the solution, known as a Kuznetsov-Ma soliton, can be described as an isolated peak which comes and goes repetitively as the signal propagates down the fiber. In other words, the Akhmediev breather is located in space and periodic in time; the Kuznetsov-Ma soliton is located in time and periodic in space. The intermediate case at $a = 1/2$ was first described by Peregrine [26]; its mathematical form can be simplified to a rational expression

$$A(T, Z) = \sqrt{P_0} \left[1 - \frac{4(1 + 2iZ)}{1 + \omega_c^2 T^2 + 4Z^2} \right] \exp(iZ). \qquad (9.63)$$

This solution is known as a Peregrine soliton, and is shown in Figs. 9.23 and 9.24. The attention of experimentalists has only recently been attracted to the Akhmediev

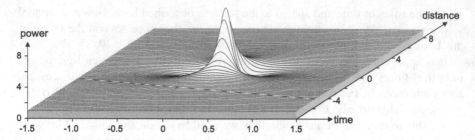

Fig. 9.23 Evolution of the Peregrine soliton. Time in units of the period ($T\omega/2\pi$), power in units of the constant background ($|A^2|/P_0$), and distance in nonlinear lengths ($Z = z/L_{NL}$). The evolution from $Z = -8$ to $Z = +8$ is shown; a solitary peak appears at $Z = 0$. This figure was drawn with help from Christoph Mahnke

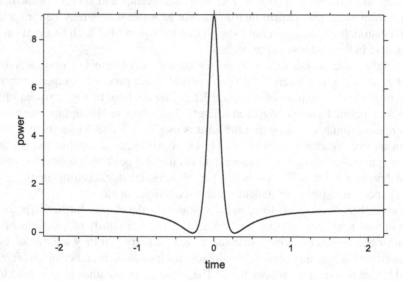

Fig. 9.24 Cross section of the Peregrine soliton in Fig. 9.23 at its culmination point at $Z = 0$. Units as in Fig. 9.23

breather [13], the Kuznetsov-Ma soliton [19], and the Peregrine soliton [18]. As the same equation pertains to optical pulses in fibers and to certain water surface waves, this entire family of solutions—all of them describe pulses on a continuous wave background—also has its analogue in water waves. Historically, the Peregrine and Kuznetsov-Ma structures received the name of 'solitons', which is a bit unfortunate because this family of solutions can in no way be understood as a mere superposition of a continuous background plus proper solitons [22].

The peak of the Peregrine soliton is localized in both domains: it is a single, large pulse which appears once and in one position, then disappears. That property seems to fit the description of a rogue wave. However, one should always be cautious about analogies. This model pertains to a single spatial dimension while the ocean surface has two dimensions; that will make a difference. In any event, several interesting hypotheses have been advanced about the mechanism responsible for the generation of rogue waves, and at this point it is not finally clear which one is best. It is entirely possible that depending on circumstances several mechanisms may be capable of producing rogue events, so that there is no unique correct explanation. The scientific debate continues.

9.6   More $\chi^{(3)}$ Processes

In our approach to derive a wave equation, we have assumed a monochromatic wave. Then we found self-phase modulation, an effect by which a monochromatic wave or, in extension, a more or less narrowband light pulse modifies itself. But this is not the only consequence arising from third-order susceptibility $\chi^{(3)}$. It also gives rise to the following effects:

- *Cross-phase modulation.* In the presence of an intensive wave of frequency ω, a wave of frequency $\omega + \Delta\omega$ gets phase modulated.
- *Frequency tripling.* A new wave of frequency 3ω arises from a wave of frequency ω.
- *Four-wave mixing.* The three fields involved in a $\chi^{(3)}$ process may, in the most general case, all have different frequencies. Then, new frequency components arise at combination frequencies.

Cross-phase modulation arises because the index modulation created by *one* wave also has an influence on the *other* wave. Frequency tripling, also known as third harmonic generation, can be understood in either one of two ways: In the wave picture, the light field acting on an atom can be written as an oscillation of the type $\sin \omega t$. In a $\chi^{(3)}$ process, there are three waves acting simultaneously. All three may have the same frequency: Then there is a term $\sin^3(\omega t) = \frac{1}{4}(3 \sin \omega t - \sin 3\omega t)$ containing the third harmonic. In a particle picture, there are three photons acting on the atom simultaneously. It absorbs three times the energy of a single photon. The probability of this process rises with the third power of the photon density because it takes three simultaneously arriving photons. Also, the process becomes much more probable when an atomic energy level exists at or near $E = 3\hbar\omega$.

In the same way, one can discuss four-wave mixing: In the wave picture the three irradiated waves may all have different frequencies. Then, one can use a relation between trigonometric functions of the type

$$\sin \omega_1 t \cdot \sin \omega_2 t \cdot \sin \omega_3 t =$$

$$\frac{1}{4} \left[\sin(\omega_1 + \omega_2 + \omega_3)t + \sin(-\omega_1 + \omega_2 + \omega_3)t \right.$$

$$\left. + \sin(\omega_1 - \omega_2 + \omega_3)t + \sin(\omega_1 + \omega_2 - \omega_3)t \right]. \tag{9.64}$$

There are four combination frequencies, hence the name of the process. In the particle picture, there are three photons acting on an atomic medium. Both absorption or stimulated emission can occur. If all three photons are absorbed, the atom stores an energy equal to the sum and can reradiate a wave with the corresponding frequency. The other combination tones occur when one of the photons stimulates an emission. (The loss of energy from the atom is reflected in the negative sign in the respective term of Eq. (9.64).)

For our purposes, the following remark is important: In the special case that all ω_i are integer multiples of a certain fundamental frequency ω_0, then the same is also true for all combination frequencies: Let $k, l, m \in \mathbf{N}$. Then, if $\omega_1 = k\omega_0$, $\omega_2 = l\omega_0$, and $\omega_3 = m\omega_0$, it follows that the combination frequencies are $(k + l + m)\omega_0$, $(-k + l + m)\omega_0$, $(k - l + m)\omega_0$, and $(k + l - m)\omega_0$, all integer multiples of ω_0. In the special case called *degenerate four-wave mixing*, two of the three frequencies are the same. Let, e.g., $\omega_1 = \omega_2$ and $\omega_3 = \omega_1 + \Delta\omega$. Then the fourth wave has the frequency $\omega_4 = \omega_1 - \Delta\omega$. As a result, a pair of frequencies separated by $\Delta\omega$ will produce two new frequencies. One is $\Delta\omega$ below the lower frequency, and the other $\Delta\omega$ above the higher frequency.

This consideration is simplified insofar as it makes no statements about the intensities of the generated waves. In order to assess that aspect we need to consider the following.

Let us assume that energy is transferred through some nonlinear mixing process from one wave to another. Both propagate through the material in the same direction. In general their frequencies will differ, and in the presence of dispersion they have different phase velocities. This implies that their relative phase will wander as they propagate.

As is well known from coupled oscillators, energy always flows from that with advanced phase to that with retarded phase. The energy transfer is most efficient if the phase of the driving oscillator is 90° advanced with respect to the driven oscillator. This can be generalized to traveling waves.

Right after the launch point, energy from wave A feeds wave B which at this point just emerges. The phase of wave B is automatically arranged such that an energy transfer takes place. A certain distance down the fiber, both waves have experienced a relative phase shift of 90° and energy transfer ceases. A little further on, the phase of wave B is advanced and energy is transferred back!

Instead of an unlimited increase of the energy of wave B, there is a periodic exchange of energy between both waves. This becomes noticeable at the point where the relative phase is rotated by 90° for the first time. If a most effective energy transfer is desired, one has to make this distance long; the most obvious means to do that is to make dispersion small. The technical term is *phase matching* of both waves. If, on the contrary, one wishes to thwart the energy transfer, one can arrange for strong dispersion. We will look closer into this logic in Sect. 11.2.3.

9.7 Inelastic Scattering Processes

An important class of nonlinear processes in fibers are scattering processes in which light is scattered by the medium (glass) either elastically or inelastically. In the case of elastic scattering, the energy of the light quanta, and thus the frequency, is unaltered. This puts elastic scattering into the realm of linear optics. We have already discussed *Rayleigh scattering*, a process in which scattering occurs in all directions, creating a linear loss: a loss of photons in proportion to the existing number of photons.

For inelastic scattering processes, an amount of energy δE is exchanged with the medium (either absorbed by the medium or released). Since $\delta E = h\delta\nu$, there is a frequency shift $\delta\nu$. Two types of inelastic scattering processes in fibers are distinguished: *Brillouin scattering* and *Raman scattering*. These are scattering processes either at the acoustic (Brillouin) or optical (Raman) phonon branch. In either case, in principle there can be an upshift or downshift of frequency. The irradiated wave is called pump wave; the scattered wave is called the Stokes wave in case of downshift, and the anti-Stokes wave in the case of upshift.

Almost always the downshift (Stokes wave) is much more pronounced. The medium usually consists of atoms or molecules which are in or near their energetic ground states, as given by the thermal energy and a Boltzmann distribution of occupation numbers. Then the medium can absorb, but not release energy.

Photons can be scattered spontaneously in both cases, but the rate is low. On the other hand, beginning at a certain threshold intensity a stimulated scattering process sets on. This is then called stimulated Raman scattering (SRS), or stimulated Brillouin scattering (SBS). The process can become stimulated when a sufficient number of spontaneously generated photons is already present and interacts with the pump wave. Then the polarization of the medium is driven, and above the threshold the process grows exponentially. Of course this exponential growth of Stokes or anti-Stokes wave cannot go on indefinitely: eventually the pump is depleted so that further growth is halted.

To get an idea of all this in more quantitative terms, we use the following rate equation model: Let N_s be the number of Stokes photons and N_p the number of pump photons. Then, in the stimulated process, we have

$$\frac{dN_s}{dz} = \text{const.}\, N_p(N_s + 1).$$

The "1" inside the parenthesis is for the spontaneous rate without which the process can never start (just like in a laser). Once the startup phase is over, the spontaneous rate may be neglected in comparison to N_s. Then there is a solution

$$N_s(z) = N_s(0)\exp(gIz),$$

where g is the gain coefficient for the Stokes wave.

By exponential growth even a single spontaneously occurring photon may produce a macroscopic light wave. Macroscopic means that the growth continues until the energy of the pump wave is noticeably depleted. Then a limit to growth is reached—but this does not necessarily imply a steady equilibrium. Quite to the contrary, it has been shown for Brillouin scattering in fibers that the Stokes wave has strong, irregular power fluctuations which can be interpreted as the remainder of the stochastic signature of the startup process.

The stimulated scattering process can have profound impact on a light wave propagating in a fiber. Often this is a detriment, which can in no way be neglected. On the other hand, we will see that the influence is not always unwelcome, but can also be harnessed to perform useful functions.

It goes without saying that in any scattering process, conversation of both energy and momentum must hold. We can write that as

$$\sum_{\text{in}} \omega = \sum_{\text{out}} \omega, \tag{9.65}$$

$$\sum_{\text{in}} \vec{k} = \sum_{\text{out}} \vec{k}. \tag{9.66}$$

On the LHS, there are all waves entering the interaction process, and on the RHS all waves that exit from the process.

9.7.1 Stimulated Brillouin Scattering

We begin by considering the case of Brillouin scattering in which an acoustic wave is generated.

$$\omega_{\text{p}} = \omega_{\text{s}} + \omega_{\text{a}}, \tag{9.67}$$

$$\vec{k}_{\text{p}} = \vec{k}_{\text{s}} + \vec{k}_{\text{a}}. \tag{9.68}$$

Indices p, s, and a refer to pump, Stokes, and acoustic waves, respectively.

Pump and Stokes waves oscillate at optical frequencies while the acoustic wave has a considerably lower frequency. Therefore the wave vectors for pump and Stokes wave will be similar and much larger than that of the acoustic wave. We can therefore approximate that $|\vec{k}_{\text{p}}| \approx |\vec{k}_{\text{s}}|$ and $\omega_{\text{a}} \ll \omega_{\text{p}}, \omega_{\text{s}}$. Referring to Fig. 9.25, we can then write

$$|\vec{k}_a| = 2|\vec{k}_p| \sin \frac{\theta}{2}$$

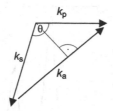

Fig. 9.25 Sketch for the relation between three wave vectors involved in Brillouin scattering and the angle θ. This is very nearly an equilateral isosceles triangle, and therefore the perpendicular halves both $|\vec{k_a}|$ and θ. Then $|\vec{k_a}|/2 = |\vec{k_p}| \sin(\theta/2)$

with θ the angle between the propagation directions of pump and Stokes waves. The pump wave propagates along the fiber, and so θ is also the angle with the fiber axis for the Stokes wave.

A wave vector equals angular frequency divided by velocity, so that

$$\omega_a = v_a |\vec{k_a}| = v_a 2 |\vec{k_p}| \sin \frac{\theta}{2}.$$

v_a is the velocity of sound, which in fiber is 5960 m/s.

We see that the Stokes shift ω_a depends on θ and disappears in forward direction ($\theta = 0$). This gives rise to conflict with energy conservation except when the energy of the forward-scattered wave vanishes. In backward direction, on the other hand, the frequency shift acquires its maximum. The physical interpretation can be given as a Doppler effect of a wave which is scattered off a grating traveling itself with the velocity of sound. Directions other than forward and backward are not relevant in fibers, and single-mode fibers in particular. (Strictly speaking, SBS does not entirely vanish in forward direction; there is a minimal forward scattering known as GAWBS (*guided acoustic wave Brillouin scattering*), which is several orders of magnitude weaker than backward scattering.)

If we rewrite in terms of natural, rather than angular frequencies and use index "B" for "Brillouin", the Brillouin shift is given by

$$\nu_B = \frac{\omega_a}{2\pi} = \frac{2 v_a |\vec{k_p}|}{2\pi} = 2 v_a \frac{n}{\lambda_p}$$

because $|\vec{k_p}| = 2\pi n / \lambda_p$. If for example $n = 1.46$ and $\lambda_p = 1.55\,\mu$m, one obtains $\nu_B = 11.2$ GHz. As a general statement, SBS produces frequency shifts on the order of 10 GHz, which is a relative shift of 10^{-4} (Fig. 9.26).

The change of Stokes power with distance along the fiber can be described by

$$\frac{dI_s}{d(-z)} = g I_p I_s - \alpha_s I_s. \tag{9.69}$$

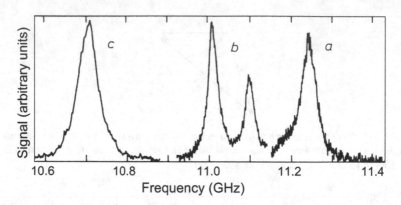

Fig. 9.26 Brillouin scattering spectrum for three different fibers. (*a*) undoped silica core, (*b*) depressed-clad fiber, (*c*) dispersion shifted fiber. In all cases the shift is near 10 GHz. From [34] with permission

Indices refer to Stokes and pump wave as before. The derivative here has been taken with respect to $(-z)$ because for SBS the scattered wave travels backward. The frequency of the acoustic wave is so much smaller than the optical frequencies that we may write $\omega_p/\omega_s \approx 1$ and $\alpha_s \approx \alpha_p$. The gain factor g for SBS is about $g_B = 20$ pm/W, slightly lower than for bulk fused silica with about $g_B \approx 50$ pm/W.

The corresponding equation for the pump wave reads

$$\frac{dI_p}{dz} = -\frac{\omega_p}{\omega_s} g I_p I_s - \alpha_p I_p. \tag{9.70}$$

It is easy to check that for the lossless case ($\alpha_s = \alpha_p = 0$), the following holds:

$$\frac{d}{dz}\left(\frac{I_s}{\omega_s} - \frac{I_p}{\omega_p}\right) = 0.$$

This demonstrates the conservation of photon number.

The first term on the RHS of Eq. (9.69) is the gain, the second, loss. Correspondingly, in Eq. (9.70) the first term is for saturation, the second, loss. We now ask for the threshold pump power for the generation of the stimulated effect. Surely, close to that threshold, the Stokes wave is still weak so that we can neglect saturation to obtain an expression for the threshold:

$$\frac{dI_p}{dz} = -\alpha_p I_p,$$

$$I_p(z) = I_{p0} e^{-\alpha_p z},$$

$$\frac{dI_s}{d(-z)} = gI_s I_{p0} e^{-\alpha_p z} - \alpha_s I_s$$

$$= I_s \left(gI_{p0} e^{-\alpha_p z} - \alpha_s \right).$$

As we integrate, the first term in parentheses yields

$$\int_0^L I_{p0} e^{-\alpha_p z}\, dz = \frac{I_{p0}}{\alpha_p} \left(1 - e^{-\alpha_p L} \right) = I_{p0} L_{\text{eff}}.$$

Now we solve

$$I_s(0) = I_{sL} \exp \left(gI_{p0} L_{\text{eff}} - \alpha_s L \right).$$

By convention, a useful criterion for threshold is that (in the absence of saturation) $I_{s,\max} = I_{p,\min}$. As initial value for the Stokes wave, one assumes a single photon inserted at the (near or far, whichever applies) fiber end to generate spontaneous scattering. Figure 9.27 clarifies to which positions the quantities are referred.

As for values of the gain coefficient, strictly speaking they depend somewhat on spectral line shape, the state of polarization of both waves, etc. However, as a reasonable order of magnitude, we may use $gI_{p0} L_{\text{eff}} \approx 20$. If one inserts the above

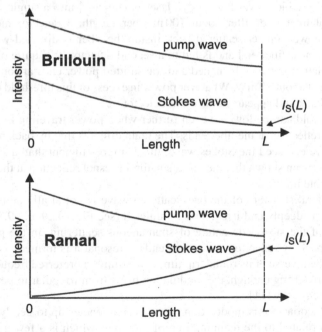

Fig. 9.27 Sketch to explain the spatial evolution of pump wave and Stokes wave in stimulated Raman and Brillouin scattering

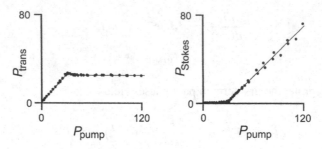

Fig. 9.28 Experimental observation of stimulated Brillouin scattering in a fiber. All powers are given in milliwatt. *Left*: Above threshold of stimulated Brillouin scattering the power of a continuous wave laser that is transmitted through the fiber is clamped. *Right*: The power "missing" in transmission appears in the backscattered Stokes wave

values of g and $L_{\mathrm{eff,max}}$, one finds a threshold of 2.5 mW in a fiber with $A_{\mathrm{eff}} = 50\,\mu\mathrm{m}^2$.

This extremely low value renders stimulated Brillouin scattering the nonlinear process with the lowest threshold. SBS gets in the way whenever continuous wave experiments are considered. One important consequence is a severe limitation of the fiber's ability to transmit power: As soon as the pump wave exceeds threshold, the excess is transferred into a Stokes wave which travels back to the light source. Figure 9.28 shows an experimental result to illustrate this point. Continuous wave light with adjustable power from a dye laser is launched into a single-mode fiber. At the distal fiber end, after about 100 m fiber length, a detector monitors the transmitted power. Power scattered back inside the fiber is diverted with a beam splitter at the near fiber end and is fed to a second detector. It is quite obvious that the linear relation between launched and transmitted power ends at some point, in this example at about 20 mW. Whatever power in excess of this threshold is launched is scattered back and appears at the other detector.

The threshold can be lowered even further when power traveling in the fiber is recycled by reflection at the fiber ends. The main effect is that by back reflection, a coherent wave can seed the Stokes wave; this is more efficient than a spontaneous photon. It has been shown that the minute natural Fresnel reflection at the fiber ends has appreciable influence.

The temporal structure of the backscattered wave is not at all continuous. The Stokes wave is deeply and irregularly modulated (see Figs. 9.29 and 9.30). This is a signature of the stochastic nature of spontaneous scattering. In the presence of optical feedback by reflection at the fiber ends, a resonator is formed; its round trip frequency (the inverse of its round trip time) constitutes a preferred frequency. If the reflections are strong enough, the modulation turns from irregular to periodic with nearly this frequency [11].

Without feedback, this modulation contains frequencies up to nearly 100 MHz. This limit is related to the damping rate of phonons, which is a few nanoseconds and which also sets the spectral width of the Stokes wave. Therefore, the Brillouin

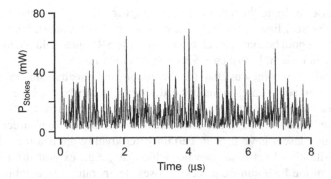

Fig. 9.29 The backscattered wave (the Stokes wave) has a deep and irregular temporal modulation, a result of the origin of the Stokes wave in spontaneous scattering processes

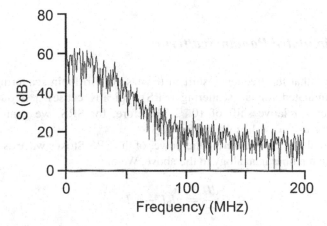

Fig. 9.30 The Fourier spectrum pertaining to the data shown in Fig. 9.29. The modulation of the Stokes wave extends up to tens of megahertz. This bandwidth is related to the damping rate of the acoustic wave in the nanosecond regime

line width (SBS line width) is about $\Delta \nu_B \approx 10\,\text{MHz}$, corresponding to a relative width of 10^{-3}.

There is an important conclusion with regard to some fiber applications here. It is often the case that in laser-based materials processing the laser power must be delivered from a bulky laser head to various positions on the workpiece. Fiber would provide perfect flexibility here, but SBS poses a severe limitation, and renders the idea to transmit sheer power useless unless extra measures are taken.

One such measure can be to avoid near-monochromatic pump light. For broadband pump light (band width $\Delta \nu_p$), the effective Brillouin gain is reduced according to

$$\tilde{g}_B = g_B \frac{\Delta \nu_B}{\Delta \nu_B + \Delta \nu_p}.$$

For short pulses of light, the threshold is then higher because the pulse is spectrally wider than the SBS line width. The consequence is that for short, i.e., broadband, pulses, the threshold becomes much higher than the SRS threshold when the pulses have picosecond width. In that case SBS loses its importance.

We should also point out that the Brillouin gain mechanism, most often a nuisance, can in some cases actually be desirable. It can be exploited to build a Brillouin laser which can provide laser oscillation on a frequency offset from the pump by one Brillouin shift. Since the latter is within the reach of direct electronic detection, such lasers have uses in certain heterodyning applications. Moreover, there are Brillouin effect-based sensors which can, e.g., exploit the temperature dependence of the Brillouin frequency to assess temperature. In combination with (long) pulses and an evaluation of the temporal structure, one can even have a position-resolved measurement. Fiber-optic sensors are treated in Chap. 12.

9.7.2 Stimulated Raman Scattering

We have seen that the frequency shift in the case of Brillouin scattering is about 10 GHz. Stimulated Raman scattering (SRS) typically causes a frequency shift of 10 THz or a relative shift of 10^{-1}. Therefore, for SRS, we cannot use the approximation $\omega_p/\omega_s \approx 1$ as we did for SBS.

Other than that we can describe the power of the SRS Stokes wave as a function of position in the fiber in analogy to the above. We obtain

$$\frac{dI_s}{dz} = gI_pI_s - \alpha_sI_s. \tag{9.71}$$

The gain factor g is about $g_R = 0.1\,\text{pm/W}$. The corresponding equation for the pump wave is

$$\frac{dI_p}{dz} = -\frac{\omega_p}{\omega_s}gI_pI_s - \alpha_pI_p. \tag{9.72}$$

Again we convince ourselves that in the lossless case ($\alpha_s = \alpha_p = 0$), the photon number is preserved:

$$\frac{d}{dz}\left(I_s + \frac{\omega_s}{\omega_p}I_p\right) = 0.$$

In analogy to the treatment above we find the threshold from

$$\frac{dI_p}{dz} = -\alpha_pI_p,$$

$$I_p(z) = I_{p0}\,e^{-\alpha_pz},$$

$$\frac{dI_s}{dz} = gI_sI_{p0}\,e^{-\alpha_p z} - \alpha_s I_s$$

$$= I_s\left(gI_{p0}\,e^{-\alpha_p z} - \alpha_s\right).$$

By integration, we conclude

$$\int_0^L I_{p0}e^{-\alpha_p z}\,dz = \frac{I_{p0}}{\alpha_p}\left(1 - e^{-\alpha_p L}\right) = I_{p0}L_{\text{eff}}$$

with the effective interaction length L_{eff} introduced in Sect. 9.1. Now we solve

$$I_s(L) = I_{s0}\exp\left(g_R I_{p0}L_{\text{eff}} - \alpha_s L\right).$$

Similarly as above, threshold is reached when without saturation $I_{s,\text{max}} = I_{p,\text{min}}$ holds. The gain term is roughly the same for SBS and SRS and comes to $gI_{p0}L_{\text{eff}} \approx 20$. By reinserting the values given above for g_R, g_B, and $L_{\text{eff,max}}$, one obtains a threshold power in a fiber with $A_{\text{eff}} = 50\,\mu\text{m}^2$ for SRS of about 500 mW, many times the value for SBS. Raman scattering becomes the dominant scattering process only when quite short pulses are used so that the Brillouin threshold rises considerably. Pulse durations on the order of 10 ps or less are required for this.

The frequency dependence of the Raman gain was first measured in [31] (Fig. 9.31); later on researchers also considered how it consists of several contributions with different temporal response [32, 33].

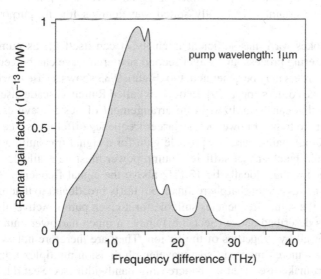

Fig. 9.31 The frequency dependence of Raman gain. The maximum of the Raman gain spectrum is reached at a detuning between pump and signal of about 13 THz, but even at smaller detunings there is an appreciable gain. After [31] with kind permission

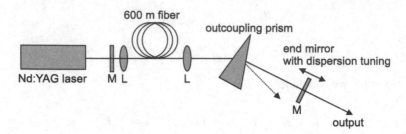

Fig. 9.32 The first tunable Raman laser 1977 consisted of a fiber which was pumped by a modelocked Nd:YAG-Laser (1064 nm) with pulses of 200 ps duration. The average pump power was 1.1 W and the repetition rate 100 MHz. A prism served to separate pump and signal waves; a moveable end mirror provided tuning. The tuning range extended from 1101 to 1125 nm; an average power up to 20 mW was generated. The threshold was at 0.7 W pump power, and the slope efficiency was 60 %. After [21] with kind permission

Like SBS, SRS is suitable for use in amplifiers and lasers; these are then called fiber Raman amplifiers or lasers, respectively. Figure 9.32 shows an experiment in which a tunable Raman laser was built [21].

Raman amplification of a signal wave by the energy taken from a pump wave was shown in several experiments; a gain of, e.g., 30 dB was obtained. An important consideration for such amplifiers is the frequency difference (detuning) between both waves as dictated by the Raman gain spectrum of fibers. The gain factor acquires its maximum near a detuning of 13 THz. Nd:YAG pump lasers emit either at 1.06 or at 1.32 μm; this is suitable for signal wavelengths of 1.12 or 1.40 μm, respectively—certainly not ideally suited wavelengths for the purposes of fiber optics.

If the Stokes wave has sufficient intensity, it can itself act as pump wave for another scattering process; this way a second scattered wave can be generated, and even higher orders may be generated, too. Figure 9.33 shows a case in which no less than five Stokes orders appear [9]. In devices called Raman cascade lasers as shown in Fig. 9.34 this can be utilized in an arrangement of nested cavities for several Stokes orders to transfer power across larger frequency differences (see, e.g., [12]).

When Raman gain is used to provide gain for a signal transmitted through the fiber, it should be clear that sufficient pump power must be available and that the pump frequency must ideally be 13 THz above the signal frequency. Fortunately, the Raman process is not sharply resonant but fairly broadband so that there is some tolerance in the signal frequencies suitable for a given pump. Active fibers like Er-doped fibers described above (Sect. 8.8.1) have a much narrower gain band, fixed once and for all by properties of the Er ion. They are therefore not as universally applicable. As increasingly massive wavelength division multiplex transmission is employed to make use of an ever-increasing bandwidth (see Sect. 11.1.5), the Er gain band begins to be a limitation for state-of-the-art systems. Raman amplifiers therefore attract more attention again recently.

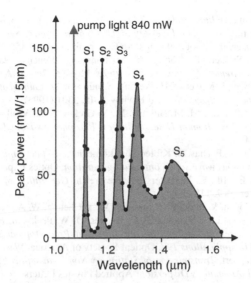

Fig. 9.33 This Raman scattering spectrum was obtained by pumping a fiber with a Nd:YAG laser at 1064 nm. The Stokes wave acts as a pump for the next order Stokes wave. This way five orders of Raman scattering are generated in this example. From [9] with kind permission

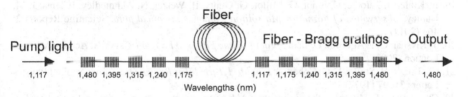

Fig. 9.34 A Raman laser using a cascade of several Raman orders for the generation of light with longer wavelength. Several selective reflectors (fiber-Bragg gratings) form nested cavities which support the pump wave, the targeted Stokes order, and all intermediate orders. The numbers shown in the example refer to the case described in [15] where fifth-order Raman scattering transfers power from a pump wave at 1117 nm to a new wavelength of 1480 nm. After [15]

References

1. G. P. Agrawal, *Nonlinear Fiber Optics*, 5th ed., Elsevier Academic Press, Oxford (2013)
2. N. N. Akhmediev, A. Ankiewicz, *Solitons. Nonlinear Pulses and Beams*, Chapman & Hall, London (1997)
3. N. Akhmediev, A. Ankiewicz, M. Taki, *Waves that appear from nowhere and disappear without a trace*, Physics Letters A **373**, 675 (2009)
4. N. N. Akhmediev, V. I. Korneev, *Modulation instability and periodic solutions of the nonlinear Schrödinger equation*, Theor. Math. Phys. **69**, 1089–1093 (1986)
5. L. Allen, J. H. Eberly, *Optical Resonance and Two-Level Atoms*, John Wiley & Sons, New York (1975), Reprint by Dover Publications (1987)
6. M. Böhm, F. Mitschke, *Soliton-Radiation Beat Analysis*, Physical Review E **73**, 066615 (2006)
7. M. Böhm, F. Mitschke, *Soliton content of arbitrarily shaped light pulses in fibers analysed using a soliton-radiation beat pattern*, Applied Physics B **86**, 407 (2007)

8. M. Böhm, F. Mitschke, *Solitons in Lossy Fibers*, Physical Review A **76**, 063822 (2007)
9. L. G. Cohen, Ch. Lin, *A Universal Fiber-Optic (UFO) Measurement System Based on a Near-IR Fiber Raman Laser*, IEEE Journal Quantum Electronics **QE-14**, 855 (1978)
10. J.-M. Courty, S. Spälter, F. König, A. Sizmann, G. Leuchs, *Noise-Free Quantum-Nondemolition Measurement Using Optical Solitons*, Physical Review A **58**, 1501 (1998)
11. M. Dämmig, G. Zinner, F. Mitschke, H. Welling, *Stimulated Brillouin Scattering in Fibers with and Without External Feedback*, Physical Review A **48**, 3301 (1993)
12. E. M. Dianov, I. A. Bufetov, M. M. Bubnov, M. V. Grekov, S. A. Vasiliev, O. I. Medvedkov, *Three-Cascaded 1407-nm Raman Laser Based on Phosphorous-Doped Silica Fiber*, Optics Letters **25**, 402 (2000)
13. J. M. Dudley, G. Genty, F. Dias, B. Kibler, N. Akhmediev, *Modulation instability, Akhmediev Breathers and continuous wave supercontinuum generation*, Optics Express **17**, 21497 (2009)
14. P. A. Franken, A. E. Hill, C. W. Peters, G. Weinreich, *Generation of Optical Harmonics*, Physical Review Letters **7**, 118 (1961)
15. S. G. Grubb, T. Erdogan, V. Mizrahi, T. Strasser, W. Y. Cheung, W. A. Reed, P. J. Lemaire, A. E. Miller, S. G. Kosinski, G. Nykolak, P. C. Becker, D. W. Peckham, *High-Power 1.48 μm Cascaded Raman Laser in Germanosilicate Fiber*, in *Proc. Topical Meeting on Optical Amplifiers and Their Applications 197*, Optical Society of America, Washington DC (1995)
16. A. Hasegawa, F. Tappert, *Transmission of Stationary Nonlinear Optical Pulses in Dispersive Dielectric Fibers. I. Anomalous Dispersion*, Applied Physics Letters **23**, 142 (1973)
17. International Tsunami Information Center: *Massive Tsunami hits Indian Ocean Coasts*, Animation by Kenji Satake. See http://ioc.unesco.org/itsu
18. B. Kibler, J. Fatome, C. Finot, G. Millot, F. Dias, G. Genty, N. Akhmediev, J. M. Dudley, *The Peregrine soliton in nonlinear fibre optics*, Nature Physics Letters **6**, 790 (2010)
19. B. Kibler, J. Fatome, C. Finot, G. Millot, G. Genty, B. Wetzel, N. Akhmediev, F. Dias, J. M. Dudley, *Observation of Kuznetsov-Ma soliton dynamics in optical fibre*, Scientific Reports **2**: 463 (2012)
20. Y. Kivshar, G. P. Agrawal, *Optical Solitons. From Fibers to Photonic Crystals*, Academic Press, London (2003)
21. Ch. Lin, R. H. Stolen, L. G. Cohen, *A Tunable 1.1 μm Fiber Raman Oscillator*, Applied Physics Letters **31**, 97 (1977)
22. Ch. Mahnke, F. Mitschke, *Possibility of an Akhmediev breather decaying into solitons*, Physical Review A **82**, 033808 (2012)
23. F. M. Mitschke, L. F. Mollenauer, *Experimental Observation of Interaction Forces Between Solitons in Optical Fibers*, Optics Letters **12**, 355 (1987)
24. L. F. Mollenauer, R. H. Stolen, J. P. Gordon, *Experimental Observation of Picosecond Pulse Narrowing and Solitons in Optical Fibers*, Physical Review Letters **45**, 1095 (1980)
25. A. C. Newell, J. V. Moloney, *Nonlinear Optics*, Addison-Wesley, New York (1992)
26. D. H. Peregrine, *Water waves, nonlinear Schrödinger equations and their solutions*, J. Austr. Math. Soc. B **25**, 16 (1983)
27. K. Porsezian, V. C. Kuriakose (Eds.), *Optical Solitons. Theoretical and Experimental Challenges*, Springer, New York (2003)
28. J. S. Russell, *Report on Waves*, 14th Meeting of the British Association for the Advancement of Science 311 (1844)
29. J. Satsuma, N. Yajima, *Initial Value Problems of One-Dimensional Self-Modulation of Nonlinear Waves in Dispersive Media*, Suppl. Progress of Theoretical Physics **55**, 284 (1974)
30. D. R. Solli, C. Ropers, P. Koonath, and B. Jalali, TITLE, Nature **450**, 1054 (2007)
31. R. H. Stolen, C. Lee, R. K. Jain, *Development of the Stimulated Raman Spectrum in Single-Mode Silica Fibers*, Journal of the Optical Society of America B **1**, 652 (1984)
32. R. H. Stolen, J. P. Gordon, W. J. Tomlinson, H. A. Haus, *Raman Response Function of Silica-Core Fibers*, Journal of the Optical Society of America B **6**, 1159 (1989)
33. R. H. Stolen, W. J. Tomlinson, *Effect of the Raman Part of the Nonlinear Refractive Index on Propagation of Ultrashort Optical Pulses in Fibers*, Journal of the Optical Society of America B **9**, 565 (1992)

34. R. W. Tkach, A. R. Chraplyvy, R. M. Derosier, *Spontaneous Brillouin Scattering for Single-Mode Optical-Fibre Characterization*, Electronics Letters **22**, 1011 (1986)
35. N. Zabusky, M. D. Kruskal, *Interaction of Solitons in a Collisionless Plasma and the Recurrence of Initial States*, Physical Review Letters **15**, 240 (1965)
36. V. E. Zakharov, A. B. Shabat, *Exact Theory of Two-Dimensional Self-Focusing and One-Dimensional Self-Modulation of Waves in Nonlinear Media*, Soviet Physics JETP **34**, 62 (1972)
37. V. E. Zakharov, S. Wabnitz (Eds.), *Optical Solitons; Theoretical Challenges and Industrial Perspectives*, Springer, Berlin (1999)

Chapter 10
A Survey of Nonlinear Processes

10.1 Normal Dispersion

10.1.1 Spectral Broadening

Self-phase modulation (SPM) broadens the frequency spectrum. This effect is not pronounced as long as the peak nonlinear phase shift remains below π or so. At a few π, however, the spectrum begins to develop strong undulations as shown in Fig. 10.1.

The figure is based on numerical calculations based on the nonlinear Schrödinger equation. This prediction is borne out well by experiment, as shown in Fig. 10.2. Intense light pulses in sufficiently long fiber easily achieve $\phi_{nl} \gg \pi$. Then the spectrum takes a nearly rectangular shape (Fig. 10.3), mostly an effect of the linear chirp across the central part of the pulse.

This spectral broadening may be desirable, e.g., to filter out different frequency components simultaneously. Sometimes, one wishes to generate a spectral continuum over a certain frequency range. Our main interest at this point, however, is that a broad spectrum is an important prerequisite for the generation of shorter pulses. In other words, strong self-phase modulation is a step toward pulse compression.

10.1.2 Pulse Compression

Let us assume that a light pulse has assumed a broad spectrum by self-phase modulation. Then, there will be a strong chirp in its temporal evolution.

Now, if all this happens in the presence of normal dispersion, the different spectral components of the pulse will be stretched out temporally. Then, the pulse will take on a nearly rectangular temporal profile with a very nearly linear chirp when the components responsible for the flat central part of the spectrum are rearranged in time.

© Springer-Verlag Berlin Heidelberg 2016
F. Mitschke, *Fiber Optics*, DOI 10.1007/978-3-662-52764-1_10

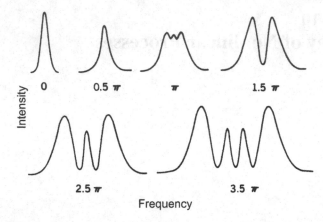

Fig. 10.1 Calculated spectral broadening by SPM; the maximum nonlinear phase shift is given as a parameter. From [18] with kind permission

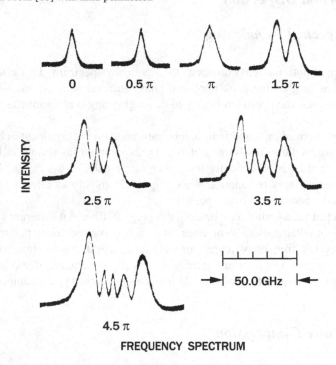

Fig. 10.2 Observed spectral broadening by SPM; the maximum nonlinear phase shift is given as a parameter. From [18] with kind permission

By using a diffraction grating, one can generate an opposite (anomalous) dispersion which can recompress the distorted pulse to the shortest duration compatible with its spectral width. A pair of gratings (see Fig. 10.4) is more convenient to handle. A combination of a fiber and a grating pair as sketched in Fig. 10.5 is

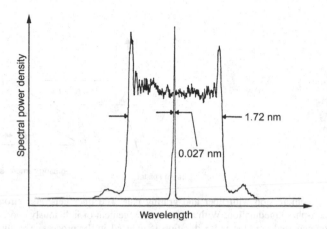

Fig. 10.3 Spectral broadening by self-phase modulation can, in extreme cases, give a nearly rectangular spectrum. Here the pulses were taken from a frequency-doubled Nd:YAG laser (532 nm) and had a duration of 35 ps. After [7] with kind permission

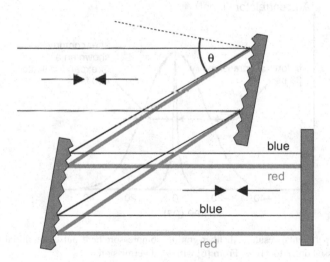

Fig. 10.4 Schematic representation of dispersion from a pair of diffraction gratings. Shorter wave light (labelled as '*blue*') takes a shorter path than longer wavelength light ('*red*') (Color figure online)

available commercially as a pulse compressor. It works in the wavelength regime where the fiber is normally dispersive (which is useful for light from dye lasers or Nd:YAG lasers), and it can considerably reduce the pulse duration, as shown in Figs. 10.6 and 10.7.

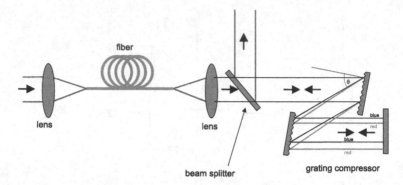

Fig. 10.5 Pulse compression with fiber and grating pair. The fiber generates strongly chirped pulses due to self-phase modulation. With a grating arrangement of judiciously chosen dispersion the chirp is compensated, and the pulse duration is reduced in the process. The figure shows a setup with double pass through a grating pair and output coupling from a beam splitter (partially reflecting mirror)

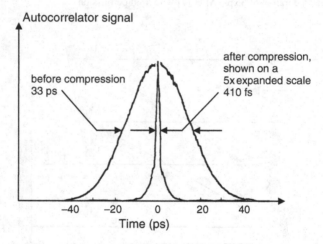

Fig. 10.6 Experimental result with fiber-grating compressor: here pulses of initial width 33 ps were compressed down to 410 fs. From [6] with kind permission

10.1.3   Chirped Amplification

There are now laser systems capable of generating peak powers of more than 1 PW. They rely on an oscillator–amplifier concept: Pulses generated by an oscillator are amplified and brought well into the terawatt regime and above. Such light sources are important tools for basic physics research.

The technical difficulty is that optical components of the amplifier must withstand the enormous intensities and thus are subject to a damage hazard. This can be avoided by the concept of "chirped pulse amplification" or CPA [16], which has its origin in radar technology and is a method to avoid high peak powers acting

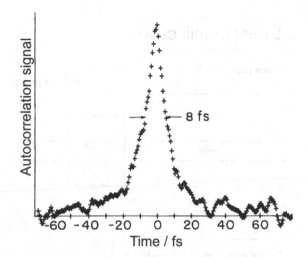

Fig. 10.7 Extreme pulse compression down to 8 fs. For some time in the 1980s, this result represented the world's shortest pulses. From [8] with kind permission

on components. One inserts a dispersive element—either a fiber or a grating—after the oscillator to produce a strong dispersive broadening of the pulse, with the accompanying reduction in peak power. The spectral components of the pulse then do not occur at the same time but sequentially. This predistorted pulse is fed to the amplifier where the highest intensity peaks are now reduced by the broadening factor. This can amount to several orders of magnitude, and the damage risk is drastically reduced. After amplification, all Fourier components are shifted together again by sending the pulses through another dispersive element which has the same absolute value of dispersion, but the opposite sign. An example is shown in Fig. 10.8 where the first dispersive element is a fiber, the second, a grating.

CPA is now the method of choice to produce petawatt powers in several laboratories around the world. To put this into perspective, consider that all electric power generated in the USA is below 1 TW. Of course, the petawatt level is maintained only for a split second, indeed, a few hundred femtoseconds. A pioneering experiment at the Lawrence Livermore Laboratory 1999 [16] demonstrated pulses with peak power ≥ 1 PW, 680 J energy, and a duration of 440 fs. Pulses were stretched 25,000-fold before amplification. Recompression had to be performed in vacuum due to the enormous field strength of the final pulse. It exceeded by three orders of magnitude those typical field strengths by which electrons are bound to nuclei in most atoms; any material would instantly break down. When focused, an intensity of 10^{25} W/m^2 was obtained at an energy density of 30 PJ/m^2; this is a lot more than inside stars.

In this case, however, fibers were not used but rather a combination of gratings. While it is true that fibers provide more dispersion, there are also contributions from higher-order dispersion that make it difficult to undo the chirp completely. Also, on a grating one can distribute the power over a larger area, thus reducing intensity and risk of damage. Therefore, fibers are preferentially found in systems that do not aim at the ultimate power limit but that are intended to work as a handy laboratory tool. Commercial CPA systems are available.

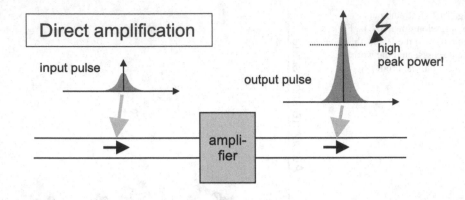

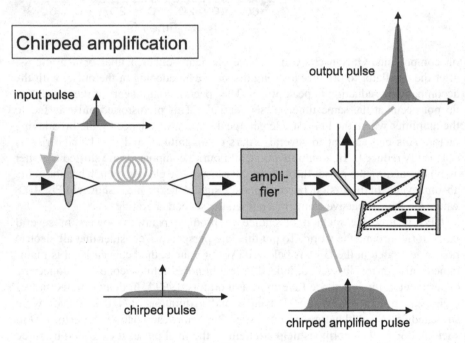

Fig. 10.8 *Top*: When a light pulse is amplified to very high energy, excessive peak powers may damage the gain medium. *Bottom*: This is circumvented by broadening of the pulses with a dispersive element prior to amplification, and a restoration of the initial pulse width with an oppositely dispersive element after amplification. As dispersive elements, either fibers as in Fig. 10.5 or gratings can be used; for the compression gratings are usually preferred

10.1.4   Optical Wave Breaking

We have seen above that through strong self-phase modulation, pulses acquire an almost rectangular spectrum. In the presence of normal dispersion, the spectral components are pulled apart temporally, so that there is an almost linear chirp in the central section of the pulse, and the temporal profile also approximates a rectangular. Both the leading and trailing slopes are fairly abrupt.

If one then keeps increasing the amount of self-phase modulation by increasing the power, there is a phenomenon called "optical wave breaking" [20]. Fig. 10.9 shows the progression of events: The portion of the pulse with the highest frequency is delayed from column (a) to (b) so that eventually in column (c) it falls behind the background in the far pulse wings. At the same time, the part with the lowest frequency passes the background. That is, at the positions of the slopes the pulse "folds over" and interference phenomena arise [17]. Then oscillations appear in the wings of the temporal profile and in the spectral profile as well.

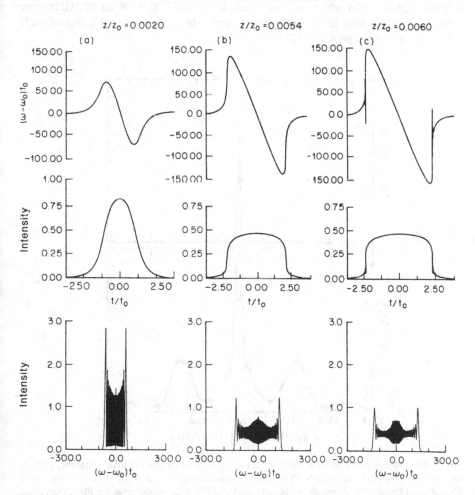

Fig. 10.9 Optical wave breaking. *Top row*: Evolution of the instantaneous frequency profile. There is a considerable nearly linear chirp. *Center row*: Evolution of the power profile. The pulse shape becomes nearly rectangular. *Bottom row*: The corresponding power spectra. After sufficiently long propagation (*right column*) the wave breaks; interference fringes arise. From [20] with kind permission

10.2   Anomalous Dispersion

10.2.1   Modulational Instability

We have seen in Sect. 9.4.1 that noise or tiny perturbations are subject to gain when the dispersion is anomalous and that the gain prefers certain frequencies (typically on the order of 1 THz). This gain can be utilized for the generation of signals in its preferred frequency range. This comes in handy because it is not trivial to generate signals at frequencies around 1 THz; there are not many alternative methods. Figure 10.10 shows the first experimental proof that modulational instability (MI) sidebands grow from noise; in this example an oscillation of ca. 450 GHz was generated [19]. In fiber lasers one can now generate continuous oscillation of such sidebands, at least in principle [2].

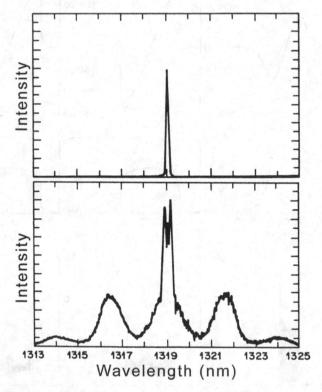

Fig. 10.10 The first observation of terahertz signals generated by modulational instability [19]. A sequence of 100 ps pulses from a Nd:YAG laser with an average power of 7.1 W is launched into a fiber of 1 km length. The figure shows power spectra of the pulse sequence at the fiber input (*top*) and at the fiber output (*bottom*). The newly generated sidebands, indicative of the modulation, are separated from the seed by 2.6 nm or ca. 450 GHz. From [19] with kind permission

10.2.2 Fundamental Solitons

Solitons exist due to the simultaneous presence of both group velocity dispersion and Kerr nonlinearity. This can be tested by a very simple experiment: Pulses of a given duration (in this example, 560 fs) and wavelength (here, 1.5 μm) are sent through a variable attenuator into a fiber (Fig. 10.11). At the distal fiber end pulse shape and duration are monitored. As long as the power remains weak, nonlinearity does not yet play any role. Due to dispersion the pulses will broaden out, here to ca. 50 ps, which is a 100 times their initial width. As power is gradually increased, the pulse duration at the fiber end is significantly reduced. When ca. 6 mW average power is reached, the initial pulse shape is faithfully reproduced at the fiber end. This is the power level at which the pulse propagates without any change of shape. We have found the fundamental soliton! Its pulse shape is stable. Actually, it is stable even in the sense that mild deviations from the right shape will automatically be reduced.

If the power is further increased, the pulse undergoes a net compression; the dispersive broadening is overcompensated. But the pulse duration does not fall monotonously. This is particularly clear where the fiber length happens to be an integer multiple of z_0: Then, at power levels equal to 4, 9, 16, etc. times the fundamental soliton power, the initial pulse duration is reproduced again. This repetition is of course due to higher-order solitons, but it is difficult to observe this cleanly because a multitude of effects gets in the way of an exact reproduction of the initial shape. On the other hand, the reproduction of the duration and shape by the fundamental soliton is quite robust and straightforward to observe experimentally.

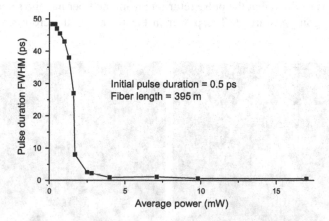

Fig. 10.11 Observation of pulse broadening by dispersion and pulse compression by nonlinearity, and the equilibrium of the two. A sequence of light pulses with an initial width of 0.5 ps is sent through a 395-m long fiber. At low power the pulses broaden out dispersively to ca. 50 ps. As the power is increased, nonlinearity counteracts dispersion and mitigates the broadening. At ca. 6 mW average power, the initial pulse width and shape are reproduced at the fiber end; indeed this is the power at which for the parameters of the fiber used here, the fundamental soliton is expected

10.2.3   Soliton Compression

As described in Sect. 9.4.6, higher-order solitons have an oscillating pulse width and
shape. This is a useful feature for pulse compression. One launches a pulse into a
fiber with suitable power so that it propagates as a soliton of, say, second order. If its
initial shape is reasonably close to a sech, then after a distance $L = z_0/2$ it will be
compressed in duration to 23% of its initial width. Figure 10.12 shows an example.
For even higher-order solitons, the compression is even stronger. The disadvantage
of this technique is that the resulting compressed pulses are not chirp-free, but
fortunately for some applications that is less important than the temporal duration.
If the reader compares this scheme with the fiber-grating compression described in
Sect. 10.1.2, it should be apparent that here the fiber performs all functions at once
so that additional components like gratings are not required.

10.2.4   The Soliton Laser and Additive Pulse
Mode Locking

When it comes to the generation of short laser pulses it has become common practice
by now to exploit optical nonlinearities directly. A precursor of today's Kerr lens
modelocked lasers was conceived in the mid-1980s by Linn F. Mollenauer of Bell
Laboratories when a resonator containing a piece of fiber was coupled to the laser
resonator. Both resonators were adjusted to have the same round trip time. The idea
was that the coupled system would provide a stable pulse shape when the stationarity
condition was fulfilled that the pulse returning from the fiber had the same duration
as the pulse going toward it. The power in the fiber was therefore set such that

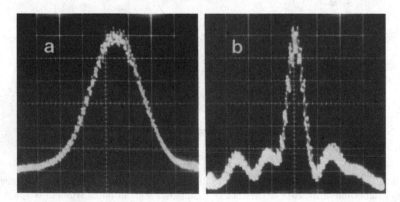

Fig. 10.12 Compression of pulses at anomalous dispersion, also known as "soliton compression."
In the example shown a pulse of 60 fs full width (**a**) is sent through a fiber of length $z_0/2$, in this
case about 50 cm. It is compressed to 19 fs full width (**b**). From [13]

solitons were formed; only solitons can maintain their shape or so the reasoning went. Indeed stable pulses were obtained even though the pulses in the fiber were closer to an $N = 2$ soliton, with the fiber length (return trip) close to z_0 [10]. The pulse durations obtained from this "soliton laser" set records in their day for the wavelengths near the third window. Directly from the laser pulses as short 60 fs were obtained; with external soliton compression 19 fs were reached. This corresponds to less than four cycles of the optical wave!

When pulses from both coupled resonators interact, they do so interferometrically. This means that the length difference of the two cavities must be maintained to within a fraction of a wavelength during operation. This can only be performed successfully with an active servo control loop as presented in [10]. The average power circulating in the fiber resonator is tapped at an otherwise unused port and measured by a photodetector. After electronic processing, it is fed to a piezoceramic transducer which serves to fine-tune the fiber resonator length. The processing involves subtraction of a manually set suitable reference value and amplification with what is known in control systems engineering as a PI (proportional-integral) characteristic (Fig. 10.13).

Later on it turned out that the concept is more general than to be restricted to the wavelength regime of anomalous dispersion in the fiber. The relevant mechanism is the Kerr nonlinearity which creates a self phase modulation. In the interference process this is translated to a modification of the pulse shape, usually a reduction of the duration. Dispersion is not really too important in this. This insight led to the concept of "additive pulse modelocking" (APM), also known as "interferential modelocking" or "coupled cavity modelocking" [14, 15]. Several different types of lasers were used in this way to generate short pulses, including Nd:YAG lasers at both 1064 and 1319 nm.

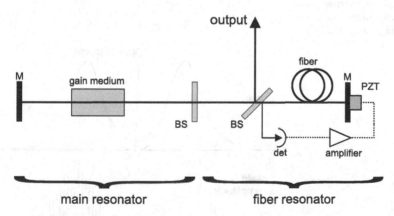

Fig. 10.13 Schematic setup of a soliton laser, a.k.a. additive pulse mode locked (APM) laser. M: mirror, BS: beam splitter (partially reflecting mirror). Both resonators are arranged to be synchronous by careful length adjustment. The servo loop consisting of detector det, amplifier, and the piezoceramic transducer PZT maintains the length with interferometric stability

10.2.5   Pulse Interaction

It is the remarkable property of a soliton that it induces a perturbation of the refractive index in the fiber which is just right to make it hang together and keep its shape. If more than a single pulse propagates down the fiber, each of them can "feel" the perturbation of the refractive index caused by its neighbor, in particular when they get into close proximity with each other. The relevant question to ask is for the relative phase of the optical field of the two pulses in their slopes where they overlap: are the fields in phase, in opposite phase, at any other phase angle? For the "in phase" situation, there is constructive interference, and each pulse "feels" a stronger index modulation on that side that faces the other pulse. This perturbation acts asymmetrically on the pulse (Fig. 10.14)!

Fig. 10.14 Interaction of co-propagating light pulses. *Upper part*: Two pulses are in phase with each other. Constructive interference will then increase the intensity in the middle, as compared to the case that the other pulse is absent. *Lower part*: Opposite phase pulses interfere destructively; the intensity profile goes down in the middle. Pulses are always attracted to the point of highest intensity (and thus, index). Then, in-phase pulses experience mutual attraction, opposite-phase pulses, repulsion

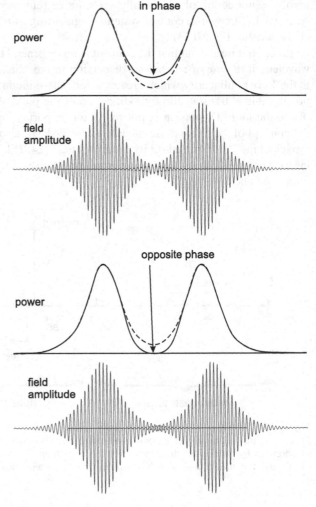

For opposite phase the fields interfere destructively, and the power in the overlap region is less than what it would be in the absence of the other pulse. Again, there is an asymmetric (one-sided) effect.

In-phase pulses will both move slightly toward their mutual center-of-mass, opposite-phase pulses will move away from each other. This interaction force was first demonstrated in [12] after a theoretical prediction had been made in [3].

In effect, there is what can be described as a force between the light pulses. Depending on the relative phase, this force can be attractive or repulsive. If one lets the separation between the pulses slide to tune the relative phase, the force will basically change in a sinusoidal fashion. Once the separation increases noticeably, the force will be reduced exponentially because the slopes of sech pulses roll down exponentially. Once pulses are separated more than five or seven or so pulse widths, the interaction force becomes negligible.

The first experimental proof [12] is shown in Fig. 10.15. Time measurements on a femtosecond scale are only feasible by way of the autocorrelation technique (see Appendix 18). In this experiment the interaction was easily measured. It was also found that in the case of attraction the pulses move toward each other, but they do not collide: this is surprising because collisions would be expected both intuitively and by the nonlinear Schrödinger equation. However, higher-order effects perturb the pulses as they get increasingly close to each other so that eventually they actually fly apart [9].

The concept of attraction and repulsion can be extended to the case of chirped pulses where it is not so straightforward to speak of in-phase or opposite-phase pulses, see [5].

10.2.6 Self-Frequency Shift

One might be forgiven for adopting the following simple-minded approach to pulse propagation in optical fiber: While it is possible that the pulse shape is corrupted by

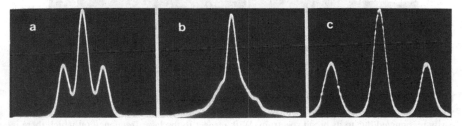

Fig. 10.15 Experimental proof of pulse interaction forces. Autocorrelation traces of the pulse pair from the source (**a**) represents a double pulse, consisting of two humps each 0.9 ps wide and separated by 2.33 ps. (About the interpretation of autocorrelation traces see Appendix 18.) If the pulses have the same phase, after traveling down 340 m of fiber they have moved toward each other so far that they are no longer resolved (**b**); for opposite phase, they have moved away from each other (**c**). From [12]

influences like dispersion and self-phase modulation, the optical center frequency remains unaffected. However, the exact opposite is true: Dispersion and self-phase modulation combine in such a way that in the case of solitons the pulse shape is preserved; the optical center frequency, however, is shifted. This latter fact was first discovered experimentally [11] and is easily explained by considering the effect of Raman scattering [4].

The Raman gain spectrum is broad and, as Fig. 9.31 shows, begins at very small frequency detunings. Therefore, there is Raman self-pumping even within the bandwidth of a single pulse: The high-frequency slope acts as a pump for the low-frequency slope. As a result, the spectral center-of-mass of the pulse shifts continuously toward lower frequencies. If the pulse is a soliton, then its inherent robustness lets it hang together as an entity; pulses of inferior structural stability are likely to decay in the process (Figs. 10.16 and 10.17).

The amount of frequency shift depends strongly on the pulse duration: As the pulses get shorter, peak power grows quadratically, and the spectral width linearly. The Raman gain curve grows approximately linearly for small detunings (see Fig. 9.31). Taking all this together, the frequency shift is proportional to the inverse fourth power of pulse duration [4] (see Fig. 10.18). For 1 ps pulses the effect is so weak as to be noticeable only after very long distances; for 10 ps it may be safely neglected in almost all cases. On the other hand, for subpicosecond pulses, the frequency shift becomes a dominant effect: A pulse of less than 100 fs

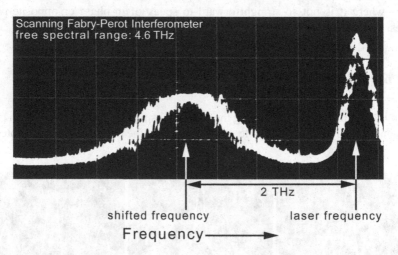

Fig. 10.16 First observation of the soliton self-frequency shift. The figure shows the power spectrum at the far fiber end when short laser pulses are launched into the near end. The soliton is easily recognized due to its broad spectrum. With respect to the laser frequency (at which there is a narrower peak), the soliton is shifted downward in frequency. The amount of shift fluctuates with the laser power because the power defines the soliton width (this was explained in Sect. 9.4.4); power fluctuations during exposure of this photographic picture result in a "flat rooftop" of the soliton. From [11]

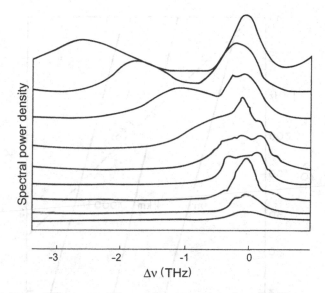

Fig. 10.17 Soliton self-frequency shift as a function of launch power. Here power was varied over a wide range with a modulator; its control voltage also produced the offset between traces. It is clearly visible that the soliton's spectral width increases with increasing power; so does the spectral shift. Only above a certain threshold does the soliton spectrum become visible as a separate structure. From [11]

duration is shifted considerably after only 1 m propagation distance in standard fiber. The shift can reach large values, amounting to a noticeable fraction of the optical frequency. However, even a strong shift slows down to a halt once the pulse is shifted by hundreds of nanometers toward longer wavelengths because at the longer wavelength the fiber probably has much higher dispersion, and also is no longer a low-loss medium. These modifications conspire to reduce the peak power and increase the duration so that the frequency shifting rate comes down.

10.2.7 Long-Haul Data Transmission with Solitons

Fundamental solitons are the natural units, or bits, for the transmission of information over optical fiber. They are more robust and stable than any other pulse because they embrace Kerr nonlinearity in the first place, and therefore do not get perturbed by it. They exist at anomalous dispersion; it is fortunate that the wavelength regime of anomalous dispersion in fiber coincides with the wavelength regime of lowest losses. This is why solitons lend themselves to applications in long-haul data transmission. Chapter 11 is devoted to a more detailed discussion of this aspect.

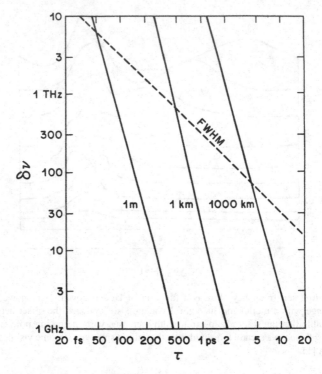

Fig. 10.18 Expected amount of self-frequency shift. The shift basically scales with the inverse fourth power of pulse duration. From [4] with kind permission

10.2.8   Supercontinuum Generation

A novel application of nonlinear effects in optical fibers is called Optical Supercontinuum Generation. The name refers to the generation of broadband light; while the terminology is not entirely well-defined, most authors consider a spectrum 'broad' when its width is about an octave or more: $\nu_{upper} \geq 2\nu_{lower}$. Here, frequencies ν describe the upper and lower limit of the spectrum notwithstanding the complication that there is ambiguity at which point exactly to read the limit. In many instances, spectra of several octaves have been generated between points where the spectral power density had fallen to -20 dB from the highest value.

It seems ironic that one would consider to create a wide spectrum by starting from laser light which for most people epitomizes the most narrow-band light source. Why not take an ordinary light bulb—it surely generates broadband light more economically?

The catch is how much spectral power density can be obtained, and can be coupled into some structure. Thermal light sources are subject to the Planck limit which dictates that in the best case (for a 'perfectly black surface') the spectrum is given by

$$I_\nu = \frac{2\pi h\nu^3}{c^2} \frac{1}{e^{\frac{h\nu}{kT}} - 1}.$$ (10.1)

Here, I_ν is the emitted power per spectral interval and cross sectional area in W/(Hz m^2), taking two orthogonal polarizations into account. h is Planck's constant, k is Boltzmann's constant, and T is temperature in Kelvin units. At any desired frequency, the only parameter available to maximize power is T. However, one cannot indefinitely raise the temperature of a thermal emitter. Lamp filaments are made from tungsten, the metal with the highest melting point (3422 K). A few solid state materials with melting points above that of tungsten exist: diamond, and some carbides, but none of these is suitable for making lamp filaments.

In the interest of reasonable lifetime operating temperatures must stay well below melting temperature; accordingly, light bulbs operate near 2800 K (in the case of halogen light bulbs one can go a tad higher due to the tungsten recirculation process provided by the halogen filling). If one demands higher temperature, the only option is to use arc discharges which require much maintenance effort and are expensive to operate.

On the other hand, non-thermal light sources like lasers and LEDs are not subject to the Planck limit and can reach much higher spectral power density, but only over their narrow spectral range. The idea about supercontinuum generation is that relatively powerful but narrowband light from a laser source is coupled into a fiber which is selected for having a large nonlinearity coefficient γ—which usually means a small modal area A_{eff}. Nonlinear processes in the fiber then redistribute the power in the frequency domain, and thus create a broad spectrum. It turns out that in this way the above definition of supercontinuum can be met without much difficulty [1]. Indeed, there are now commercial supercontinuum sources available which routinely generate spectra spanning much of the visible and near-infrared portion of the spectrum with spectral power densities far above the Planck limit.

Let us sketch what happens: Typically, holey fibers (see Chap. 4.7) are used in this context because they can confine the light to a much smaller cross section than conventional fibers. In most approaches a short light pulse of high peak intensity is launched into this fiber, although in a few cases even continuous wave light has been used successfully. The light pulse may be thought of at least approximately as a soliton of very high order, $N \gg 1$; higher-order solitons, in turn, are a combination of several fundamental solitons all centered right on each other (see Sect. 9.4.6). However, higher-order solitons are prone to decay because after the most minute perturbation there is no restoring force that would nudge them back. Therefore, higher-order solitons easily decay into their constituent fundamental solitons; the process has been called soliton fission.

In order to get an estimate how rapidly the fission takes place, it helps to note that the propagation distance to the decay point is usually well described by the distance to a first massive soliton compression. The high peak powers occurring there help create the perturbation that breaks the N-soliton apart. This distance is a fraction of z_0; as an approximation

$$z_{\text{fission}} = \frac{z_0}{N}$$

may be used [1]. As a quick glance at Table 9.1 shows, this predicts a distance of the order of centimeters to a few meters at most for sub-picosecond pulses when the soliton number is of order $N \approx 10$. In a fiber longer than that, fundamental solitons will propagate individually, each of them subject to their mutual interactions as well as to other perturbations—most notably by the Raman shift.

On the other hand, for many-picosecond pulses the fission distance becomes hundreds of meters, which is longer than most laboratory-experiment fibers. Fission is therefore not expected; nevertheless, the N-soliton decays. Obviously there is a different mechanism at work for 'long' pulses: and that is modulational instability. If one thinks of a 'long' pulse as an approximation to a continuous wave, one can apply the logic of Sect. 9.4.1; an exponential growth leads to a structure which is similar to a sequence of pulses plus some background. Again, other perturbations like Raman shift perturb this structure enough to set individual pulses into motion independent of the others.

In either case, after the initial breakup of the launch pulse into many subpulses and usually also some background radiation, during further propagation several interaction mechanisms between these pulses and radiation go on. As the Raman shift is power-dependent, different soliton pulses experience different amounts of frequency shift which implies that they can collide. In collisions, they can exchange energy, etc. The process may be quite complex in detail, but all exchange of energy between various subpulses contributes to redistribution of power in the spectrum, so that after some distance a wide spectrum is obtained. Its total power is, of course, given by the input power (with some correction for coupling and other losses). In the desired (but not easily obtained) case that the generated spectrum is flat, the power density is easily calculated as total power divided by spectral width.

Note that here the output light has the perfect spatial coherence of the fiber mode, but the temporal coherence T_{coh} has been reduced from whatever the laser source may have had to a much smaller value which may be estimated as the inverse of the spectral width, $T_{\text{coh}} \approx 1/\Delta\nu$. For a width of about one octave or more, the temporal coherence length $c\,T_{\text{coh}}$ is roughly the (center) wavelength. Short coherence length is exactly what is needed in some applications, like coherence tomography. That is a type of interferometry where light is focused onto a sample under study (e.g. a piece of biological tissue) to obtain spatially resolved information in three dimensions. While the positional resolution is given by the spot size and can be reduced to about the wavelength, depth resolution is obtained thanks to the short temporal coherence length, also of the order of micrometers. As a result, the technique provides microscopic resolution in the specimen in three dimensions, as far into the material as the light penetrates.

References

1. J. M. Dudley, G. Genty, S. Coen, *Supercontinuum generation in photonic crystal fiber*, Review Modern Physics **78**, 1135 (2006)
2. P. Franco, F. Fontana, I. Christiani, M. Midrio, M. Romagnoli, *Self-induced Modulational-Instability Laser*, Optics Letters **20**, 2009 (1995)
3. J. P. Gordon, *Interaction Forces Among Solitons in Optical Fiber*, Optics Letters **8**, 596 (1983)
4. J. P. Gordon, *Theory of the Soliton Self-Frequency Shift*, Optics Letters **11**, 662 (1986)
5. A. Hause, H. Hartwig, M. Böhm, F. Mitschke, *Binding Mechanism of Temporal Soliton Molecules*, Physical Review A **78**, 063817 (2008)
6. A. M. Johnson, R. H. Stolen, W. M. Simpson, *80x Single-Stage Compression of Frequency Doubled Nd: Yttrium Aluminum Garnet Laser Pulses*, Applied Physics Letters **44**, 729 (1984)
7. A. M. Johnson, W. M. Simpson, *Tunable Femtosecond Dye Laser Synchronously Pumped by the Compressed Second Harmonic of Nd: YAG*, Journal of the Optical Society of America B **2**, 619 (1985)
8. W. H. Knox, R. L. Fork, M. C. Downer, R. H. Stolen, C. V. Shank, *Optical Pulse Compression to 8 fs at a 5-kHz Repetition Rate*, Applied Physics Letters **46**, 1120 (1985)
9. Y. Kodama, K. Nozaki, *Soliton Interaction in Optical Fibers*, Optics Letters **12**, 1038–1040 (1987)
10. F. M. Mitschke, L. F. Mollenauer, *Stabilizing the Soliton Laser*, IEEE Journal Quantum Electronics **QE-22**, 2242 (1986)
11. F. M. Mitschke, L. F. Mollenauer, *Discovery of the Soliton Self Frequency Shift*, Optics Letters **11**, 659 (1986)
12. F. M. Mitschke, L. F. Mollenauer, *Experimental Observation of Interaction Forces Between Solitons in Optical Fibers*, Optics Letters **12**, 355 (1987)
13. F. M. Mitschke, L. F. Mollenauer, *Ultrashort Pulses from the Soliton Laser*, Optics Letters **12**, 407 (1987)
14. F. Mitschke, G. Steinmeyer, M. Ostermeyer, C. Fallnich, H. Welling, *Additive Pulse Mode Locked Nd:YAG Laser:An Experimental Account*, Applied Physics B **56**, 335 (1993)
15. F. Mitschke, G. Steinmeyer, H. Welling, *Coupled Nonlinear Cavities: New Avenues to Ultrashort Pulses*, in: *Frontiers in Nonlinear Optics – The Sergei Akhmanov Memorial Volume*, H. Walther, M. Koroteev, M. Scully (Eds.), IOP Publishing, Bristol (1993)
16. M. Perry, *Crossing the Petawatt Threshold* (1996). See www.llnl.gov/str/Petawatt.html
17. J. E. Rothenberg, D. Grischkowsky, *Observation of the Formation of an Optical Intensity Shock and Wave Breaking in the Nonlinear Propagation of Pulses in Optical Fiber*, Physical Review Letters **62**, 531 (1989)
18. R. H. Stolen, Ch. Lin, *Self-Phase-Modulation in Silica Optical Fibers*, Physical Review A **17**, 1448 (1978)
19. K. Tai, A. Hasegawa, A. Tomita, *Observation of Modulational Instability in Optical Fibers*, Physical Review Letters **56**, 135 (1986)
20. W. J. Tomlinson, R. H. Stolen, A. M. Johnson, *Optical Wave Breaking of Pulses in Nonlinear Optical Fiber*, Optics Letters **10**, 457 (1985)

Part V
Technological Applications of Optical Fibers

Laying optical fiber cables—here within sight of the author's house—is not nearly as spectacular as the performance of fiber during operation.

Part V
Technological Applications of Optical
Fibers

Chapter 11
Applications in Telecommunications

11.1 Fundamentals of Radio Systems Engineering

We first present a brief introduction to essential concepts of telecommunications engineering, insofar as they are relevant for our topic.

11.1.1 Signals

The central concept of all communication engineering is that of a *signal*. In the most general case, it is left open what this is physically; it suffices to state that it is a scalar, real-valued function of time. We assume that the signal contains information which is meant to be taken from some transmitter to some receiver. The signal may be represented by some physical quantity such as an electric voltage, the position of an indicator needle, or the brightness of a light source; one common realization would be that at each instant, the value of the quantity is proportional to that moment's value of the signal.

We must first distinguish *continuous-time signals* and *discrete-time signals*. The latter have a defined value only at certain instants in time, or in other, more mathematical words consist of a sequence of Dirac pulses (delta functions), each weighted in accord with the signal value. One can obtain a discrete-time signal from a continuous-time signal by *sampling*. Very frequently one chooses to take samples at a fixed rate, i.e., in equal time steps. Below we will assume a fixed sampling rate, or clock frequency, throughout.

The other fundamental distinction is between *analog signals* and *digital signals*. An analog signal has a continuous range of values, i.e., can take any intermediate value within the interval of possible values. In contrast, a digital signal has a finite number of possible states known as its *alphabet*.

A thermocouple yields a voltage proportional to temperature; this is an example for an analog continuous-time signal. A dynamic microphone is another example of

© Springer-Verlag Berlin Heidelberg 2016
F. Mitschke, *Fiber Optics*, DOI 10.1007/978-3-662-52764-1_11

the same. A sequence of results when dice are thrown or the roulette wheel is turned would represent a discrete-time digital signal. Is is quite often the case that digital signals are also discrete-time.

An important subclass of digital signals are *binary* signals. For these the alphabet has just two symbols which, depending on context, may be called "zero" and "one", "high" and "low", "true" and "false", or "plus" and "minus", etc.

11.1.2   Modulation

Only in the simplest cases a signal is transmitted just as is. There are many benefits if one uses a *carrier* oscillation, a wave on which the signal is "impressed". This is familiar from broadcast signals: At the receiver one selects the carrier frequency of the desired program. This way several different programs can be transmitted simultaneously and independently.

The impression of a signal onto a carrier is known as *modulation*. It can be done in a variety of ways: If the carrier is a harmonic periodic function (this is a very common situation), which can be written as

$$A = A_0 \cos(\Omega t + \varphi) \tag{11.1}$$

with amplitude A_0 and angular frequency Ω, one has the options of subjecting either A_0, Ω, or φ to the signal. The result is then referred to as amplitude modulation, frequency modulation, or phase modulation, respectively.

11.1.2.1   Amplitude Modulation

Let us assume for simplicity that the range of values of the signal $S(t)$ is restricted to the interval $-1 \leq S(t) \leq +1$. This can always be achieved by proper normalization. One can then make the amplitude signal-dependent by letting $A_0 = \hat{A} (1 + S(t))$ in Eq. (11.1) to obtain amplitude modulation (Fig. 11.1).

Again for the sake of simplicity, we consider the simplest possible signal, a sinusoidal oscillation with angular frequency ω:

$$S(t) = \sin \omega t. \tag{11.2}$$

We now let $A_0 = \hat{A} (1 + M \sin \omega t)$ where $0 \leq M \leq 1$ is called modulation depth. We insert in Eq. (11.1) and obtain

$$A = \hat{A} (1 + M \sin \omega t) \cos(\Omega t + \varphi). \tag{11.3}$$

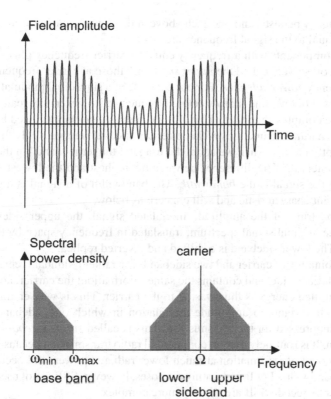

Fig. 11.1 Sketch to explain amplitude modulation. *Top*: The modulation of a carrier wave with frequency Ω with a signal $\omega = 0.05\,\Omega$ is shown at a modulation depth of $M = 0.5$. *Bottom*: Modulation generates two sidebands above and below the carrier, at the distance of the signal frequency. The figure suggests a signal occupying the band $\omega_{min} \leq \omega \leq \omega_{max}$ (baseband). The sidebands have the same width. Note that the lower sideband is inverted

Using the well-known relation between harmonic functions

$$\sin x \cos y = \frac{1}{2}\left[\sin(x-y) + \sin(x+y)\right],$$

this then yields

$$
\begin{aligned}
A &= \hat{A}\left[\cos(\Omega t + \varphi) + M\sin\omega t \cos(\Omega t + \varphi)\right]\\
&= \hat{A}\left[\cos(\Omega t + \varphi) + \frac{M}{2}\left[\sin(\omega t + \Omega t + \varphi) + \sin(\omega t - \Omega t - \varphi)\right]\right]
\end{aligned}
\tag{11.4}
$$

This result contains terms of three different frequencies: The first term on the RHS at Ω corresponds to the carrier. The second and third terms have frequencies $\Omega \pm \omega$. Amplitude modulation (AM) generates new frequency components: the one at the carrier frequency persists, and one each above and below the carrier frequency by a difference equal to the signal frequency are new.

Fourier components with a frequency equal to carrier frequency plus signal frequency are collectively called the *upper sideband*; those at carrier frequency minus signal frequency, *lower sideband*. For our example of a sinusoidal modulation, these "bands" consist of only one sharply defined frequency. However, any realistic signal will be more complex than that of Eq. (11.2), and may be decomposed by Fourier analysis into harmonic functions within a certain spectral interval. Only then are the sidebands aptly named, when we adopt the usage of the term "band" in the sense of "frequency interval." The difference between the highest and the lowest frequency occurring in the signal is the *bandwidth*. The bandwidth of a signal is arguably its most important characteristic and will concern us below.

In the spectrum of the amplitude modulated signal, the upper sideband is a replica of the original signal spectrum, translated in frequency space by the carrier frequency. The lower sideband is a shifted and inverted replica.

The combination of carrier and two sidebands is a rather redundant representation of the signal. Each sideband contains the same information; the carrier, none. Much of the transmitted energy is thus wasted to the carrier. This is why engineers have come up with variants to amplitude modulation in which one sideband and the carrier are suppressed for the transmission. This is called *single sideband*, or SSB, transmission. It is routinely used in commercial radio transmission because it carries the same amount of information at much lower radiated power, and occupies only half of the bandwidth. For broadcasting purposes, however, SSB is not used because the receivers to decode SSB are slightly more complex.

11.1.2.2 Angle Modulation

Instead of imposing the signal onto the amplitude in Eq. (11.1), one can make either Ω or φ signal dependent. (Of the three quantities, two remain constant in each case). Then one obtains frequency or phase modulation, respectively (Fig. 11.2). In both cases it is the phase angle of the carrier which is acted upon, so that both cases are collectively called angle modulation. Mathematically, in both cases, there is a term of the form

$$\sin\left(a + b\sin(\Omega t)\right) \quad (a, b \text{ are constants}), \tag{11.5}$$

and "sine of sine" produces Bessel functions (see Chap. 15).

Angle modulation creates sidebands, too. The difference is that even in the case of a purely sinusoidal signal, more than a single Fourier component appears both above and below the carrier. Their frequency differences from the carrier are equal to integer multiples of the signal frequency; their amplitudes can be evaluated using said Bessel functions. We will not pursue this context any further here; instead, the interested reader is referred to texts on communications engineering (e.g., [45]).

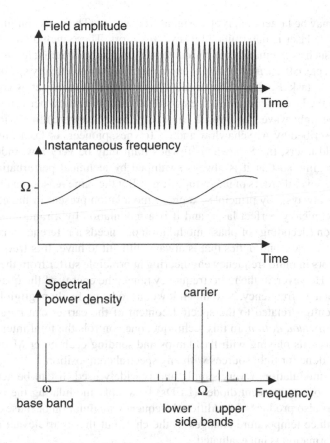

Fig. 11.2 Frequency modulation. *Top*: The modulated oscillation in the time domain. *Center*: The instantaneous frequency follows the signal. *Bottom*: Several sidebands are generated both above and below the carrier

11.1.2.3 Intensity Modulation

The types of modulation described so far are applicable when a monochromatic carrier wave is available. In the realm of optics, only lasers can provide monochromatic waves or an approximation thereof.

Unfortunately, in most cases, it is not guaranteed that the laser emission is truly single frequency. This is certainly not the case for lasers operating on several modes simultaneously. Not all types of lasers can easily be operated in a single mode, and for many laser types used in optical telecommunications this is indeed difficult to achieve. If, however, a laser operates on a multitude of modes simultaneously, the modulation formats described above are not applicable.

Moreover: even when a laser runs in single-mode operation, strictly speaking the oscillation is not monochromatic. Rather, it covers a narrow frequency band (the *emission line width*) which may be small compared to optical frequencies, but at the same time may be larger than typical signal frequencies. The emission line width of a single-mode laser is determined by several factors. These include technical considerations such as fluctuations of parameters (vibration of components, temperature fluctuations, etc.); these may be removed in principle, but practically speaking that is a very difficult task. But then there are also fundamental limits as set by spontaneous emission in the laser medium. Each emission act brings about a perturbation of the phase of the light wave. As a result there is a finite (nonzero) line width, which was first described by A. Schawlow and C. Townes, pioneers of laser physics [73]. In real-world lasers, the Schawlow–Townes limit may be very low, indeed in the millihertz regime so that it is always swamped by technical perturbations which typically are several orders of magnitude larger. But the same reasoning also implies that there is always—by principle—a phase modulation present in the emission of even the technically perfect laser, and it is a modulation by a random signal. For demodulation (decoding) of phase modulation one needs a reference phase, and in the context of lasers and optics that is always difficult to have. It is true, of course, that oscillators in radio frequency engineering in principle suffer from the same line width limit. However, in the radio frequency range, the energy of the quanta, which is proportional to frequency, is so much lower as to be perfectly negligible.

All difficulties related to the spectral content of the carrier can be avoided by using *intensity modulation*. In this technique, one controls the total intensity in the same way as kids playing with flashlamps and sending each other Morse signals. This can be done for light sources with any spectral composition.

Intensity modulation is very simple and is widely used. It can be achieved for laser diodes or luminescent diodes (LEDs) by simply modulating the operational current. This also produces an additional frequency modulation because changes in current produce temperature changes in the chip, but that is irrelevant as long as spectral information is not evaluated.

Applications which demand the highest data rates and/or longest distances of transmission are sensitive to dispersion and thus to spectral composition. In such cases single-mode lasers (e.g., of the distributed Bragg type, see Sect. 8.9.3.3) are preferred light sources; in the interest of keeping the emission frequency stable, one keeps the current constant and applies the modulation with an external modulator.

11.1.3 Sampling

Digital transmission formats are today by far the most successful formats. The signal to be transmitted is digitized, i.e., reduced to a finite number of values in the process of sampling, almost always at a certain fixed rate, the sampling rate. Speaking in mathematical terms, the original signal is multiplied with a periodic sequence of delta functions (a "picket fence"). The continuous signal is thereby replaced with a sequence of delta functions with weight factors corresponding to the respective signal value. This must be done in such a way that the relevant information contained in the signal is represented by the sequence.

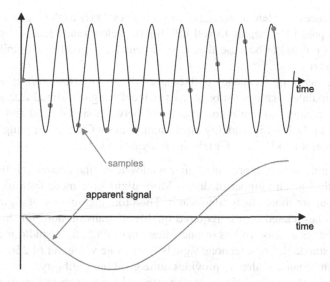

Fig. 11.3 If the sampling rate is only marginally larger than the signal frequency, a beat note at the difference frequency appears in the sampled data. This spurious contribution is called *aliasing* signal

It is therefore important to select a suitable sampling frequency. It should be obvious that the sampling frequency must be higher than the highest signal frequency of interest; one can hardly represent an oscillation with fewer sample points per period than just one. If the sampling rate is too low, another complication arises: Fig. 11.3 demonstrates that in the sampling process, certain new frequency components are generated which were not present in the original signal. The reason is that the sequence of delta functions can create beat notes with Fourier components of the signal,[1] so that difference frequencies between sampling frequency and some signal frequencies appear. These undesired additions to the signal are called *aliasing* signals.

Aliasing signals are, of course, highly undesirable because they prevent a faithful reconstruction of the original signal at the receiving station. They can be avoided by the following precautions:

1. The signal bandwidth is strictly limited with steep-slope low-pass filters to a certain maximum frequency. For this limit, one selects the highest frequency deemed necessary for the transmission in terms of reproduction quality. For high-fidelity music formats as used for CD recording, this limit is chosen as 20 kHz, i.e., the highest frequency audible to a human ear under the most favorable

---

[1] According to the convolution theorem of Fourier transforms, the spectrum of the sampled signal is found as the product of the spectra of original signal and picket fence. It contains the infinite series of harmonics from the sequence of delta functions, each with an upper and lower sideband from all Fourier components of the signal.

circumstances. For telephone signals one chooses 4 kHz as the highest frequency (the low-pass filter begins to roll off slightly below that because "brick wall" filters do not exist) because that is sufficient for a good intelligibility of the spoken word.

2. Then the sampling frequency is fixed according to the *sampling theorem* [71], which stipulates that it must be at least twice the highest signal frequency. This way it is guaranteed that no overlap exists between signal band and alias band (see Fig. 11.4). For high-fidelity music signals as on CD the sampling frequency is chosen as 44.1 kHz, and for telephone signals, 8 kHz.

Each sample is then represented after analog-to-digital conversion by a binary number with a given number of digits. More digits give more faithful amplitude resolution but are more costly to transmit. Therefore the number of digits (the "bit resolution") is dictated by the required quality of transmission. For high-fidelity music, one takes at least 16 bits or one value out of 65,536; in studio recording 24 bits is now standard. For telephone signals 8 bits (one value out of 256) is deemed good enough because it already provides quite good intelligibility.

In this way, the original signal is represented by a stream of binary digits, i.e., zeroes and ones. They represent a discrete-time, discrete-amplitude version of the signal. The bit rate is obtained from sampling rate and bit resolution. For sampling with 8 kHz and at 8 bits resolution one has 64 kbit/s; during each time slot of $1/64{,}000\,\mathrm{s} = 15.625\,\mu\mathrm{s}$, one bit is transmitted. This is the value used in telephony worldwide. For music in the CD format, there are 44,100 samples of at least 16 bits each, and twice that for stereo. On top of the signal proper, a CD contains test bits, track information, etc. The standardized SPDIF format of the digital signal stream in CD players contains as many as 64 bits at each sample point, used for two stereo channels plus overhead. Then the total data rate is 2.8224 Mbit/s.

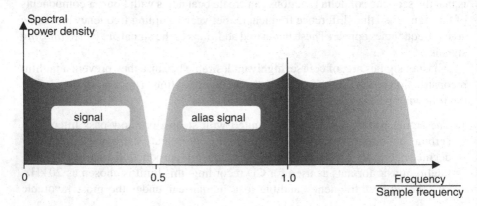

Fig. 11.4 Beat notes between Fourier components of the signal and the sampling frequency generate sidebands to the sampling frequency called aliasing bands. If prior to sampling the signal bandwidth is clipped with a low-pass filter at a frequency below one half of the sampling frequency, signal band and aliasing band cannot overlap. This is the prerequisite for faithful signal reconstruction

11.1.4 Coding

The sampled signal, a sequence of zeroes and ones, can now be transmitted, at least in principle. At the receiver, the first task is to recover the clock rate from the received bit sequence; only then can the bit stream be decoded by deciding which time slots contain a zero and which a one. To make sure that decoding can be done error-free, it is advantageous to re-code the bit stream before transmission. The objective is

- that no long strings of consecutive equal symbols can occur. A long string of zeroes makes it difficult to regenerate the clock rate.
- that the numbers of zeroes and ones, both presumably of equal probability in the long run, get equilibrated as quickly as possible. The advantage is that the demodulated signal then does not contain a DC component; this simplifies receiver construction.
- that sensitivity toward perturbations is reduced. One possibility is to transmit test bits along with the data which allow a parity check and possibly some error correction.

For example, the so-called 5B/6B code uses a lookup table by which each block of 5 bits is replaced by a 6-bit block. The table is set up such that no more than three consecutive zeroes can ever occur. This makes for a low DC component and allows easy clock regeneration. The additional bit serves as a parity check bit and helps in error correction. Of course, the data rate is increased by a factor of $6/5 = 1.2$, and correspondingly more bandwidth is required.

In the CMI format (coded mark inversion) each "zero" is replaced by the sequence "zero–one", and each "one" alternatingly by "one–one" and "zero–zero". It is obvious that this eliminates the DC component and completely avoids long strings of equal symbols. On the other hand, the price to pay is that the effective data rate is doubled, and twice the bandwidth is required.

11.1.5 Multiplexing in Time and Frequency: TDM and WDM

No single data source can generate the enormous data rates successfully transmitted today over a single fiber. The fiber can carry terabits per second! Such rates are only obtained when data from many sources, possibly an entire country, are combined. To compose separate data streams into one can be done by two methods and by combinations thereof:

TDM: *Time division multiplex* is an interleaving of bit streams in time. For long-haul transmission this is universally done to increase the rate to typically 10 Gbit/s, or more recently to 40 Gbit/s. At this speed, even fast electronic circuitry comes to its limits. Also, at that rate, errors due to polarization mode dispersion become noticeable and are difficult to keep in check.

WDM: *Wavelength division multiplex* is the transmission of independent bit
streams at different optical frequencies. This is the equivalent of different radio
stations transmitting on different frequencies: Different programs are modulated
to carriers of different frequencies and can easily be separated at the receiver by
selective means. With WDM, several bit streams can be launched into a fiber
simultaneously so that the available (low loss) spectral range can be utilized,
more or less. However, WDM is expensive: For each WDM channel a complete
set of hardware including laser diodes is required. Therefore an economic
incentive exists to first increase the bit rate as far as possible by TDM; this "only"
requires some fast electronics.

Figure 11.5 shows both variants: The right part depicts the spectrum obtained for
the combined signal. In the final analysis, TDM and WDM use the same amount of
bandwidth for the transmission of the same amount of data per unit time.

It is common engineering practice to first combine many telephone channels
with TDM to the highest frequency, which can still be conveniently worked with.
Resulting data rates are not exactly multiples of 64 kbit/s but slightly more due to an
overhead from additional bits required for controlling the decoding. Unfortunately,
different countries started using different numbers of telephone channels for TDM
(24 in the USA, 30 in Europe), so that on the transmission lines different data rates
existed. In order to assure smooth international traffic, a standardization became
inevitable.

First the USA created a standard called SONET, for *synchronous optical
network*. The fundamental clock rate is 51.48 Mbit/s and is referred to as OC-1 (as
in *optical carrier*). Integer multiples of this clock rate may be used; in particular,
OC-3 at three times that rate (155.52 Mbit/s) and OC-12 at 12×51.84 Mbit/s $=$
622.08 Mbit/s are being used.

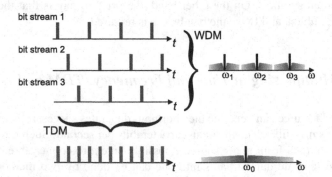

Fig. 11.5 Comparison of time division multiplex (TDM) and wavelength division multiplex
(WDM) formats. For TDM several bit streams are interleaved temporally; the resulting bit rate
is the sum of the individual bit rates. For WDM each bit stream is coded onto its own carrier. The
right half of the figure shows the spectral composition of both formats; for this example we assume
amplitude modulation. All told, both formats occupy the same bandwidth in frequency space

By international standardization SDH or *synchronous digital hierarchy* was created. The fundamental rate is 155.52 Mbit/s; data packets according to this standard are referred to as STM-1 (as in *synchronous transport module*). Note that OC-3 and STM-1 share the same clock rate.

On long distances OC-48 signals, or STM-16, have been common for several years; they have ca. 2.5 GBit/s. For intercontinental traffic many commercial systems use OC-192 (STM-64) at ca. 10 GBit/s. OC-768 or STM-256 at ca. 40 GBit/s was introduced ca. 2008. An increase in steps of factor-of-four was considered good business practice as it presents four times the payload at something like two and a half times the hardware cost; on top of that there are space savings in comparison to four OC-192 sets of hardware. However, beginning at 40 GBit/s problems from polarization mode dispersion arise that remain negligible at lower rates. This is because the relevance of the effect is determined by the relative propagation time scatter, i.e. the scatter *in units of the clock period*. Shorter pulses have a proportionally wider spectrum and thus 'feel' more of the dispersion. On the other hand, the clock period shrinks inversely with clock rate. The relative propagation time scatter then grows quadratically with clock rate. For OC-768 signals it is 16 times as large as for OC-192 signals. Polarization mode dispersion causes a random fluctuation of the state of polarization of the received optical signal which translates to level fluctuations. Quite complex compensation apparatus has been introduced to assure glitch-free operation. Nevertheless, after 40 GBit/s the industry did not take another leap of a factor-of four; rather, somewhat reluctantly 100 GBit/s was introduced beginning ca. 2011.

11.1.6 On and Off: RZ and NRZ

The physical representation of a bit value—a zero or a one—in an optical format is usually obtained by intensity modulation of a light wave. Again, there are basically two options; the relative advantages and disadvantages have been under discussion for many years.

Discrete-time signals have a certain clock rate which defines the time slots for the individual bits. To assign a binary value, zero or one, to a time slot one may

- either turn the intensity off or on during the entire duration of the time slot; or
- place a short signal pulse inside the time slot for a one, and no pulse for a zero.

Figure 11.6 these variants are compared. In the first case, the intensity remains the same during the entire time slot of duration T_c or, in the event of several ones or zeroes in a row, for several clock periods. In the second case, the intensity is always zero when one time slot is over and the next begins. Hence the names *no return to zero* or NRZ for the first case and *return to zero* or RZ for the second.

There are two relevant practical differences. When both zeroes and ones are statistically equally probable, the average for NRZ is ½, and for RZ close to zero. This plays a role in the construction of receivers where an AC coupling is usually employed to get rid of 1/f noise and drift.

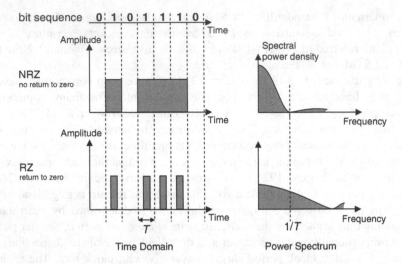

Fig. 11.6 Comparison of coding binary data in the NRZ and RZ formats. For NRZ the pulse occupies the entire time slot of the clock period T_c, for RZ just a fraction of it. (In principle the pulse might be much shorter than the time slot; for practical considerations, it is not very much shorter. In the figure, it is about one half.) Shorter pulses have wider bandwidth, so RZ occupies more bandwidth

More relevant is the difference in usage of frequency space: RZ uses more bandwidth because shorter pulses are spectrally broader. Keeping in mind that bandwidth is a nonrenewable resource, this is not economical. On the other hand, an RZ data stream contains a strong Fourier component at the clock frequency, which makes the design of clock regeneration circuits in the receiver easy. In the case of NRZ less bandwidth is occupied, but in the spectrum of an NRZ signal there is a null at the clock frequency. This is easy to see: for each rising slope in the signal there is also a falling slope. Both types of slopes occur equally often. They therefore introduce Fourier components of the same magnitude but opposite phase at the clock frequency which mutually cancel out. The absence of a strong Fourier component at the clock frequency makes its regeneration more difficult. It can be done by first differentiating the signal to emphasize the temporal positions of the slopes with a narrow spike, then rectifying the result to make all spikes positive-going. This way one obtains a strong spectral component at the clock rate which can easily be filtered out.

11.1.7 Noise

Noise is the collective term for all kinds of external influences that can hamper signal transmission. They include *man-made*, *natural*, and *fundamental* perturbations. The term "noise" must be taken in a broad sense here to denote any type of extraneous material imposed on the signal, be it coherent or incoherent, etc.

Man-made noises include emissions from machinery which find their way into the transmission channel. The reader may have experienced a radio crackling when a car with inadequate radio frequency noise suppression drove by. In the case of wavelength division multiplexing the emissions may arise from other channels: This is then referred to as channel crosstalk.

Natural noises may be caused by electric storms (lightning flashes), solar storms, etc.

Fundamental noises include quantum noise. Any signal is quantized because the basic physical constituents are: Electric currents consist of a certain number of electrons flowing per second and this number is subject to fluctuations. Similarly, any detected light power consists of a certain number of photons received per second; this number, too, fluctuates. The fluctuations constitute the quantum noise. Fundamental noise sources also include thermal noise as it occurs in any electronic circuit. At any temperature other than absolute zero, all constituents of matter including electrons undergo a random motion due to their thermal energy; this produces a noise voltage and a noise current in any real impedance.

Thermal noise can be derived directly from Planck's distribution formula for radiation [72]; this is indicative of its fundamental nature. One needs to consider the spatial and spectral density (power per frequency interval and per volume element) of a one-dimensional perfect emitter (what physicists call a "black body") at temperature T. Planck's distribution in one dimension, written as a function of frequency v (in Hertz),[2] is

$$I_v \, dv = \frac{2hv}{e^{\frac{hv}{kT}} - 1} \, dv. \tag{11.6}$$

Here c is the speed of light in vacuum as usual and Boltzmann's constant $k = 1.38 \times 10^{-23}$ J/K converts temperature T to energy units. $h = 6.6256 \cdot 10^{-34}$ Js is Planck's constant and hv is the energy of an individual photon.

In the "radio engineering limit" the quantum energy is much smaller than the thermal energy: With $hv \ll kT$ we obtain

$$I_v \, dv = 2kT \, dv;$$

from this one can deduce the thermal noise as described by Johnson and Nyquist [25, 41], with a "white" spectrum

$$\tilde{P} = 4kTB, \tag{11.7}$$

where $\tilde{P}$ is the product of open circuit voltage and short circuit current which produces noise in a bandwidth B.

Above that frequency at which $hv = kT$, the Nyquist–Johnson formula is no longer valid. At standard ambient temperature around 300 K this limit is in the

---

[2]Most textbooks describe Planck's law for three-dimensional emitters; for the connection with electronic noise we need the one-dimensional case.

far infrared. Therefore, it is perfectly justified that electronics engineers disregard quantum noise entirely and deal with thermal noise. In the visible and near-infrared optical range, however, quantum noise has the upper hand and quantum effects present very real limits.

In either case one deals with noise which approaches a Gaussian amplitude distribution and a "white" spectrum; the latter means that the noise's correlation time is shorter than all correlation times of the signal.

Noise establishes a fundamental limitation to the transmission. In the most favorable case that all technical and thermal sources of noise are negligible, there is quantum noise left. Let us assume binary coding and estimate the limit of reach. We will keep in mind that in terms of realistic systems the following is far too optimistic; we are after the ultimate limit.

As discussed in Sect. 5.4, there must be at least a single photon received for a signal to be detectable. (We are serious about the ultimate limit!) The photon energy is $E = h\nu \approx 6.6 \times 10^{-34}\,\text{Js} \times 200 \times 10^{12}\,\text{Hz} \approx 10^{-19}\,\text{J}$ in the near infrared. The average launch power is limited to around $1\,\text{W}$ so that thermal damage to the fiber is avoided, this corresponds to 10^{19} photons/s. Then an attenuation of no more than $1/10^{19}$ or $190\,\text{dB}$ is admissible when we assume for simplicity the ridiculous bit rate of 1 bit/s. We also accept that due to the statistical nature of the photon number in some cases, zero photons will be detected instead of one; this would constitute a transmission error, and we will come to that. For a fiber with $0.2\,\text{dB/km}$, this gives a maximum distance of $950\,\text{km}$.

> Due to energy loss, optical fibers are quantum limited in their reach. Even in an unrealistically optimistic estimate the maximum distance is less than $1000\,\text{km}$.

Distances spanned in practical systems are much shorter than that; hence the requirement of optical amplifiers.

Of course it is not possible to detect a signal consisting of a single photon without error, due to the statistical nature of both their generation and their detection. Therefore it is useful refine our estimate as follows: We set an upper limit to the bit error probability which is deemed sufficient for practical purposes, and calculate how many photons on average must be contained in a light pulse to accommodate that limit. For the distribution of photon numbers we may assume Poisson statistics. At an average photon number N, the probability to have the value n (do not confuse this symbol with the refractive index!) is given by

$$p(n) = N^n\,\text{e}^{-N}/n!.$$

Then, the probability to erroneously measure a logical "one" when indeed a "zero" was sent is

$$p(1) = 0^1 e^{-0}/1! = 0.$$

Zeroes are detected error-free! This is no surprise because when zero photons are sent, and all other noise sources are excluded, the arriving number of photons got to be zero. For the "ones" it is different: The probability to measure a "zero" when in fact a "one" was sent is

$$p(0) = N^0 e^{-N}/0! = e^{-N}.$$

In the telecommunications industry it is common to set the maximum allowed bit error rate in telephony to 10^{-9}. We insert this value and solve for N. On average, there are as many "zeroes" as there are "ones", but "zeroes" are detected error-free. Then we can admit an error of 2×10^{-9} for the "ones". It follows that

$$N_{min} = \ln 2 \times 10^{-9} = 20.03.$$

In an ideal situation it would suffice to have 20 photons for a "one":

> In order to detect a signal with a bit error rate below 10^{-9}, photon statistics dictates that a logical symbol on average must contain at least 10 photons.

In any practical context other sources of noise and error will also contribute; therefore even the best available detectors require at least ten times as many photons, and typical decent detectors maybe a hundred times as many. Detectors that are uncompromisingly optimized for highest speed may require even more than that.

11.1.8 Transmission and Channel Capacity

Now we consider the compound signal coded in one of the formats described above: RZ or NRZ; TDM and/or WDM. This signal is eventually fed to a receiver. The idea is that this occurs across a certain distance; this implies that over the distance there is some suitable transmission medium, like a cable. The medium acts as a *channel*.

External noises and perturbations also act on the channel; as a result, what arrives at the receiver is a mix of the signal proper and some noise. It is the task of the receiver to reconstruct the signal without error and to disregard the noise. That may or may not be possible. This is the topic of communications theory, a field which was started by a seminal work by Claude Shannon [56].

Shannon's work shows that one of the most relevant parameters is the bandwidth available for the transmission. Assuming that the channel can provide the bandwidth B, transmission can take place with a data rate R as long as R remains smaller than the *channel capacity* C. The latter is defined by

$$C = B \cdot \log_2 \left(1 + \frac{S}{N} \right). \tag{11.8}$$

Here S is the signal power and N the noise power. Noise is assumed to be Gaussian white noise. Shannon showed that provided $R < C$, a coding can be found such that the bit error rate can be made arbitrarily small. It is well possible that the coding gets increasingly complex as R approaches C (from below), but virtually error-free transmission is possible. If, on the other hand, transmission with $R > C$ is attempted, the bit error rate can no longer be kept down.

According to Eq. (11.8), the dominant factor determining the channel capacity is the bandwidth. One might think, then, that an infinite amount of data can be transmitted when B is allowed to grow indefinitely. This is not so. The catch is that the noise also depends on bandwidth. If one restricts the discussion to white noise, the noise power is proportional to bandwidth. Then one can write $N = N_0 B$, with $N_0 = $ const. the spectral noise power density which is constant. In the limit

$$\lim_{B \to \infty} C \neq \infty,$$

it follows that

$$\lim_{B \to \infty} C = \frac{S}{N_0} \cdot \log_2 e.$$

In reality, of course, the available bandwidth cannot grow indefinitely anyway but is bounded by physical considerations.

Transmission through optical fiber, in comparison to electric cables, enjoys the benefit of a wide spectral region of low loss. If we take the regime of the third window generously as 1400–1600 nm corresponding to a frequency interval of 214–188 THz, the bandwidth is 26 THz. With the best available fibers, one may be able to utilize an even more extended range of (optimistically) 1250–1650 nm; this corresponds to 240–180 THz implying a bandwidth of 60 THz. Of course, toward the end points of this interval, losses are much higher than in the middle so that for long-distance transmission one may be tempted to return to a less optimistic estimate. In any event, realistic estimates produce bandwidths on the order of 50 THz.

The *spectral efficiency*

$$\eta = \frac{R}{B} \qquad \text{Bits/s/Hz} \tag{11.9}$$

indicates how well the data rate makes use of the available bandwidth. For binary signals only two values are used: off and on, or zero and some power at least equal to noise power. This can be formally introduced into Eq. (11.8) by letting $S = N$; then we obtain

$$C = B \qquad \text{channel capacity for binary transmission.} \qquad (11.10)$$

In other words, for binary transmission the maximum rate is 1 bit/s in 1 Hz of bandwidth.

11.2 Nonlinear Transmission

In long-haul transmission, it is unavoidable that the fiber's nonlinearity becomes noticeable. Nonlinearity is special because in electrical cables both attenuation and dispersion are well known, but a phenomenon corresponding to the Kerr nonlinearity in fiber does not exist. This may be why engineers trained in electronics instinctively considered nonlinearity as an impediment and an utter nuisance for a long time. The way to avoid nonlinearity, in this logic, is to use large-mode area fibers to reduce the nonlinear coefficient and to use low power signals. This approach can go a long way. Indeed, remarkable progress has been obtained, and data transmission rates of several TBit/s over long distances have been demonstrated (see Sect. 11.4.2). Only a few researchers pointed out as early as in the 1980s that nonlinearity also presents an opportunity to counteract dispersion's detrimental effects and thus to improve the transmission system as a whole. Both approaches are being pursued, and only the future can tell which one will ultimately be better.

Before any commercial system can be deployed, there are years of extensive research experiments and laboratory tests, and finally field trials. Lab tests are not done in actual long-distance optical cables but in closed fiber loops which can be set up in a laboratory. Signal degradation with distance can then be assessed in detail by just letting the signal go around the loop for more and more turns; Fig. 11.7 shows what insiders tongue-in-cheek call a carousel.

Much research deals with that paradigm of transmission in the presence of Kerr nonlinearity, the fundamental (i.e., $N = 1$) optical soliton. Solitons are the natural units (*bits*) for transmission of data over optical fibers because they are more robust than any other type of pulse. They require anomalous fiber dispersion; it comes in handy that the spectral regime of lowest loss coincides with the anomalous dispersion regime.

According to the latest research, the best results are obtained not with *pure* solitons but rather with a certain generalization of the soliton concept. A number of subtle effects become noticeable on truly long distances on the order of thousands of kilometers, some in an individual wavelength channel and others only in the case of WDM. These effects make the situation a little more complex; it is then a matter of taste whether one still calls the modified pulses by the name of solitons or by some other name. A few books have recently become available that are devoted to solitons in optical fiber [7, 20, 36].

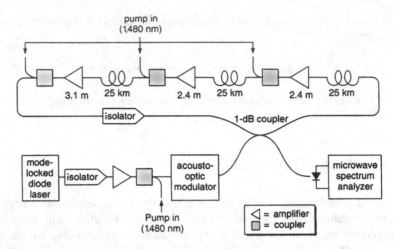

Fig. 11.7 A typical laboratory experiment for the study of long-distance transmission. The long distance is here represented by multiple round trips in a fiber ring. In this example, the ring has 75 km circumference and contains three amplifiers. From [35] with kind permission

11.2.1 A Single Wavelength Channel

In spite of their extraordinary stability, solitons do not enjoy eternal life. Perturbations arise from energy loss, Raman scattering, and by mutual interactions of pulses (see above). Combined, they eventually destroy even solitons [11, 31].

Energy loss may be compensated by optical amplifiers (of the Raman type or with Er-doped fiber) at least on average. The first question to ask is at which intervals L_{amp} one should insert amplifiers into the fiber. It turns out that the condition $L_D \gg L_{amp}$ must be maintained in order to avoid a resonant perturbation of the solitons [10, 18, 19, 36]. For typical standard fiber and picosecond pulses, L_D is a few to a few tens of kilometers. Here is a quick estimate:

$$L_D = \frac{T_0^2}{|\beta_2|} = \frac{(10\,\text{ps})^2}{20\,\text{ps}^2/\text{km}} = 5\,\text{km}.$$

It would be awkward to insert amplifiers at distances shorter than this. If dispersion is reduced to ca. 1 ps/(nm km), however, L_D becomes about 100 km. Then, very reasonable intervals between amplifiers on the order of tens of kilometers are possible. A useful side effect of low-dispersion fiber is that the soliton's energy is also scaled down so that less power is required for their generation. Also, the combined power of possibly a hundred WDM channels is kept low so that handling live fibers does not pose a health hazard to a service crew. On the other hand, one should not push dispersion reduction too far because there is also a signal-to-noise issue when the soliton energy goes down too much.

11.2.1.1 Gordon–Haus Effect

Amplifiers, by their nature, cause additional noise due to spontaneous emission; this noise degrades signal integrity in a subtle manner. It modifies the pulse energy, its optical phase, and its temporal position. Modifications of amplitude, phase, and position of solitons are not a big worry. However, frequency deviations spell trouble. They arise from asymmetric components of the noise with respect to the spectral center of the pulse (see Fig. 11.8). In the presence of dispersion, frequency changes produce changes in the pulse arrival time, which, after a long distance, add up to a considerable random pulse jitter. If the jitter becomes too large (i.e., comparable to the clock period), the signal is rendered unreadable. This phenomenon is called Gordon–Haus jitter [16] in honor of James P. Gordon and Hermann A. Haus who predicted it. They showed that the jitter grows with the third power of distance.

The existence of the Gordon–Haus jitter was clearly shown experimentally (Fig. 11.9). In the experiment, it was necessary to average over many consecutive

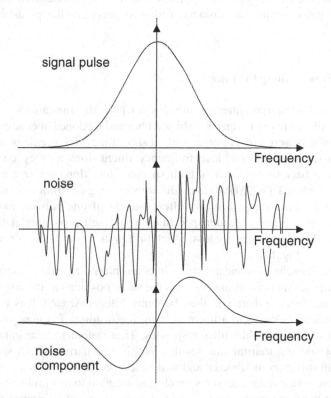

Fig. 11.8 With respect to the pulse spectrum (*top*), noise (*center*) may have asymmetric components (*bottom*). If noise is then added to the pulse, the spectral center-of-mass (i.e., the center frequency) is shifted ever so slightly. Due to dispersion in the fiber, this results in a modified time of arrival. These random fluctuations of arrival time are called Gordon–Haus jitter

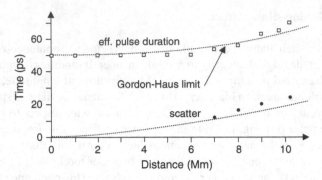

Fig. 11.9 Experimental demonstration of Gordon–Haus jitter. The apparent increase of pulse width with increasing distance is in perfect agreement with the prediction. From [33] with kind permission

pulses so that the arrival time jitter appeared like a pulse broadening. The apparent increase of pulse width with distance followed precisely the prediction of the jitter [33].

11.2.1.2   Filters Along the Line

Insight about Gordon–Haus jitter was the reason why in the transatlantic cables TAT-12 and TAT-13 (see below) dispersion shifted fiber and Er-doped fiber amplifier were used, but solitons were not. However, only a short time later a remedy was found: Wavelength-selective filters reduce frequency fluctuations as they continuously nudge solitons back to the center of their spectral slot. Moreover, in a wavelength division multiplexed (WDM) system, differences in gain from one wavelength channel to the next are equalized by filters because if one pulse is momentarily too powerful, it acquires a broader spectrum through self-phase modulation; at the next filter, it then suffers greater loss which brings its power back to normal [34]. This is shown in Fig. 11.10.

Of course the scheme mandates that filters are placed at certain intervals along the line; for practical reasons one chooses the same positions as the amplifiers. On a transoceanic distance there will then be many filters cascaded. It is well known that for cascaded elements in a linear system, the resulting frequency response is the product of the individual filter responses. That statement here implies that a very narrow spectral transmission results. Within this narrow width spontaneous emission can still grow unhindered and will pose a problem.

Again a very simple idea presents an elegant solution to this problem. All filters do not have the same center frequency; rather, the center frequencies are sliding along the line. Then there is no single wavelength for which spontaneous emission can transit the whole distance because in the product of filter responses, there is always one factor practically zero for any frequency. For linear signals, a *sliding*

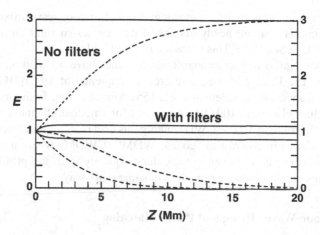

Fig. 11.10 Computer simulation of the equalization of power variations of solitons by filters. If a pulse has more than its normal power, it will be shorter in time and broader spectrally. In the selective filter, it experiences extra loss which will overcompensate the gain and reduce the power toward the equilibrium value. In this example, three wavelength channels are considered. Stable propagation is only obtained with the use of filters. From [27] with kind permission

filter system is opaque! For solitons it is different: Solitons are creatures of the nonlinear realm. They can adjust their shape and center frequency at each filter and thus pass through the entire system without any problem. Such a discrimination between signal and noise has no correspondence in linear systems!

11.2.2 Several Wavelength Channels

The maximum data rate of an individual wavelength channel is basically limited by the speed of electronic components as they are available. This limit is optimistically at 100 Gbit/s, and more realistically in the tens of Gbit/s. The rate may be increased somewhat by *optical time division multiplexing* when two or more bit streams are first converted into a sequence of pulses with fast electronics, then interleaved by using optical delays of half the clock time. 100 Gbit/s have been obtained routinely, but that is still a far cry from the 50 THz or so bandwidth which the fiber offers. That tremendous spectral range can only be utilized by parallel operation of a multitude of wavelength channels, i.e., by wavelength division multiplex or WDM. The first question arising is at which spacing to place the spectral channels: should they be equidistant?

Several valid points in favor of equidistance can be brought forward:

- It represents better conceptual clarity.
- It is in accord with common use in radio and TV transmission.
- Filters with equidistant transmission frequencies are particularly easy to construct (Fabry–Perot filters).

The counter argument is that the detrimental effect of four wave mixing is most pronounced in this case: all newly generated frequencies sit right on top of some other channel (see Sect. 9.6). This is shown in Fig. 11.11.

Nevertheless, international standardization bodies have adopted an equidistant channel grid. The ITU[3] grid uses a reference frequency of 193,100 GHz, which corresponds to a vacuum wavelength of ca. 1552.5 nm. Starting from this frequency, there is one channel at every 100 GHz increment. Intermediate channels on a 50 GHz or even 25 GHz grid may be used. WDM using this grid is also called "dense WDM" or DWDM. This is in contrast to "coarse WDM" (CWDM) where a much wider spacing of 20 nm is used. Given the regular frequency grid, the problem arising from four wave mixing must be remedied in some other way.

11.2.2.1   Four-Wave Mixing and Phase Matching

The amount of degradation caused by four-wave mixing is also determined by the degree of phase matching of this process (Fig. 11.12). In a fiber without any dispersion, perfect phase matching of both the generating and the generated wave would be guaranteed, and mixing products and thus signal perturbation would reach a maximum. Therefore, dispersion is definitely helpful in this context. Even small amounts of dispersion reduce the efficiency of four wave mixing noticeably.

11.2.3   Alternating Dispersion ("Dispersion Management")

An invention conceived for a different purpose is the solution to the four-wave mixing problem. Engineers had tried to solve the problem of dispersive pulse broadening by tinkering with the fiber's dispersion. Their idea was to basically compensate the dispersion and make it zero at least as a path average by inserting segments of dispersion-compensating fiber. The latter designates dispersion-shifted fibers (see Sect. 4.5.5) which have dispersion β_2 of the opposite sign as the main fiber. It turned out that a full compensation to zero is not at all desirable. In the (unavoidable) presence of fiber nonlinearity, a certain residual dispersion was found beneficial. This is, of course, rooted in soliton formation.

In order to reduce the detriment of four-wave mixing by the introduction of phase mismatch, it suffices to have strong *local* dispersion (see Sect. 9.6); even for zero-path average dispersion this end would be achieved. The idea then is to optimize dispersion along the path by judicious *dispersion management* (DM) to create a *dispersion map*. The dispersion map typically consists of a periodic alternation of fibers with different dispersion (Fig. 11.13); typical DM period lengths are a few tens of kilometers.

---

[3] International Telecommunication Union, a United Nations agency for information and communication technology issues.

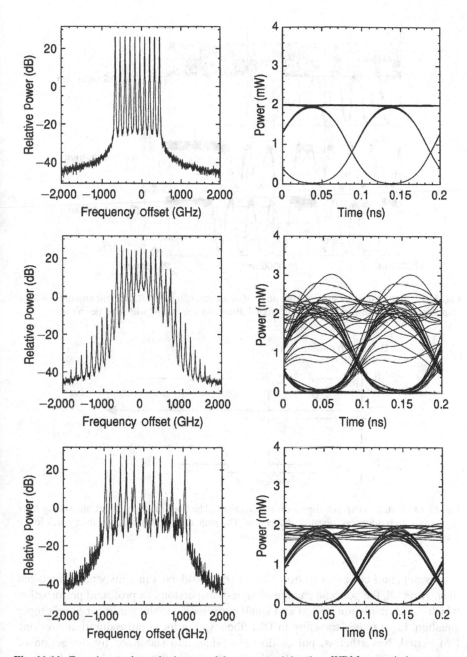

Fig. 11.11 Experiment about the impact of four wave mixing in a WDM transmission system. *Left*: spectra, *right*: eye diagrams (these are explained in Sect. 11.3.2). *Top row*: Ten equidistant channels are launched simultaneously into a fiber. *Center row*: At the fiber end numerous mixing products ("combination tones") have been generated. The eye diagram indicates severely degraded signal integrity. *Bottom row*: If nonequal channel separations are chosen, both the number and the strength of mixing products are reduced, and the eye diagram indicates good signal integrity (the "eye" is completely open). From [14] with kind permission

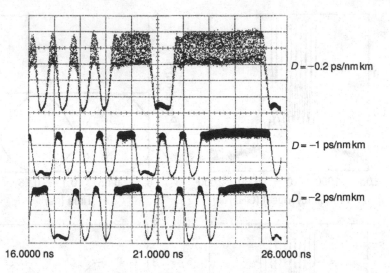

Fig. 11.12 Impact of four-wave mixing on a bit sequence, compared at different amounts of fiber dispersion. The less dispersion, the more signal distortion. From [14] with kind permission

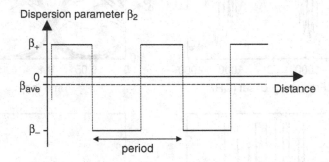

Fig. 11.13 Sketch to explain dispersion management. The fiber line consists of alternating fiber segments with positive and negative dispersion. The path average dispersion is then much lower than the local dispersion and may be close to zero

It is not at all clear that soliton-like pulses would exist in a dispersion managed fiber. After all, the periodic change of sign of dispersion is a profound perturbation which can certainly not treated as a small perturbation to the nonlinear Schrödinger equation. Light pulses traveling in DM fibers vary a lot in pulse duration over one DM period. Nevertheless, pulses do exist which are stabilized by the action of nonlinearity [39, 40]. This stabilization implies that after a complete DM period, the original pulse shape and width are restored. In a "stroboscopic" representation, in which the pulse shape is only shown at a particular position within the DM period, one sees a stably propagating pulse again. In a certain generalization of the concept of solitons such pulses are referred to as *DM solitons*. Their pulse shape is different

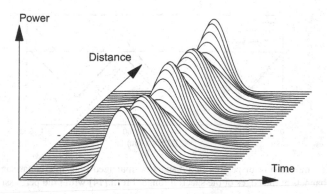

Fig. 11.14 A DM soliton in a computer simulation. Two periods of the dispersion map are shown

from the sech shape of conventional solitons, as can be seen from Fig. 11.14. Indeed, it more closely resembles a Gaussian. On a log power scale one even discerns undulations in the wings.

The repetitive variation of dispersion brings about a further benefit. Since the pulse shape "breathes" over one dispersion period, the pulse's phase profile breathes, too: underneath the envelope there is a chirp bending back and forth. Where neighboring pulses overlap, the phase relation varies rapidly so that interaction is mostly washed out. Moreover, for long stretches of the path, the peak power is reduced and with it the effective nonlinearity. To make up for that, the power of the DM soliton is higher than in the comparable case of a fiber with constant dispersion equal to the path average value [57]. This so-called DM power enhancement [57, 68] provides advantages in terms of signal-to-noise ratio and also in the context of Gordon–Haus jitter [60].

Due to higher-order dispersion, there is a different β_2 value at the center frequency of each wavelength channel. Different channels thus experience different dispersion, both for local and path average values. This is illustrated in Fig. 11.15 where the propagation of signals in neighboring WDM channels with different dispersion is compared.

As a consequence, in different WDM channels the pulse streams have different power. If we take the unperturbed soliton of the nonlinear Schrödinger equation as a reference, its energy is found as

$$\hat{P}_1 = \frac{|\beta_2|}{\gamma T_0^2} \quad \text{and} \quad E_1 = 2\hat{P}_1 T_0 = \frac{2|\beta_2|}{\gamma T_0}.$$

The energy is proportional to dispersion. As Fig. 11.16 shows, this relation carries over to the relation between energy and path average dispersion of DM solitons [34]. Meanwhile fibers with an inverse trend of dispersion (inverse β_3) have been suggested in order to obtain a flat resulting dispersion so that power differences are equalized.

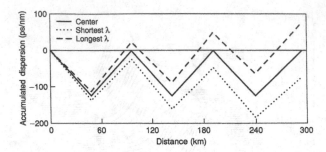

Fig. 11.15 Accumulation of dispersion in a dispersion-managed fiber in comparison of a central channel and channels at the edges of the spectral range. From [14] with kind permission

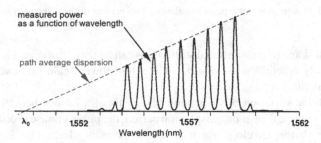

Fig. 11.16 Nine WDM channels of a DM system are operated with signals of the same bit rate. After a sufficiently long distance, the power in each channel adjusts itself according to the path average dispersion of that channel. This is the same scaling behavior as known from "ordinary" solitons. From [34] with kind permission

In a similar fashion, gain must be equalized, too. The spectral gain curve of Er-doped fibers as shown in Fig. 8.20 is not at all flat. By tweaking fiber design, an essentially flat range of more than 80 nm has been demonstrated. A more broadband alternative is to use gain by means of the Raman effect; see e.g., [42].

At the turn of the millennium, researchers had succeeded to transmit several terabits per second over a single fiber. Engineers describe these systems as *chirped RZ*, i.e., they realize that the pulses have the chirp that a soliton acquires in a dispersion-managed fiber, yet they typically avoid to speak of solitons. This seems to be a case of two cultures which, meaning the same thing, call it a by different name.

For many years, the telecommunications industry has been driven by ever-increasing demands for transmission capacity. It is a fact of life that fiber is a nonlinear transmission medium; therefore one only has the choice of either avoiding the impact of nonlinearity by using wide area fibers and low signal power, or to embrace it and accept nonlinear chirp—whether one calls that format "chirped RZ" or "soliton" is of lesser importance. Fiber nonlinearity will not go away, and only soliton-based coding takes it fully into account. Will soliton-related data formats become a standard in the future? It is always difficult to make predictions, especially when they are about the future—as the famous quote goes (it is variously attributed to Winston Churchill, Niels Bohr, George Bernard Shaw, and many others).

11.3   Technical Issues

11.3.1   Monitoring of Operations

In commercial service, a permanent monitoring of system integrity is mandatory. This is done in the following way: The intermediate amplifiers along the line are combined with so-called *loopback* modules. These consist of four fiber couplers as shown in Fig. 11.17. In this arrangement the signal stream can pass, but a minor portion of it is branched off, attenuated by 45 dB, and sent back toward where it came from. This weak signal does not interfere with other data streams.

Somewhere among the multitude of WDM channels a pseudo-random sequence is transmitted. By way of correlation measurement, the return signal can be detected in spite of being weak. Such monitoring allows to detect additional losses due to damage or whatever cause. The damage can also be localized because the return signals from different loopback modules arrive with different delay.

In addition, loopback modules are built such that backpropagating light from Rayleigh scattering can bypass the amplifiers so that OTDR measurements may be performed. For extremely long distances, one combines OTDR with coherent detection to increase sensitivity. With these measures the entire fiber length can be monitored, in part during life data traffic on a reserved channel, and with improved sensitivity during a routine maintenance interval. Figure 11.18 shows an example of monitoring a fiber of about 4400 km length [28].

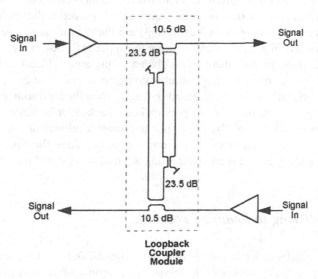

Fig. 11.17 A *loopback* module serves to monitor a dual-fiber line during operation with life traffic. After [70] with kind permission

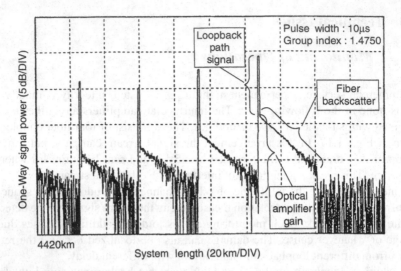

Fig. 11.18 OTDR measurement of a very long fiber line including several amplifiers. From [13] with kind permission

11.3.2   Eye Diagrams

One of the simplest and most efficient ways to test the quality of transmission is to inspect the so-called *eye diagram*. The name refers to an oscilloscope display of the bit stream such that the horizontal deflection is synchronized with the clock rate. In principle, all slopes (rising as well as falling) are then at the same position on the screen; in between, in principle there is either the upper or the lower level. Therefore, in the middle of the picture, there should be an empty area, which is referred to as the "eye." All kinds of signal impairments conspire to close the eye (Fig. 11.19): The slopes of the pulses may be smeared out, e.g., when the light source or detector have insufficient bandwidth, fiber dispersion is unchecked, or by timing jitter. Then the eye is narrowed horizontally. The upper and lower level may not be maintained, or there may be excessive noise or channel crosstalk: Then the eye is narrowed vertically. A wide open eye is an instant indication of good signal integrity.

11.3.3   Filtering to Reduce Crosstalk

Intersymbol interference can lead to channel crosstalk and can be a severe perturbation, but by judicious choice of the frequency response of the transmission chain it can be nearly eliminated. The reader is reminded that most filters, in particular those with steep slopes of their spectral transmission, have a nonmonotonous step response. The trick then is to select a frequency response such that in the distorted

Fig. 11.19 An eye diagram is obtained by displaying the bit stream over time with synchronization to the clock rate. The *upper* and *lower* levels of the binary signal and the steep slopes at the beginning and end of the bit slot can be assessed quickly and conveniently; if the eye is wide open, one may expect error-free reception

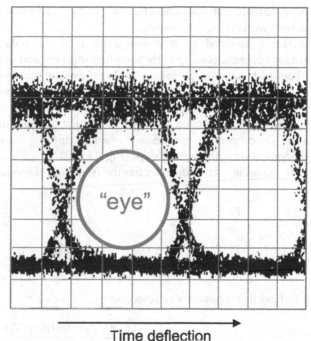

Time deflection

filtered temporal shape there are zeroes at multiples of the clock period T_c. If this is the case, then the crosstalk is eliminated. For example, a filter with step response

$$u(t) = \frac{\sin(\pi t/T_c)}{\pi t/T_c} = \text{sinc}\left(\frac{t}{T_c}\right) \qquad (11.11)$$

has the desired property. Here we have used the sinc function (*sinus cardinalis*), which is defined as

$$\text{sinc}(x) = \frac{\sin(\pi x)}{\pi x}.$$

With this step response, the frequency response is

$$U(f) = \begin{cases} T_c & : \ f < \dfrac{1}{2T_c}, \\[2mm] 0 & : \ f > \dfrac{1}{2T_c}. \end{cases}$$

(T_c appears here from a normalization $\int_{-\infty}^{+\infty} U(f) = 1$.) Such filter with "vertical" slope, usually referred to as a brick wall filter, cannot be built because it would require an infinite number of selective elements. Also, a filter with time-symmetric response defies causality (any effect takes place only *after* the cause) and thus cannot be built with ordinary hardware.

This is too bad because such a filter would make it possible to optimize bandwidth use according to the sampling theorem and also because it would reject all out-of-band noise. So can one still make use of these ideas?

One can approximate the step response of the acausal filter quite well if one does not perform the filtering in real time but accepts a certain latency. It turns out, fortunately, that a latency of only a few clock periods suffices. The infinite "brick wall" slope becomes unnecessary when using a *raised cosine* filter, which is its generalization with rounded slope (Fig. 11.20).

If one arranges that the filter has the frequency response

$$U(f) = \begin{cases} T_c & : \quad 0 \le |f| \le \dfrac{1-\beta}{2T_c}, \\[4mm] \dfrac{T_c}{2}\left[1 + \cos\left(\dfrac{\pi T_c}{\beta}f - \pi\dfrac{1-\beta}{2\beta}\right)\right] : \dfrac{1-\beta}{2T_c} \le |f| \le \dfrac{1+\beta}{2T_c}, \end{cases} \tag{11.12}$$

it follows that it has the step response

$$u(t) = \frac{\sin(\pi t/T_c)}{\pi t/T_c}\,\frac{\cos(\beta\pi t/T_c)}{1-(2\beta t/T_c)}, \tag{11.13}$$

which in turn is a generalization of a sinc function. By tweaking the parameter β, one can fine-tune the sharpness of the transition from pass band to rejection band. Such filters can be made to a good approximation and allow to have a well-defined pass band to keep noise in check. At the same time, they cancel channel crosstalk from intersymbol interference. Pulses shaped such that they fulfil this description are known as Nyquist pulses.

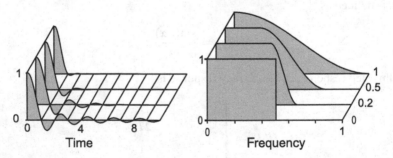

Fig. 11.20 Explanation of the *raised cosine* filter. Parameter β (increasing from front to back) sets the sharpness of transition from pass band to rejection band. The zeroes in the temporal response remain centered on the positions of the adjacent bits so that channel crosstalk is eliminated

11.4 Telecommunication: A Growth Industry

Nothing displays the increasing globalization as clearly as the increasing demand for long-distance communications lines. For many years, there has been an annual growth by some 20–30 %. The race to keep up with the rising demand is fueled by the incentive of money to be made; some of that money is well invested in research to advance the technology involved. We now give a historical sketch of the development of telecommunications.

11.4.1 Historical Development

1851: The first undersea cable commences service. It crosses the English Channel, connecting Dover and Cape Gris Nez, and it will work well for 24 years.

1858: The first transatlantic cable begins operation, but it is broken after only 1 month.

1866: With the largest ship of its time, the Great Eastern, the first successful transatlantic cable is deployed. It operates in Morse code.

1927: The first transatlantic telephone connection is inaugurated. It is based on radio transmission in SSB mode between New York and London.

1956: The first transatlantic telephone cable ("TAT-1") takes up service. Its coaxial cable can accommodate 48 simultaneous telephone channels in *analog* format. The amplifiers use electron tubes (transistors are invented only shortly before and are not mature yet). In a sophisticated scheme, even the silent intervals in natural conversation are used for transmission of other channels, a scheme called TASI for "time-assignment speech interpolation." This technology is very successful so that a few years later more cables follow, which use the same basic technology but an increasing number of channels. The seventh of these (TAT-7, with 4000 telephone channels) is the last of its kind. It is commissioned in 1983 and can handle 4200 simultaneous telephone channels. TAT-1 is decommissioned in 1978, TAT-7 in 1994.

1962: Telstar I, the first active telephone satellite, is launched.

1965: Intelsat I ("Early Bird"), a greatly improved telephone satellite, is launched.

1966: Kao and Hockham predict the possibility of making fibers with loss of not more than 20 dB/km.

1970: The prediction comes true: The first fiber with less than 20 dB/km loss is introduced. Only a few years later, even 0.2 dB/km are reached.

1976: A first system experiment in the transition from research to commercial use is started by Bell Laboratories in 1976 in Atlanta. Two cables, made by Western Electric Co., having 640 m length and containing 144 fibers each, are laid in existing ducts. Each fiber transmits 44.7 Mbit/s corresponding to 672 telephone channels. The strands are hooked up in series to create a longer effective distance. The performance is virtually error-free over about 11 km. Including 11 repeaters, even 70 km transmission is successfully demonstrated. The trial shows that the fibers survive intact all the bending and pulling involved in placing the fiber, certainly a much harsher treatment than in a laboratory.

1977:   Other countries follow suit. The first comparable experiment in Germany takes place 1977 in Berlin. A cooperation of AEG-Telefunken, Standard Elektronik Lorenz, Siemens, and TeKaDe places a 4.3-km cable between Assmannshauser Straße and Uhlandstraße. In the same year, England and Japan perform similar tests.

1985:   The first fiber-optic undersea cable, Optician 1, connects the Canary Islands of Tenerife and Gran Canaria. There are initially problems with fiber damage by shark attacks; additional steel strength members avoid that problem.

1988:   TAT-8 constitutes the beginning of a new era: that of *optical* transatlantic data transmission (Fig. 11.21). This cable operates in the second window at 1.3 μm and is the first to transmit in a digital format. Its two pairs of fibers, each with a capacity of 280 Mbit/s, allow it to transmit 40,000 telephone channels. The cost per channel is thus dramatically lowered by two orders of magnitude. A steel cladding of the cable is used to provider electrical energy as a supply for the repeaters. At a constant current of 1.6 amperes a voltage of 7500 V is required, the return is through the ocean water. One year later a transpacific cable TPC-3 and a connection between mainland USA and Hawaii, HAW-4, follow in the same technology. These cables form the first generation of fiber-optic intercontinental cables. TAT-8 is decommissioned in 2002.

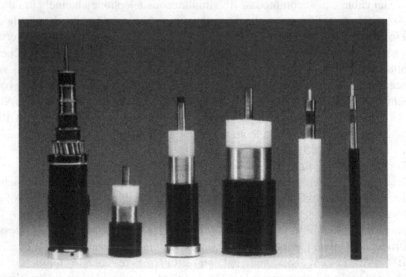

Fig. 11.21 Six generations of data transmission cables: In the 1950s a cable (far *left*) could transmit 36 telephone channels, the optical fiber cable from the early 1990s (far *right*) handled 40,000. Since then, capacity has risen to several million telephone channels without any major change in outside appearance

1991:   Fiber-optic cables surpass telephone satellites in terms of number of transmitted calls. 33 million km of fiber have been laid out. Half of it, 16 million km, is in the USA. Europe has 9 million km, the Pacific Rim 8 million km.

1992:   The first cable of the second generation, TAT-9, starts in March. The wavelength is now in the third window at $1.55\,\mu m$. Advanced components like DFB lasers and APD diodes are used; the transmission format is NRZ. At 565 Mbit/s per fiber in two fiber pairs, 80,000 telephone channels are transmitted simultaneously. The cable is 9310 km long. It costs 450 million US$ and is owned by a consortium from 35 international telecommunications companies. It links USA and Canada on one side with England, France, and Spain on the other. For Spain, it is the first fiber-optic direct link to the USA; before, they had to be content with TAT-5 with 845 telephone channels. Italy, Greece, Turkey, and Israel are connected via Spain.

In 1992/1993, the same technology is used in the Pacific for TPC-4. The next cables (TAT-10 between USA and Germany/the Netherlands) and TAT-11 are configured in a new topology: Instead of a line between two points a "ring" is used, basically a pair of independent cables between the same two points. The rationale is that if any damage occurs at any position, one can route all data traffic around the damage location. The idea is to ultimately have nets, or webs, that can better survive damage. In view of the enormous data traffic, it is clear that any service interruption immediately leads to considerable financial damage. Like the earlier fibers before, TAT-11 is switched off in 2004.

1994:   Unification of Germany has created a new market for telecommunication because in communist Eastern Germany telephones had been available only to a narrow privileged class. Meanwhile in the 1980s, Western Germany had fallen behind other countries in making the transition to fiber optics because the responsible ministry favored copper cables. In the late 1980s this course was reversed, and Western Germany invested heavily in fiber optics. After unification 1990, this situation led to the inspired decision to immediately go for the most advanced technology as the country's telecommunications infrastructure got an overhaul. Within a few years, the existing 111 lines between both Germanies were replaced with several tens of thousands. In mid-1994, Deutsche Telekom had the world's most close-meshed fiber-optic network. At a total length of 80,000 km, the fiber network exceeded the highway network. Also, Deutsche Telekom started early to put fibers all the way to the subscriber, an activity which is now described with several new acronyms: FTTH is for *fiber to the home*, FTTC for *fiber to the curb*, and FTTP for *fiber to the premises*. These acronyms can be wrapped up under FTTx for *fiber to the whatever*.

1995:   The third generation begins with TAT-12; TAT-13 follows in late 1996, and TPC-5 and TPC-6 soon thereafter. Again important technical novelties have been introduced. Dispersion shifted fiber is used; Erbium-doped fiber amplifiers make it possible to increase the distance between repeaters. Now RZ is used as the transmission format. The data rate is 5 Gbit/s equivalent to 1,228,800 telephone channels. Meanwhile, a good fraction of the total traffic is no longer traditional telephone voice communication ("POTS" or plain old telephone service), but also fax and data transmission between computers. Cost per telephone channel has again come down from TAT-8 levels by more than an order of magnitude. TAT-8 through TAT-11 are decommissioned 2002 and 2003. Technically they still work well, but the more recent cables are so much superior that it does not make any business sense to keep them alive. Some of the decommissioned cables have later been used for research purposes.

2001: The transatlantic cable TAT-14 takes up service in May. It has been built for 1.5 billion US$ and can handle 640 Gbit/s (corresponding to 8 million telephone calls). In October, a competing consortium opens Flag Atlantic-1 on the same route; this cable has six fiber pairs with a combined capacity of 4.8 Tbit/s. The telecommunications industry thrives on short return-on-investment time and gigantic growth figures. Many competitors join the industry to lay and operate fiber-optic cables. For the first time in the history of telecommunication, there is an excess capacity: supply surpasses demand. As a consequence the prices come further down, and revenues of all involved parties plummet. There is a string of insolvencies, some of which are quite spectacular (2002: Global Crossing, WorldCom). This is at the same time that the internet bubble bursts, and the two upheavals are related. Euphoria from the late 1990s dissipates very quickly, and recovery to normal business takes several years.

Also in 2001, Lucent Technologies rolls out a new DWDM system called Lambda Extreme for use on long-haul and ultralong-haul segments. It is based on dispersion-managed soliton transmission with Raman amplification, and is specified for 128×10 Gbit/s wavelengths (1.28 Tbit/s) up to 4000 km or 64×40 Gbit/s wavelengths (2.56 Tbit/s) up to 1000 km, at a bit error rate of better than 10^{-16} [1].

2002: This year heralds the start of commercial soliton transmission systems to carry actual life traffic. Lucent's Lambda Extreme technology is deployed between Tampa and Miami (both Florida). Existing fiber designed for only 10 Gbit/s and owned by Verizon is used over a distance of 500 km to transmit 100 Gbit/s signals. In Germany, Deutsche Telekom conducts trials over 4000 km with Lucent's 128-channel version of Lambda Extreme [3]. While there are several sales of this soliton-based system over the next few years, Lucent does not publicly disclose any details, and available information is spotty.

British equipment manufacturer Marconi Solstis deploys an all-optical network based on solitons which takes up operation at the turn of 2002/2003. This ultralong-haul optical DWDM system, operated by the Australian carrier IP1, consists of a 2900 km all-optical connection (without signal regeneration) between Perth on the west coast of Australia and Adelaide on the south coast. It uses standard single-mode fiber; solar-powered amplifiers are typically spaced 90 km along the link. It is configured to use 40 out of possible 160 channels of 10 Gbit/s each for later upgrade capability. The system works well in technical terms. Unfortunately, at a time when the telecom industry is forced to release their workforce by the tens of thousands it does not work equally well in business terms, so that it gets decommissioned after only a few years.

Also in 2002, improved fibers are introduced by major fiber manufacturers. Due to increased purity they avoid the OH absorption peak near 1.4 μm so that in effect the second and third transmission window are merged into one that stretches from ca. 1280–1625 nm.

2003:   The dotcom bubble and ensuing economic woes have haunted the telecom industry for several years, but business is gradually coming back to life. New cables are being installed all the time (e.g., Apollo on the North Atlantic route in 2003), but certainly not at the same hectic pace as before. During the crisis, research is also trimmed back in the companies involved because they cannot generate the revenue that it takes to run large labs. One of the major telecom equipment providers, Lucent Technologies with its famous Bell Laboratories, is sold in 2006 to the French company Alcatel. Two years later, Alcatel-Lucent is pulling out of basic science, material physics, and semiconductor research and will instead turn its focus on more immediately marketable areas such as networking, high-speed electronics, wireless, nanotechnology, and software.

2007:   In March, the record data transmission rate over a single fiber reaches 26 Tbit/s, at a span length of 240 km [15]. At that rate, this entire book could be transmitted in under 1 ms. This is not a soliton system, but it makes use of all the tricks that are there in "linear" systems. It uses 160 WDM channels and polarization multiplexing. For coding, an RZ format and differential quaternary phase shift keying is used (see below in Sect. 11.4.2); this achieves an impressive 3.2 bits/(s Hz) of spectral efficiency. Distributed Raman amplification balances the losses. Just 5 years earlier, before the introduction of OH peak-free fibers, this signal would have come close to reaching the limit of the available bandwidth.

11.4.2   Continued Growth, Approaching the Limit

About 10 years into the new millennium it became quite clear that some entirely new approach became necessary in order to cope with further increase of demand. Binary coding of the 'OOK' (*on-off keying*) type ran into the limit imposed by the Shannon theorem (see Sect. 11.1.8). Lab experiments had almost reached the limit, and commercial systems were not far behind.

As explained in Sect. 11.1.5, data streams from a multitude of sources are combined in order to make good use of the available capacity. Individual bitstreams are interleaved by TDM to obtain 10 Gbit/s, 40 Gbit/s or, after pushing the speed of electronic circuitry further, 100 GBits/s bit rates; the latter after climbing the learning curve about how to deal with polarization mode dispersion. Then, several such bit streams are combined by WDM, and the resulting several-THz combined bandwidth signal is launched into a fiber. The phrase *massive wavelength division multiplex* (MDWM) is used when indeed large numbers (on the order of a hundred) channels are used. In some systems, nonlinear effects were taken into account from the outset by using solitons, but in most systems the power level was kept low so that nonlinearity was avoided. In any event, tremendous increase in transmission rate has occurred over the years through technical improvements such as Erbium-doped fiber amplifiers and dispersion managed fibers in combination with MWDM. The entire useful spectral transmission window of the fiber is nearly exhausted. Therefore, a further increase of the clock rate would only shift the transition point from TDM to WDM, without making more aggregate capacity available.

To simply deploy more fibers was also not a good option as that is a very expensive way of increasing data-carrying capacity.

11.4.2.1   Coding Schemes Beyond Binary

Fibers offer a useful bandwidth of ca. 50 THz. Shannon's theorem predicts a channel capacity of 50 Tbit/s for binary signals when a spectral efficiency of 1 bit/s/Hz assumed. Once the best published experiments were approaching that limit, there were basically has two options to accommodate the relentlessly growing demand for ever more data volume:

Either install more fibers, or find better coding for more data handling capacity of existing fibers.

It seems straightforward to simply add more fibers. Never mind that new fiber was already deployed at amazing speed. More relevantly, if one keeps using essentially the same technology, one keeps the cost per transmitted bit more or less constant. That is not a sustainable proposition in the face of exponentially growing volume. Incidentally, also the energy consumption per bit would be held roughly constant and the total energy consumption would rise exponentially; that has meanwhile grown to a non-negligible factor.

Another close look at the Shannon theorem is a good starting point to find better ways of using existing fibers. There are several reasons why the Shannon limit as quoted in Sect. 11.1.8 above requires some modification in the context of fiber optics:

1. Shannon's theorem holds for a *linear* channel but the fiber is inherently nonlinear.
2. A single mode optical fiber actually supports two polarization modes. This holds potential to increase the data rate by another factor of 2.
3. By taking phase modulation with coherent detection into consideration, the data rate can be further increased considerably.

This suggests that better coding schemes are possible that would allow to transmit more than one bit per clock period. That would be most welcome; however, the role of nonlinearity needs to be carefully considered.

Ad 1.   Shannon had realized that even in a continuous amplitude range one can only discern levels when they are different by at least as much as the ever-present noise dictates. Assuming that the lowest nonzero amplitude level used for transmission can not be lowered any further due to noise constraints, the addition of more amplitude levels increases the signal power and hence the impact of nonlinearity. Nonlinearity then causes stronger channel crosstalk, and this leads to increased impairment from four-wave mixing. Mitra and Stark [32] found from simulations that an increase is possible only up to 4 bit/s/Hz. With a somewhat different approach, J. Tang [63] arrives at similar conclusions.

Ad 2. Light waves are polarized: They can have either one out of two orthogonal states of polarization, or a combination thereof. Polarization multiplexing is therefore an option that can, in principle, contribute a factor-of-two to capacity. However, as discussed in Sect. 4.6, in ordinary fiber both polarization modes are not entirely independent from each other. Due to crosstalk, the state of polarization is scrambled after perhaps a few hundred meters. While polarization-preserving fibers do exist, they are more expensive and more lossy than standard telecom fiber, and the existing cables are not of this type anyway. The saving grace is that while the state of polarization may be scrambled, it is scrambled for both orthogonal components in the same way. That means that long after the state of polarization has changed, both components remain very nearly orthogonal. Therefore they can still be separated and decoded after kilometers. Only over very long distances will crosstalk between both states render the signal's bit error rate unacceptably high. Polarization multiplexing can therefore contribute moderately to an enhancement of capacity (a factor-of-two at best). In a less ambitious scheme, adjacent wavelength channels were successfully used at alternating state of polarization, to reduce channel crosstalk, without actually using polarization multiplexing. In one case 1.6 bit/s/Hz was obtained with binary data [58]. Another case combining polarization multiplex with differential phase shift keying (this is explained in the next paragraph) for a 3.2 bit/(s Hz) efficiency [15] was already mentioned above; also, a similar scheme was used in [69]. In such schemes polarization coding contributed indirectly by mitigating adverse effects from close spectral proximity of adjacent channels.

Ad 3. Information can be coded onto a carrier wave by modulating its phase; this was not considered in Shannon's original work. The fact is well known from radio engineering, and it also holds for optical carriers. One possible approach is to use *pure* phase modulation: All signal pulses have the same amplitude so that they tend to be distorted by nonlinear effects in the same way [23], but their phases can take one value out of a discrete set of different phases. This is known as *phase shift keying*, or PSK.

Of course there must be a phase detector, to decode the message, at the receiver. As is customary in optics, phase is detected through interference with a known reference wave of defined phase. In this context, that is called *coherent detection*. There is one fundamental difference between electronics and optics, though: In the best of cases, lasers do not emit pure sine waves with well-defined frequency and phase. Rather, they are subject to a random phase diffusion (the Schawlow-Townes effect described above in Sect. 11.1.2.3). On top of that, many technical noise sources conspire to let the phase fluctuate at random. Finally, the optical Kerr effect in the fiber will cause phase modulation in dependence of the signal amplitudes (see Sect. 9.2). A local oscillator providing a phase reference at the receiver will also have a randomly fluctuating phase. This makes straightforward phase detection impossible, but luckily, there is a very simple way around the problem.

In the absence of an absolute phase reference, the phase of each pulse is detected in reference to the previous pulse. It is tacitly assumed, and indeed holds very well, that in the short time of one clock period the random phase evolution of the laser can be neglected. In technical terms, pulses are encoded not with the value that pertains to this time slot, but with the difference to the previous time slot. To retrieve the message at the receiver, the previous pulse is delayed by one clock period and made to interfere on the detector with the current pulse. In other words, a Mach-Zehnder interferometer with a differential delay of one clock period is inserted in front of the photodetector. This modulation format is called *differential phase shift keying*, or DPSK [67].

Figure 11.22 illustrates different modulation formats. In the traditional amplitude modulation (AM) at left, a single degree of freedom is used for coding: here, the amplitude of the pulse. (The horizontal axis is merely for illustrative purposes and does not signify any physical quantity). As Shannon pointed out, amplitude levels can only be distinguished at the receiver when they are spaced by at least as much as what occupied by the noise. The noise is symbolized here by fuzzy bands. Binary on-off keying would use just two states, one of them being at zero. PSK-type modulation is conveniently represented in the complex plane as real and imaginary part of the complex amplitude. Distance from center signifies amplitude, and the angle stands for the phase. In engineering one also speaks of the in-phase and in-quadrature (with respect to some reference phase) components of the amplitude, or as 'the quadratures' for brief. In the center of Fig. 11.22, *quadrature phase shift keying* (QPSK) is shown: Four symbols have the same amplitude, but four different phase values in 90° increments in a shamrock (lucky 4-leaf clover) arrangement. This QPSK concept, adopted from radio engineering, constitutes a quaternary, rather than binary, coding: two bits are transmitted in each clock period. In extension of this scheme, more than four different phases may be used at the same amplitude.

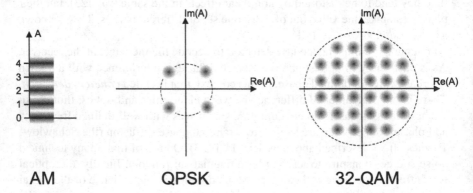

Fig. 11.22 Schematic representation of the configuration spaces of AM, QPSK, and 32-QAM. See text

Note that at the receiver each symbol is contaminated with some noise in both amplitude and phase; this uncertainty is here indicated by a fuzzy area around it, the 'noise ball'. Noise sets a limit to how many symbols can be placed on the circle: as their number increases, so does the risk that their noise balls begin to overlap, with ensuing bit errors. Power loss during propagation lets the configuration space shrink, but not the noise balls (see Fig. 11.23). If there is too much loss, it becomes difficult to tell them apart without error.

Kerr nonlinearity, and self phase modulation in particular, deforms the noise balls from spherical to banana shape because larger amplitude experience more phase rotation. A further consequence of nonlinearity is the creation of mixing products between different WDM channels which makes error-free detection even more difficult.

The quest for more bits per clock period led to consideration of even more intricate modulation formats known from electronic engineering. Among these there is a mixed phase-and-amplitude modulation called *quadrature phase and amplitude modulation* (QAM). The right part of Fig. 11.22 shows a specific example, reflecting the 32-QAM, 5 bits per clock scheme reported in [54]). With different amplitudes involved in this coding format, it is mandatory that the maximum amplitude be limited so that all symbols safely stay within the linear regime (delimited schematically by the dashed circle). Loss would move the symbols closer together, and s nonlinearity would again create self phase modulation as in Fig. 11.23, compounding the risk of overlap. In other words, QAM is a linear coding scheme which would suffer from nonlinearity, so the amplitudes need to be kept low enough that nonlinearity can be neglected. In this situation the majority of researchers and engineers in the field have decided that the best way to avoid penalties from nonlinearity is to keep the power of the data stream low enough so that over the transmission distance there is only a very moderate and nearly negligible amount of nonlinear chirp. This does nothing to eliminate the signal-to-noise issue at the receiver if one goes for very long distance.

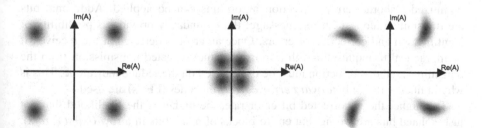

Fig. 11.23 Schematic representation of distortions in configuration space. Note that detected signals include detection noise which is independent of the signal. *Left:* After a short distance, QPSK symbols are clearly separated. *Center:* After a longer propagation distance (assuming absence of nonlinearity) loss has shrunk the intersymbol separations. *Right:* Same as *left*, but taking self phase modulation into account. It introduces a phase shift in proportion to intensity; the shear renders the 'noise balls' into 'banana' shapes

Nonetheless, remarkable success has been obtained with such formats: The added complexity of coding and decoding is rewarded with a remarkable increase of data rate: QAM has been demonstrated to support up to 11 bits of information per clock period [9]. The area within the maximum-amplitude circle is then densely packed with symbols, with the minimum intersymbol distance set by the size of the noise balls. Such advances significantly increase the spectral efficiency.

For further increase of capacity, either the noise balls must get smaller. That implies a limited reach, because with attenuation over distance, signal power decreases while noise (e.g. detector noise) stays the same. Or one accepts a certain amount of overlapping noise balls; then the bit error rate will go up. Generally speaking, the reach of QAM signals is shorter than that for less advanced coding, and compromises need to be made on bit error rate. Given that the relative size of the noise balls grows with distance (when the signal is attenuated), one pays the price for the increased capacity in either reduced reach, or increased bit error rate (or both).

It seems that when nonlinearity is avoided through low power operation one cannot simultaneously maximize capacity, reach, and data integrity. Maybe high-number QAM will be helpful at interoffice, perhaps intercity distances, but may be less suitable for the transoceanic long haul.

11.4.2.2   Error Correction

The level of bit error rate which is still deemed acceptable depends in part on the type of data being transmitted, and is more demanding for financial data than for voice mail, for example. In any event, the recent coding schemes designed to transmit several bits per clock period tend to have higher bit error rates than the more traditional on-off keying, so that some error correction needs to be performed at the receiver. The reader should understand that a consumer product like the compact audio disc also relies on error correction algorithms.

Fortunately, powerful computing has become inexpensive so that mathematically optimized elaborate error correction algorithms can be applied. Additional bits are transmitted along with the message; the redundancy provides a possibility of identification and correction of errors. Once an error is detected at the receiver, it is impractical to inquire through a return channel repeated transmission from the transmitter; rather, correction must be done based on the redundancy alone. This is why in fiber optics only *forward error correction* codes (FECs) are used.

The higher the uncorrected bit error rate, the higher is the likelihood that not just isolated bits are wrong, but entire blocks of many bits in a row (*burst errors*). Algorithms now exist by which the bit error rate can be improved significantly for both isolated and burst errors. The redundancy overhead usually makes up a few percent of the data volume, and rarely more than 20 %.

FEC has been employed for about 20 years now, with steeply increasing level of sophistication. A recent development is *soft decision based FEC*. In conventional (hard decision based) FEC the detection within a clock period amounts to a thresholding somewhere in the middle of the eye (refer to the eye diagram in Sect. 11.3.2), which results in a binary decision of either a logical *high* or a *low*. With advances in the construction of very high speed analog-to-digital converters, it becomes feasible to represent the detected level by an *n*-bit number (e.g. $n = 8$) instead of a binary value. This provides additional information on the confidence level of the *high/low* decision: if the level is high above the binary threshold, it has very high probability to represent a *high* whereas if the level is barely above the decision threshold, this is less certain, etc. Soft decision schemes therefore have the potential to perform even better, but they are very demanding in terms of hardware. Finally it should be noted that the more involved the postprocessing of the raw received data gets, the more delay occurs which adds to the link's latency, which we will comment on in the entry for 2015 in Sect. 11.4.3 below.

11.4.2.3 Nonlinearity Mitigation

Nonlinearity in fibers does not have a power threshold, or sudden onset: its impact grows continuously when the power level is raised. This implies that there is no nonzero power level at which nonlinearity is strictly absent: Even when the impact of nonlinearity is minimized by keeping power levels low, there is some level of it which creates some extra bit errors; engineers speak of a nonlinearity penalty.

In recent years there is research seeking to exploit the predictability of nonlinear impairment in fibers in order to remove this penalty; this is known as 'nonlinearity mitigation'. Mathematical models exist that are quite accurate in describing fiber propagation, with all nonlinear effects included.

The central idea in mitigation is to detect the signal at the distal fiber end, however distorted it may be, and subject it to subsequent 'backward propagation'. Backward propagation typically means that a computer takes the received data as input for a simulated propagation through a fiber with the same properties and length, but with relevant coefficients reversed [24]. As all fiber and signal parameters are known, so goes the reasoning, the inverse propagation would restore the signal shape as it has been at the launch point so that all distortions—linear (dispersion) as well as nonlinear—would be undone. In practice, a perfect cancellation is not possible, and success has been modest so far [66]. Also, the computation time adds considerably to the transmission latency—and this at a time where great efforts are made to cut down on latency (see the entry for 2015 in Sect. 11.4.3 below).

11.4.2.4 Sub-Nyquist Signaling

Nyquist pulses, as explained in Sect. 11.3.3, minimize the intersymbol interference because nulls of the pulse shape fall on the center positions of all adjacent time slots. It was proposed as early as 1975 [30] that if the achieved bit error rate is better than actually required, one might sacrifice some of it by placing pulses closer to each other in time; the increase of bit rate would then come without any relevant penalty in bit error rate. It turns out that as long as closer packing is done with modesty, the impact on neighboring time slots is predictable so that with some smart techniques it may be mitigated with good success. In practice, the increase of bit error rate is quite benign, and with some smart error correction a successful sub-Nyquist signaling can be implemented [8].

11.4.2.5 Beyond-Binary Soliton Format

Solitons were always thought of as a RZ format that is limited to on-off keying. Several years ago it was shown, however, that in dispersion-managed fibers bound states of solitons exist [59]. It was now demonstrated in a proof-of-principle experiment that such 'soliton molecules' may be used for quaternary coding in a soliton-derived format [48, 49]. As nonlinearity is already a contributing factor for this format, there is no need to keep amplitudes very low, and a larger reach can be hoped for. Also, as the degree of freedom exploited here is signal shape, a combination with polarization or phase multiplexing seems possible (but not with amplitude modulation). There have been no field trials yet, though.

11.4.3 Historical Development, Continued

2009: The major wire-line equipment providers never found back to their strength of the 1990s. The telecom business is now driven in large part by nifty end-user devices like wireless handheld sets loaded with features such as email and Internet access. Meanwhile, in the wake of the US housing bubble another economic recession and financial crisis has arrived and will linger on for several years. To give just one example, Canada's Nortel Networks Corporation was worth 250 billion US$ a decade ago, but now initiates bankruptcy proceedings. The global market for telecom services is estimated at 1.7 trillion US$, but an increasing share of this amount goes to wireless operators and handheld providers [22].

Also in 2009, a first commercial 100 Gbit/s system goes operational. It uses dual-polarization and is intended for rapid data exchange in the financial industry [2]. A new record transmission over a single fiber is reported in May: 32 Tbit/s over a distance of 580 km. This was achieved with 320 WDM channels at 25 GHz spacing with a combination of polarization multiplex and phase shift keying [69].

2012: Multicore fibers are demonstrated to allow enormously increased data rates as the aggregate value is essentially proportional to the number of cores. In two publications, transmission experiments with just over one Petabit (10^{15}) per second are reported: In [47], the fiber had 12 single-mode cores and two few-mode cores, and was 3 km long. The aggregate spectral efficiency was 109 b/(s Hz). In [62] there were 12 single-mode cores, and the fiber was 52 km long. The aggregate spectral efficiency was 91 b/(s Hz).

In a fiber loop with circulating signals (compare Fig. 11.7), 20 Tbit/s transmission was demonstrated over an effective distance of 6860 km. Sub-Nyquist Channel Spacing was used on 198 WDM channels with 100 Gbit/s each [12].

2014: The ever-increasing demand for more bandwidth is driven by popular new internet services to share pictures and video. The concept of the 'internet of things', where every appliance would have the capability to communicate via its own internet address, will—if it comes—create a further surge in demand of transmission capability. Both the manufacturing of fiber and the deployment of new cables are struggling to keep up with rising demand.

In a study assessing the quality of research in higher learning institutions of the UK, it is noted that a transmission system based on dispersion-managed solitons was originally conceived by British researchers, had been marketed under the name of Marconi MHL3000, and was first deployed in Australia (see the 2002 entry above). While the Marconi company has been acquired by Ericsson in 2006, the product is still marketed and generates sales of about 100 million US$ annually; more than 100 employees are involved [6].

A public debate has emerged after it was disclosed that intelligence services of several countries take advantage of the massive concentration of telecommunications data in fiber-optic cables such as TAT-14. While official bodies claim that tapping the data streams and scanning them for suspicious context is in the public interest in order to counter terrorism, civil rights advocates take offense at the massive breach of privacy.

2015: For the first time in more than 10 years, a new transatlantic fiber-optic cable goes into service. 'Hibernia Express' links Herring Cove, Nova Scotia (Canada) with Brean, Somerset (UK) over 4600 km. With six fiber pairs operating with DWDM at 100 Gbit/s, its aggregate capacity is designed to be 53 Tbit/s. The route was specifically chosen to approximate the most direct path between financial centers London and New York along the earth's great circle in order to keep the travel time of light signals, the *latency*, to the minimum. In our time of high-frequency trading at stock exchanges, a split second advantage can make an enormous difference.

As a consequence of global warming, shipping routes through Arctic waters become feasible. Plans have emerged and are being actively pursued to route fiberoptic cables through the Arctic Ocean. One such project, 'Arctic Cable', will run from UK across the North Atlantic, then through the Hudson Strait and North-West passage, following the Northern shore of the Canadian mainland (crossing the Boothia peninsula over 50 km of land), around Alaska and on south towards Japan. The challenges to lay cable in that hostile environment are considerable; much of the route is only accessible during brief annual ice-free seasons. On the other hand, once the cable is in place, the same ice will protect it from damage from man-made risks: internationally, 40 % of cable breaks are due to fish trawling and 26 % from anchors. Again, the rationale is that this route is geographically shorter than existing alternatives whereby the latency would be cut down by tens of milliseconds. It is anticipated that a faster link between financial centers of London and Tokyo will well be worth the considerable expense [21].

Similar plans under the name of 'Polarnet' have been in the making for several years. A cable will run from Japan past Kamchatka through the Arctic waters of Russia's Northern shore, to Murmansk and on to the UK. Stalled due to financial concerns for a while, the project now seems to move forward as ROTACS (*Russian Optical Trans-Arctic Cable System*).

A publication [64] claiming to eliminate the limits of nonlinearity mitigation attracted quite some attention. The authors argue that as long as individual lasers are used to generate the carrier waves of the various WDM channels, there are unavoidable fluctuations of the relative frequencies. This in turn gives rise to random propagation velocity variations, and random walk-off timing between signals in a WDM stream. They trace the inability to remove nonlinear impairment to this random effect. They suggest to use frequency combs, rather than individual lasers, for the carriers in the WDM system.

For many years now laser physicists have worked on frequency combs [17]. These are signals which consist of many carriers on a regular frequency grid, so that all carriers are in fixed phase relation to all others. In other words, the relative frequency fluctuations between individual frequency components are reduced by several orders of magnitude. Indeed, such frequency combs are used in metrology, e.g. in the definition of the second. In a demonstration experiment with a frequency comb the authors of [64] could show considerable improvement over the case of independent individual lasers (see also [65]).

A new record for fiber capacity is reported in [46]: A specialty fiber with 22 single mode cores of 31 km length was used with polarization-multiplexed 64-QAM transmission. With ca. 400 WDM channels, each core carried about 100 Tbit/s, and the aggregate value is then above 2 Pbit/s (a Petabit is 10^{18} bits). The carriers for the channels were generated from a frequency comb.

A first field trial of sub-Nyquist signaling was performed between Sidney and Melbourne; results are reported in [44].

In experiments, multicore fibers of ca. 225 μm diameter have reached 30 cores [50]; for an even thicker fiber with heterogenous (unequal index) cores, 36 [52].

2016:   Again a new cable takes up operation on the North Atlantic route: AEConnect. It was originally conceived under the name Emerald Express, but the operating company could not raise sufficient funding. Now the cable is completed under a new name; the company advertises "130 WDM channels of 100 Gbit/s per fiber pair". Fig. 11.24 shows the fiber-optic cables currently in use in the North Atlantic.

According to an estimate by network equipment provider Cisco [5], the worldwide internet traffic will reach a volume of 1 Zbyte this year. A Zettabyte equals 10^{21} bytes; the figure corresponds to an average traffic of 250 Tbit/s around the clock 24/7/365.

Figure 11.25 attempts to graphically represent the 'hero' experiments up to the present. Achieved data rates (horizontal axis) and the pertaining distances (vertical axis) are shown for published reports which are identified through the number; the key can be found at [4]. Shading suggests the approximate limits on distance (as

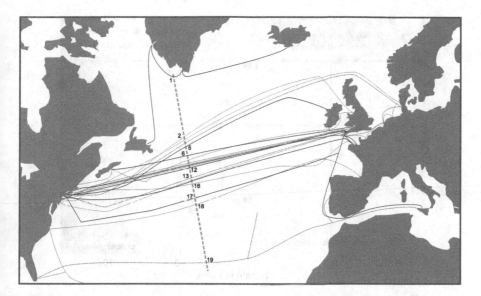

Fig. 11.24 Numerous fiber-optic cables crisscross the oceans of the world. This selection shows currently active cables between North America (USA, Canada) and Europe. Along the dashed auxiliary line they are from North to South 1: Greenland Connect, 2: Hibernia Atlantic (n) 3: TAT-14 (n), 4: AC-1 (n), 5: AEConnect, 6: Hibernia Express, 7: Hibernia Atlantic (s), 8: TGN Atlantic (n), 9: TGN Atlantic (s), 10: Flag Atlantic (n), 11: Apollo (n), 12: Yellow, 13: AC-1 (s), 14: TAT-14 (s), 15: Flag Atlantic (s), 16: Apollo (s), 17: Flag Europe Asia, 18: WASACE, 19: Columbus III. (n) and (s) indicate the northern and southern route of cable rings, respectively. AEConnect was formerly known as Emerald Express. TGN Atlantic was formerly known as VSNL and before that as TYCO Global Network Transatlantic. Yellow is also known as AC-2. TAT is for Transatlantic Telephone, AC for Atlantic Crossing. Earlier cables TAT 8 - TAT 13, Cantat-3, Gemini, and PTAT-1 are out of service. Geographic positions are approximate. The author has made an effort to represent all active cables as of March 2016

given by the size of our planet) and data rate (as given by Shannon's theorem for binary coding, with some leeway for polarization multiplexing etc.). Only single-core, single-mode fiber cases are taken into consideration, with the exception of the diamond-shaped symbols which indicate the single-core data rate (i.e. the published aggregate data rate of a multiple-core fiber divided by the core count).

11.4.4 Beyond the Single-Mode Fiber

With ever increasing demand for data volume brought about by services like streaming video etc., efforts are made to find entirely new degrees of freedom from spatial degrees of freedom. These arise when one exploits the transverse spatial dimensions, i.e. the location of the power transport within the cross section of the fiber. Such concepts abandon the time-proven single-mode fiber with its well-defined guiding mechanism. After a suggestion in [37], two related approaches have

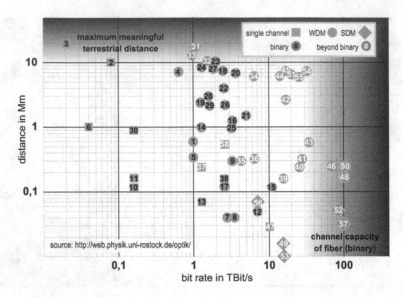

Fig. 11.25 Record experiments for high data rate transmission over a single fiber, shown here in the data rate–distance plane. For the key to the data points and detailed information see [4], a compilation maintained by Dr. Alexander Hause, Rostock University. The figure does not explicitly distinguish between different coding formats. For the sake of clarity, data are shown only for single-mode, single-core fiber transmission; diamond-shaped symbols represent single-core data rates reported from multiple-core fiber experiments. The figure represents the status as of early 2016

been pursued in recent years:

- Mode division multiplexing (MDM) uses several modes of a conventional multi-mode fiber [38].
- Space division multiplexing (SDM) uses specialty fibers with several cores side by side, distributed over the cross section [51].

Terminology has not been quite uniform from the beginning; occasionally SDM was used as a generic term for both formats. Sometimes several single- and multimode cores are combined into one fiber, effectively creating a mixed MDM / SDM arrangement [47].

Multimode Fibers: One uses fibers which carry more than a single mode; a set of several modes is used so that each carries its own independent information [38]. A detector with spatial resolution is trained to recognize the different modes individually, then uses refined algorithms to disentangle the data streams in the superpositions of various modes. The former step is repeated on the slow time scale over which channel properties drift, the latter at the clock rate. It remains to be seen how well modal crosstalk can be suppressed for successful transmission over long distances, and whether multipath interference turns out to be a problem; maybe this technology is best used for short distances (intra- and interoffice traffic).

Multicore fibers: Specialty fibers can be manufactured to have several separate cores distributed across the cladding. Issues like connectorizing etc. must be addressed, of course, but it is quite obvious that separate cores allow to transmit individual data streams, with a correspondingly increased aggregate data rate. As mentioned above, in 2012 both a team from NTT and other partners [62], and a team from NEC and Corning [47] each reported that they had crossed the 1 Pbit/s barrier. Taking distance also into account, recently the Ebit/(s km) mark was reached [55].
Crosstalk between the cores due to the proximity of the cores exists, in a way similar to that in a coupler (see Sect. 8.7.1). While it is less severe than between the modes in a multimode fiber when the number of cores is not too large, it still poses a challenge for further development of this technique.

Or the degree of freedom may be the guiding material basis:

Hollow Fibers: In yet another approach, photonic-crystal fibers with extra-wide hollow core have been designed. In these designs the core area covers many unit cells of the cladding structure. Such fibers guide light such that an extremely small fraction of power actually travels in glass; most travels in air. As a benefit, nonlinearity is greatly reduced so that linear transmission schemes may be used at elevated powers. One should also note that the propagation velocity, on account of the effective refractive index being close to unity, is almost equal to that of light in vacuum. These fibers are therefore advertised as guiding light 'at the speed of light in vacuum' [43] (about implications for telecommunications, see the comment on latency in the 2015 entry above). A combination of this type of fiber with modal multiplexing is also investigated [26]. As the interaction with the glass is reduced, the available spectral window is also larger. It remains to be seen, however, how these fibers fare with respect to bending loss, connectorization, and other aspects.

The central issue to be addressed in these formats is cross talk between data streams. In SDM the number of cores is limited by crosstalk considerations due to proximity of cores and long interaction lengths. Researchers at Fujikura have suggested a new, larger fiber diameter of 225 μm, to accommodate about 12 cores [29]. Presently the record stands at 36 cores [52]. In MDM there is the natural coupling between modes which are orthogonal only in mathematical abstraction; in reality any minimal deformation of the fiber readily causes coupling of light into other modes. Therefore one must conclude that even with these schemes one does not escape the fundamental dilemma that simultaneous optimization of capacity, reach, and data integrity implies conflicting requirements.

Maybe an even larger challenge is that both approaches, MDM and SDM, share the dramatic disadvantage that legacy fiber cannot be used. The existing network of fibers all around the globe represents a staggering financial investment. With ca. one million km of fiber-optic cables deployed on ocean floors alone (as of 2015/2016), and considering that the installation of a major transoceanic link costs on the order of 1 billion US $, the total expenditure for the fiber-optic network (oceanic and terrestrial) strung around the planet probably stands somewhere around

1 Teradollar (10^{12} US\$). This tremendous asset of legacy cables is useless for these novel formats. Moreover, before specialty fibers can be deployed on a large scale, their mass production must be set up, and the entire periphery from couplers and splicers to connectors, splitters, and amplifiers needs to be developed—not to speak of means of tailoring the dispersion which has long become standard in conventional fibers. However, as the industry faces relentless growth of demand and runs out of other options, researchers are now very actively pursuing these ideas.

Does an n-core fiber have a convincing technical, economical, or ecological advantage over n conventional fibers running in parallel in the same cable? In the light of all the challenges, the answer is not obvious to everybody.

11.4.5 What is Next?

Many millions of kilometers of optical fibers now carry the bulk of the world's data traffic most of which is internet traffic; conventional telephone calls contribute only a fraction of 1 % now. Global fiber consumption keeps growing and has now (2016) reached somewhere between 200 and 300 million km of fiber produced annually, up from 100 million km in 2006. Consider the speed at which fiber is deployed on average: 250 million km in 1 year corresponds to ca. 8 km/s—this is much faster than sound in air, and about the velocity of a satellite in a low orbit around the earth.

So far it has always been possible to increase the data-handling capacity of the fiber—even legacy fiber!—and not resort to the trivial but costly alternative of laying more fibers, which ultimately is not sustainable. It is always smarter to upgrade transmitters and detectors as to put new fibers in the ground (or to secure the rights of way for new cables). Surely there must be an ultimate limit to what fibers can do, but it is not yet clear whether we are close or whether smart ideas will buy us more time. It is also conceivable that some coding formats—including SDM—provide connections between nodes of major data centers, where distances are short but traffic volumes are fantastic, whereas more robust schemes, not quite as ambitious in data rates, are suited to the long haul. Only one prediction is universally accepted: The demand in terms of data transmission volume will keep growing. Whether the internet of things, ultrahigh definition video, massive cloud computing, and virtual reality are with us soon we do not know for certain today, but all these and other applications which have not even been conceived yet will challenge the future capacity of the communications links.

In any event, fibers remain the most capable medium to guide information: Free space optics through the atmosphere suffers from extra loss in inclement weather conditions and thus from reduced reliability. Nonetheless, it has recently been explored again as a conduit in special niches, like between offices in upper floors of neighboring high-rise buildings. In outer space, laser beams appear to be a very promising conduit for transmission when pointing direction stability issues are solved, but certainly in vacuum which is free from both loss and dispersion much larger bandwidth is possible in principle. Whatever we have learned from fiber optics in terms of light sources, data formats, and receivers will then be of benefit, but fibers themselves will no longer be needed in space. However, it will be a while before that happens, and here on the ground fibers will stay with us for a very long time.

References

1. Alcatel-Lucent, 1625 Lambda Extreme Transport Brochure, See www.alcatel-lucent.com
2. Ciena Corporation, Press release 2009. See www.ciena.com/news/news_nysc.htm
3. www.opticalkeyhole.com/eventtext.asp?ID=23702&pd=3/20/2002&bhcp=1
4. http://web.physik.uni-rostock.de/optik/wordpress/?page_id=720&lang=en. This website identifies all data points in Fig. 11.25
5. www.cisco.com/en/US/solutions/collateral/ns341/ns525/ns537/ns705/ns827/VNI_Hyperconnectivity_WP.html
6. *The world's first terabit transcontinental optical communications system exploiting dispersion managed solitons*, Impact case study REF3b, Research Excellence Framework (2014). impact. ref.ac.uk/casestudies2/refservice.svc/GetCaseStudyPDF/37025
7. G. P. Agrawal, *Nonlinear Fiber Optics, 5th* ed., Elsevier Academic Press, Oxford (2013)
8. J. B. Anderson, F. Rusek, V. Öwall, *Faster-Than-Nyquist Signaling*, Proc. IEEE **101**, 1817 (2013)
9. S. Beppu, K. Kasai, M. Yoshida and M. Nakazawa, *2048 QAM (66 Gbit/s) single-carrier coherent optical transmission over 150 km with a potential SE of 15.3 bit/s/Hz*, Opt. Express **23**, 4960 (2015)
10. K. J. Blow, N. J. Doran, *Average Soliton Dynamics and the Operation of Soliton Systems with Lumped Amplifiers*, Photonics Technology Letters **3**, 369 (1991)
11. M. Böhm, F. Mitschke, *Solitons in Lossy Fibers*, Physical Review A **76**, 063822 (2007)
12. J.-X. Cai, C. R. Davidson, A. Lucero, H. Zhang, D. G. Foursa, O. V. Sinkin, W. W. Patterson, A. N. Pilipetskii, G. Mohs, N. S. Bergano, *20 Tbit/s Transmission Over 6860 km With Sub-Nyquist Channel Spacing*, Journal of Lightwave Technology **30**, 651 (2012)
13. D. A. Fishman, B. S. Jackson, Ch. 3 in [28]
14. F. Forghieri, R. W. Tkach, A. R. Chraplyvy, Ch. 8 in [27]
15. A. H. Gnauck, G. Charlet, P. Tan, P. J. Winzer, C. R. Doerr, J. C. Centanni, E. C. Burrows, T. Kawanishi, T. Sakamoto and K. Higuma, *25.6-Tb/s WDM Transmission of Polarization-Multiplexed RZ-DQPSK Signals*, Journal of Lightwave Technology **26**, 79 (2008)
16. J. P. Gordon, H. A. Haus, *Random Walk of Coherently Amplified Solitons in Optical Fiber Transmission*, Optics Letters **11**, 665 (1986)
17. Th. W. Hänsch, *Nobel Lecture: Passion for Precision*, Rev. of Modern Physics **78**, 1297 (2006)
18. A. Hasegawa, Y. Kodama, *Guiding-Center Soliton in Optical Fibers*, Optics Letters **15**, 1443 (1991)
19. A. Hasegawa, Y. Kodama, *Guiding-Center Soliton*, Physical Review Letters **66**, 161 (1991)
20. A. Hasegawa, M. Matsumoto, *Optical Solitons in Fibers*, 3rd ed., Springer, New York (2003)
21. J. Hecht, *Fibre optics to connect Japan to the UK – via the Arctic*, New Scientist **2856** (2012)
22. E. Heinrich: *Nortel's Nadir*, Time Magazine May 11, 2009 p. 89.
23. K.-P. Ho, J. M. Kahn, *Channel Capacity of WDM Systems Using Constant-Intensity Modulation Formats*, paper ThGG85 in: Proceedings of the Optical Fiber Communications Conference Anaheim CA (2002)
24. E. Ip, J. M. Kahn, *Compensation of Dispersion and Nonlinear Impairments Using Digital Backpropagation*, J. Lightwave Techn. **26**, 3416 (2008)
25. J. B. Johnson, *Thermal Agitation of Electricity in Conductors*, Physical Review **32**, 97 (1928)

26. Y. Jung, V. A. J. M. Sleiffer, N. Baddela, M. N. Petrovich, J. R. Hayes, N. V. Wheeler, D. R. Gray, E. Numkam Fokoua, J. P. Wooler, N. H.-L. Wong, F. Parmigiani, S. U. Alam, J. Surof, M. Kuschnerov, V. Veljanovski, H. de Waardt, F. Poletti, D. J. Richardson, *First Demonstration of a Broadband 37-cell Hollow Core Photonic Bandgap Fiber and Its Application to High Capacity Mode Division Multiplexing*, Optical Fiber Communication Conference PDP5A.3 (2013)
27. I. P. Kaminow, T. L. Koch (Eds.), *Optical Fiber Telecommunications IIIA*, Academic Press (1997)
28. I. P. Kaminow, T. L. Koch (Eds.), *Optical Fiber Telecommunications IIIB*, Academic Press (1997)
29. S. Matsuo, Y. Sasaki, T. Akamatsu, I. Ishida, K. Takenaga, K. Okuyama, K. Saitoh and M. Kosihba, *12-core fiber with one ring structure for extremely large capacity transmission*, Opt. Express **20**, 28398 (2012)
30. J. E. Mazo, *Faster-Than-Nyquist Signaling*, The Bell System Techn. J. **54**, 1451 (1975)
31. C. R. Menyuk, *Soliton Robustness in Optical Fibers*, Journal of the Optical Society of America B **10**, 1585 (1993)
32. P. P. Mitra, J. B. Stark, *Nonlinear Limits to the Information Capacity of Optical Fibre Communications*, Nature **411**, 1027 (2001)
33. L. F. Mollenauer, M. J. Neubelt, S. G. Evangelides, J. P. Gordon, J. R. Simpson, L. G. Cohen, *Experimental Study of Soliton Transmission Over More than 10 000 km in Dispersion-Shifted Fiber*, Optics Letters **15**, 1203 (1990)
34. L. F. Mollenauer, J. P. Gordon, P. V. Mamyshev in [27]
35. L. F. Mollenauer, J. P. Gordon, S. G. Evangelides, *Multigigabit Soliton Transmissions Traverse Ultralong Distances*, Laser Focus World Nov. 1991 p. 159
36. L. F. Mollenauer, J. P. Gordon, *Solitons in Optical Fibers: Fundamentals and Applications*, Elsevier Academic Press, Amsterdam (2006)
37. T. Morioka, *New Generation Optical Infrastructure Technologies: 'EXAT Initiative' Towards 2020 and Beyond.* Paper FT 4 in: Proceedings of the OptoElectronics and Communications Conference (OECC) Hong Kong (2009)
38. S. Murshid, B. Grossman, P. Narakorn, *Spatial domain multiplexing: A new dimension in fiber optic multiplexing*, Optics & Laser Technology **40**, 1030 (2008)
39. M. Nakazawa, H. Kubota, *Optical Soliton Communication in a Positively and Negatively Dispersion-Allocated Optical Fibre Transmission*, Electronics Letters **31**, 216 (1995)
40. J. H. B. Nijhof, N. J. Doran, W. Forysiak, F. M. Knox, *Stable Soliton-Like Propagation in Dispersion Managed Systems with Net Anomalous, Zero and Normal Dispersion*, Electronics Letters **33**, 1726 (1997)
41. H. Nyquist, *Thermal Agitation of Electric Charge in Conductors*, Physical Review **32**, 110 (1928)
42. A. G. Okhrimchuk, G. Onishchukov, F. Lederer, *Long-Haul Soliton Transmission at 1.3 μm Using Distributed Raman Amplification*, Journal of Lightwave Technology **19**, 837 (2001)
43. F. Poletti, N. V. Wheeler, M. N. Petrovich, N. Baddela, E. Numkam Fokoua, J. R. Hayes, D. R. Gray, Z. Li, R. Slavik, D. J. Richardson, *Towards high-capacity fibre-optic communications at the speed of light in vacuum*, Nature Photonics **7**, 279 (2013)
44. L. Potì, G. Meloni, F. Fresi, T. Foggi, M. Secondini, L. Giorgi, F. Cavaliere, S. Hackett, A. Petronio, P. Nibbs, R. Forgan, A. Leong, R. Masciulli, C. Pfander, *Sub-Nyquist Field Trial Using Time-Frequency-Packed DP-QPSK Super-Channel Within Fixed ITU-T Grid*, Optics Express **23**, 16196 (2015)
45. J. G. Proakis, M. Salehi, *Fundamentals of Communication Systems*, 2nd edition, Prentice Hall, Upper Saddle River, New Jersey (2004)
46. B.J. Puttnam, R.S. Luis, W. Klaus, J. Sakaguchi, J.-M. Delgado Mendinueta, Y. Awaji, N. Wada, Y. Tamura, T. Hayashi, M. Hirano, J. Marciante, *2.15 Pb/s Transmission Using a 22-core Homogeneous Single-Mode Multi-Core Fibre and Wideband Optical Comb*, Proc. ECOC, paper PDP.3.1 (2015)

47. D. Qian, E. Ip, M.-F. Huang, M.-J. Li, A. Dogariu, S. Zhang, Y. Shao, Y.-K. Huang, Y. Zhang, X. Cheng, Y. Tian, P. Ji, A. Collier, Y. Geng, J. Linares, C. Montero, V. Moreno, X.Prieto, T. Wang, *1.05Pb/s Transmission with 109b/s/Hz Spectral Efficiency using Hybrid Single- and Few-Mode Cores*, FW6C.3, Frontiers in Optics / Laser Science Conference (FiO/LS) XXVIII (2012)

48. P. Rohrmann, A. Hause, F. Mitschke, *Solitons Beyond Binary: Possibility of Fibre-Optic Transmission of Two Bits per Clock Period*, Scientific Reports **2**, 866 (2012)

49. P. Rohrmann, A. Hause, F. Mitschke, *Two-soliton and three-soliton molecules in optical fibers*, Physical Review A **87**, 043834 (2013)

50. K. Saitoh and S. Matsuo, *Multicore Fiber Technology*, J. Lightw. Techn. **34**, 55 (2016)

51. J. Sakaguchi, B. J. Puttnam, W. Klaus, J. M. D. Mendinueta, Y. Awaji, N. Wada, A. Kanno, T. Kawanishi, *Large-capacity transmission over a 19-core fiber*, Paper OW1I.3 in: Optical Fiber Communication Conference (OFC)/ National Fiber Optic Engineers Conference (NFOEC)2013, OSA Technical Digest, Optical Society of America, 2013)

52. J. Sakaguchi, W. Klaus, J.-M. D. Mendinueta, B.J. Puttnam, R.S. Luis, Y. Awaji, N. Wada, T. Hayashi, T. Nakanishi, T. Watanabe, Y. Kokubun, T. Takahata, T. Kobayashi, *Realizing a 36-core, 3-mode Fiber with 108 Spatial Channels*, Postdeadline Paper Th5C.2 in: Tech. Digest of The Optical Fiber Communication Conference and Exposition (OFC), Los Angeles 2015

53. B. E. A. Saleh, M. C. Teich, *Fundamentals of Photonics*, 2nd ed., John Wiley & Sons, New York (2007)

54. A. Sano et al. (21 authors), *409-Tb/s + 409-Tb/s crosstalk suppressed bidirectional MCF transmission over 450 km using propagation-direction interleaving*, Optics Express **21**, 16777 (2013)

55. A. Sano, H. Takara, T. Kobayashi, Y. Miyamoto, *Petabit/s Transmission Using Multicore Fibers*, paper Tu2J.1, Optical Fiber Communication Conference OFC (2014)

56. C. E. Shannon, *A Mathematical Theory of Communication*, The Bell System Technical Journal **27**, 379 and 623 (1948)

57. N. J. Smith, F. M. Knox, N. J. Doran, K. J. Blow, I. Bennion, *Enhanced Power Solitons in Optical Fibres with Periodic Dispersion Management*, Electronics Letters **32**, 54 (1996)

58. H. Sotobayashi, W. Chujo, K. Kitayama, *Highly Spectral-Efficient Optical Code-division Multiplexing Transmission System*, IEEE Journal of Selected Topics in Quantum Electronics **10**, 250–258 (2004)

59. M. Stratmann, T. Pagel, F. M. Mitschke, *Experimental Observation of Temporal Soliton Molecules*, Physical Review Letters **95**, 143902 (2005)

60. M. Suzuki, I. Morita, N. Edagawa, S. Yamamoto, H. Taga, S. Akiba, *Reduction of Gordon-Haus Timing Jitter by Periodic Dispersion Compensation in Soliton Transmission*, Electronics Letters **31**, 2027 (1995)

61. O. Svelto, *Principles of Lasers*, 4th ed., Plenum Press, New York (2004)

62. H. Takara, A. Sano, T. Kobayashi, H. Kubota, H. Kawakami, A. Matsuura, Y. Miyamoto, Y. Abe, H. Ono, K. Shikama, Y. Goto, K. Tsujikawa, Y. Sasaki, I Ishida, K. Takenaga, S. Matsuo, K. Saitoh, M. Koshiba, T. Morioka, *1.01-Pb/s (12 SDM/222 WDM/456 Gb/s) Crosstalk-managed Transmission with 91.4-b/s/Hz Aggregate Spectral Efficiency*, European Conference on Optical Communication (ECOC), Th 3 C. 1 (2012)

63. J. Tang, *The Shannon Channel Capacity of Dispersion-Free Nonlinear Optical Fiber Transmission*, Journal of Lightwave Technology **19**, 1104 (2001) and *The Multispan Effects of Kerr Nonlinearity and Amplifier Noises on Shannon Channel Capacity of a Dispersion-Free Nonlinear Optical Fiber*, ibid., 1110 (2001)

64. E. Temprana, E. Myslivets, B.P.-P. Kuo, L. Liu, V. Ataie, N. Alic, S. Radic, *Overcoming Kerr-induced capacity limit in optical fiber transmission*, Science **348**, 1445 (2015)

65. E. Temprana, N. Alic, B.P.P. Kuo, S. Radic, *Beating the nonlinear capacity limit*, Optics & Photonics news **3**, 30 (2016)

66. K. Toyoda, Y. Koizumi, T. Omiya, M. Yoshida, T. Hirooka, M. Nakazawa, *Marked performance improvement of 256 QAM transmission using a digital back-propagation method*, Opt. Express **20**, 19815 (2012)

67. Ch. Xu, X. Liu, X. Wei, *Differential Phase-Shift Ke4ying for High Spectral Efficiency Optical Transmission*, IEEE J. Selected Topics in Quant. Electronics **10**, 281 (2004)
68. T.-S. Yang, W. L. Kath, *Analysis of Enhanced-Power Solitons in Dispersion-Managed Optical Fibers*, Optics Letters **22**, 985 (1997)
69. X. Zhou, J. Yu, M.-F. Huang, Y. Shao, T. Wang, P. Magill, M. Cvijetic, L. Nelson, M. Birk, G. Zhang, S. T. Ten, H. B. Matthew, S. K. Mishra, *32Tb/s (320x114Gb/s) PDM-RZ-8QAM Transmission over 580km of SMF-28 Ultra-Low Loss Fiber*, Optical Fiber Conference 2009, postdeadline paper PDPB4 (Optical Society of America)
70. J. L. Zyskind, J. A. Nagel, H. D. Kidorf in [28]
71. The sampling theorem was first formulated by H. Nyquist in 1928, later C. Shannon gave a mathematical proof. See [56] or [45].
72. Planck's radiation law is treated in many textbooks on optics or lasers; see, e.g., [53] or [61].
73. The laser linewidth is treated in most textbooks on lasers; see, e.g., [61]

Chapter 12
Fiber-Optic Sensors

The development of fiber-optic technology was mainly driven by the requirements of the telecommunications industry. Nonetheless one should not overlook that telecommunications is not the only application of fiber optics. The other major application area is in metrology and data acquisition.

12.1 Why Sensors? Why Fiber-Optic?

It used to be that in any major machinery or installation, gauges were located wherever the relevant information was present: a thermometer at the boiler, a tachometer at the shaft, a fuel gauge at the tank, etc. Staff could then go to these locations and take readings. Meanwhile the trend is that data acquisition and display are separated. For example, consider an airplane: Sticking out a mercury thermometer is obviously not a good idea for measuring the outside temperature. Fuel tanks are in the wings; who would climb out there to check a level tube? Instead, all data of interest are acquired at their respective location with sensors. The sensor's response is transmitted, usually by cable, to a central monitoring station where all displays are side by side to provide an overview. In the airplane, this location is in the cockpit where the pilot can check all instruments without leaving his seat.

Industrial installations, too, have a central control room where all information comes together. It is not only time-saving when staff do not need to walk around the premises to take instrument readings, but it also minimizes risks to humans because often data are taken in hard-to-reach or dangerous places, such as inside chimneys, in high-voltage apparatus, or in numerous places inside nuclear power stations.

© Springer-Verlag Berlin Heidelberg 2016
F. Mitschke, *Fiber Optics*, DOI 10.1007/978-3-662-52764-1_12

To go by such a remote sensor concept, there are three ingredients required:

1. Sensors for any physical quantity that may be of interest. This includes temperature, pressure, stress and strain, distance, filling level, speed, force, vibration, etc. The sensors must translate such quantities into a format that can easily be transmitted.
2. Transmission lines.
3. Displays that translate the transmitted data into a format accessible to human senses, i.e., typically make them visible or audible.

Of course, the scheme also facilitates the keeping of records of relevant data; witness the "flight recorder" which is of central importance after a plane crash.

It has often been taken for granted that for the transmission one uses an electric quantity: most often a voltage, but there is at least one standard where this is a current. The lines are then usually copper cables. The advantage of this approach is that there are innumerably many suppliers, and sensors can be picked from an unfathomable variety of hardware. Also, there is an abundant supply of well-trained engineers and technicians who are knowledgeable about this technology and can use it very efficiently.

Now enter optical fiber. First, one might have the idea of using sensors that do not translate the original data into an electrical format, but rather into some optical format, like a light intensity or wavelength. There is no difficulty in converting this to a display because optical formats are easily assessed at the receiver. All it takes is a photodetector, and one is back to a voltage or current that can be displayed in a routine way. Of course, the question is: If one eventually converts to electrical anyway, why bother with optics?

The point is that during transmission, the data are in an optical format. While on its way across the distance, plenty of adverse effects can act on the transmitted signal. In the case of electric cables, one severe problem is interference from external electromagnetic fields. To avoid such difficulty, one usually provides shielding, which in the case of strong external fields is quite involved. Optical fiber, by contrast, is immune to that kind of interference.

There are some other properties of optical fibers that are advantageous in this context. As we saw earlier, they are small and lightweight. The accompanying savings in space and weight can be quite important, e.g., in vehicles, in particular in aircraft or spacecraft. Also, optical fibers withstand extreme temperatures better than electrical cables. They are also more robust in the presence of aggressive chemicals. Finally, fibers provide perfectly separated electrical potentials, a fact that is greatly appreciated, e.g., in petrochemical installations.

We see, one might have benefits from an optical technology. It is good news that a wide variety of optical sensors is available. There is hardly any physical quantity for which no optical sensor exists. New sensors are added all the time for chemical and other quantities, too.

When we look at these *fiber-optic sensors*, we need to broadly distinguish two classes (Fig. 12.1): There are sensors that are mounted in front of, next to, or in proximity of the fiber, read the quantity under investigation, and launch a corresponding

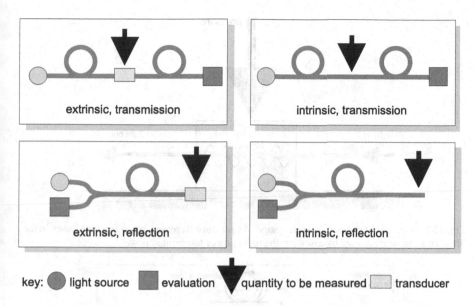

Fig. 12.1 Classification of sensor types. In extrinsic sensors (*left column*), a transducer converts the original quantity to an optical format; in intrinsic sensors (*right column*) the fiber itself is the transducer. One can also distinguish transmission sensors (*top row*) and reflection sensors (*bottom row*). The former are simpler in structure because no couplers are required to separate forward and backward traveling light. The latter are more convenient to use, though, because only one end of the fiber needs to be accessible

light signal into the fiber. In this case, the fiber is merely the transmission medium and has nothing to do with the acquisition of the original quantity. Such sensors are called *extrinsic*. In contrast, *intrinsic* sensors use the fiber itself or part of it directly to read the original quantity. Then the fiber is both sensor and cable at the same time. We will look at examples of both types. We may also distinguish the *reflective* and the *transmissive* types; while the transmission type appears more straightforward, reflective sensors have advantages in hard-to-access places.

12.2   Local Measurements

12.2.1   Pressure Gauge

The simplest type of an optical pressure gauge is shown schematically in Fig. 12.2. A fiber is placed between two corrugated surfaces; if these are pressed together, the fiber is forced into wiggles and the bending loss increases. By suitable calibration procedure, the amount of pressure can be obtained from the transmission loss. This would be an intrinsic transmission sensor. However, this very simple concept would be susceptible to errors from variation in light source output or any other influences that would effect the received power.

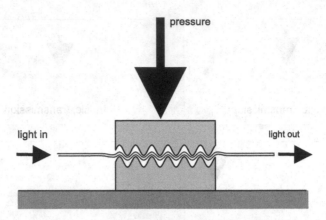

Fig. 12.2 A simple fiber-optic pressure gauge. As pressure increases, so do bending losses in the fiber. This can be monitored by assessing the reduction of transmitted power

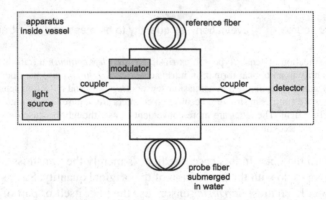

Fig. 12.3 A fiber-optic Mach–Zehnder interferometer is suitable to assess minuscule path-length variations. In this example it is employed as a hydrophone: one fiber coil is subject to pressure fluctuations under water while the other is insulated from them. Path-length changes as small as a fraction of a wavelength are easily detected; this is why such constructions can be much more sensitive than conventional microphones

12.2.2   Hydrophone

Quite often fiber-optic sensors make use of the interferometric principle to obtain an impressive sensitivity. The example shown in Fig. 12.3 consists of a Mach–Zehnder interferometer in which a light beam is split into two branches. After passing through similar but independent paths they are recombined again. Any change in the path-length difference is converted into variations of the resulting power after interference: A change of only half a wavelength provides a 100 % variation in the detected signal.

In this example one interferometer arm contains a length of fiber which is encased to insulate it from environmental effects while the other consists of the same length of fiber, wound on a hollow drum which is immersed in sea water. Sound waves, i.e., pressure fluctuations in the water, stretch the drum and the fiber with it. Again, the fiber itself is the sensing element; this is an intrinsic sensor.

The amount of stretching of the fiber is a measure of the pressure amplitude, and to the extent that the drum does not have mechanical resonances, the sensitivity is independent of sound frequency. The sensitivity can be made extremely high by using a long fiber because for the same *relative* strain the *absolute* strain increases with fiber length. Such an underwater microphone is known as a hydrophone and is extremely important, e.g., for ranging in submarines. In terms of sensitivity, the fiber-optic version is vastly superior in comparison to other technologies [7].

12.2.3 Temperature Measurement

Now we turn to an example of an extrinsic sensor. In the example depicted in Fig. 12.4, the fiber tip is coated with a layer of a thermo-sensitive phosphorescent material commonly called a phosphor (even though no phosphorous is involved).

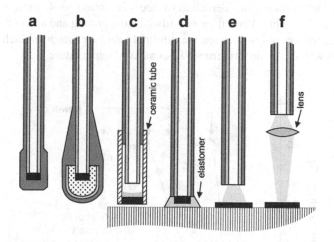

Fig. 12.4 Different versions of fiber-optic temperature sensors based on the principle of temperature-dependent luminescence decay time. The luminescent material (*black*) is deposited on the fiber tip in the standard version shown in (**a**). The sensor is coated with a protective coating (*gray*). Variant (**b**) is optimized for resilience against chemicals and oils; the luminescent material sits in a protective glass ferrule (*light gray*), which is filled with epoxy (*dotted*). In (**c**), the last 10 cm of the fiber are embedded in a tube made of aluminum oxide ceramics (*hatched*). The luminescent substance is supported by a glass bead; an air gap keeps the fiber itself away from temperature extremes. An elastic tip in (**d**) is meant to provide improved thermal contact with surfaces. (**e**) and (**f**) are versions for noncontact measurement; here the luminescent material is not applied on the fiber but directly on the workpiece. The light emerging from the fiber is reflected and captured again; in the case of extended distance a lens collimates the beam. After [10] with kind permission

This phosphor, which may actually be magnesium fluorogermanate, is optically excited by a brief flash of light (from an LED) that causes it to emit phosphorescence, i.e., luminescence light with relatively slow exponential decay. The decay time constant is a good metric for the temperature. It is a definite advantage that only ratios of intensities must be assessed to obtain the decay time, but not intensities themselves; any fluctuations due to light source instability, varying connector losses, etc. therefore cancel out. Such fiber-optic temperature gauges are commercially available, and the shape of the sensor can be chosen—depending on intended application—from a variety of several different types as shown in Fig. 12.4.

Temperature measurement with fiber-optic sensors have several advantages: The fiber has very small footprint; it has low heat capacity and conductivity and thus does not distort the heat distribution to be measured. Moreover, the sensor is fully dielectric so that measurements are not affected by the presence of strong electric, magnetic, or electromagnetic fields. Electrical thermometers (thermocouples, etc.) do not share these advantages. Figure 12.5 demonstrates a measurement of the heating process of a processed meal in a microwave oven.

Alternatively, a fiber can also be used for temperature measurement by exploiting its thermo-optic coefficient, which describes the change of refractive index of fiber with temperature changes (there is also a minor contribution from the thermal expansion). The thermo-optic coefficient of fibers is around 30–40 ppm/K [11, 13]. One can use a fiber Fabry–Perot filter or a fiber-Bragg grating and assess the spectral shift of the reflection or transmission maximum. In another variant a Mach–Zehnder arrangement would contain a reference fiber at fixed temperature.

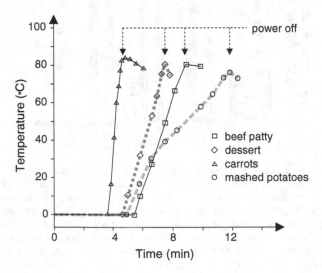

Fig. 12.5 Example for an application in which fiber-optic thermometers have a vast advantage: temperature measurement inside a microwave oven during operation. The heating of the components of a meal is shown. Now it has been proven after all that the dessert is almost boiling while the mashed potatoes are still lukewarm! After [10] with kind permission

Fig. 12.6 Calibration curve of a fiber-optic dosimeter. The additional optical loss at $\lambda = 829$ nm at room temperature was measured during an irradiation with ^{60}Co at a dose rate of $1.43 \, \mathrm{Gy(SiO_2)/s}$. The linear range extends across four orders of magnitude. Taken from [9] with kind permission

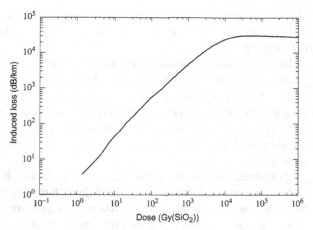

12.2.4 Dosimetry

As a final example, we look at a fiber that is subject to ionizing radiation. Dislocations are created in the glass and cause an increased optical loss which can cause a decrease of the transmitted power. The damage is essentially cumulative so that such a fiber can be used as a radiation dosimeter, i.e., an integrating gauge for the cumulative radiation received in a certain amount of time. Fiber-optic dosimeters have a wider linear range than those using other technologies (Fig. 12.6). Moreover the reading tends to be more precise because in any dosimeter, there is some degree of recovery of the dislocations after the irradiation ends. In fiber-optic versions this effect is less pronounced than in other types, and this makes a difference when an evaluation takes place only some time later.

12.3 Distributed Measurements

A more recent development may push applications of intrinsic sensors far beyond what is conceivable with electric sensors. The crucial point is this: An electric sensor measures at a point—a mathematician might say, on a zero-dimensional manifold. A fiber is extended and measures along a line, i.e., one dimensionally. As long as one is only interested in data at certain spots (a zero-dimensional manifold), this consideration is moot. However, it is important as soon as one wishes to monitor a higher-dimensional manifold. Examples for one-dimensionally distributed data acquisition are in monitoring the integrity of conduits of all types, including pipelines for oil, gas, or water, or power transmission lines, telephone lines, etc. It may be required to permanently monitor for possible temperature rises, mechanical stress and strain, vibration, etc. Similarly, surveillance of intrusion attempts at security fences calls for a sensor that reacts to perturbations anywhere along its length. It is just not practical to combine many point-like gauges in short distances because the numbers would be excessive. A single fiber can accomplish the same job.

In even higher dimensions it is easy to see: One can lay out a fiber in a zigzag pattern to cover an area. Example: A 2D pressure sensor embedded in the floor can detect whether a person is present in a certain area. This is a useful feature either for ensuring the safety of operators of dangerous machinery or for detecting intruders. Point sensors would have to be distributed in a grid pattern, implying both large numbers and high cost. As one proceeds from line to area to volume, the same logic applies.

Distributed fiber-optic sensors are available, and they provide good spatial resolution to boot! This is accomplished through propagation time effects. It should be clear how big the advantage is when one can learn in real time not only that an oil pipeline is subject to a worrying mechanical tension somewhere, but when the position of the trouble is precisely located—in some cases, within centimeters. Then, a service crew can be dispatched immediately and fix the problem before major damage occurs.

In such applications, the optical fiber itself is just about the cheapest part. It is therefore not a problem when fibers are buried in concrete when a structure is built. They can then not be replaced later on; therefore they are referred to as "lost fibers." Nevertheless, as long as their ends (or at least one end) remain accessible, the fiber embedded in a structure can provide important clues about mechanical stresses acting on the structure in real time. This idea has been applied in dams and bridges and has become quite normal now [12]. An early example was Winooski Dam in Vermont, USA, a dam with turbines providing electrical power [8]. More than 6 km fibers were embedded. Right at the first trial runs of the turbines the fiber-optic sensors showed a conspicuous resonance in the vibration spectrum of one of the turbines: The resonant frequency was at 168 Hz instead of 174 Hz as expected. Given this clue, the turbine's manufacturer could quickly fix the problem: Due to a defective component the efficiency was 81 % rather than 92 %! It would have been much more costly to discover that under load conditions during operation, and those savings alone paid for the fibers [2].

Fibers have made inroads into two-dimensional problems: Monitoring stresses on the hull of a ship is now possible in tankers as well as icebreakers. Wind turbines are another obvious field of application. In aircraft, the monitoring of structural integrity of the skin—and the wings in particular—is of the greatest importance. Using a closely knit mesh of fibers, one adds nearly nothing to overall weight but obtains the perfect means of diagnosis. Such a skin with embedded high tech is also known as "smart skin." Raman scattering types allow temperature monitoring due to temperature-dependent wavelength shifts and find applications in fire detection in tunnels, pipeline monitoring, etc. On the other hand, arrays of fiber-Bragg gratings (mostly for stress and strain detection) seem to be commercially more successful than truly distributed sensing schemes (see Figs. 12.7 and 12.8).

This is again the point where two considerations converge. Large amounts of data as produced by a smart skin need to be transmitted. There is hardly a better medium

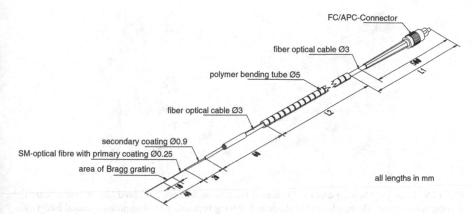

Fig. 12.7 Package of a commercial fiber-Bragg grating temperature sensor specified for the range −30 to +80 °C with 0.1 °C resolution and 0.5 °C accuracy. With kind permission by Telegärtner Gerätebau GmbH [1] and by AOS GmbH [3]

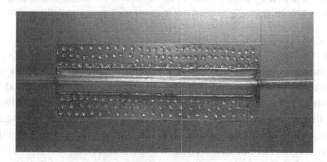

Fig. 12.8 A commercial fiber-optic strain and temperature sensor designed to be welded directly onto the metal structure to be monitored. This 15 by 40 mm size device is specified to acquire ±2500 μstrain with a resolution of 0.4 μ strain and a temperature of −170 to +150 °C with 0.05 °C resolution. With kind permission of Smart Fibres Limited, Bracknell, UK [4]

for this task than optical fiber, an excellent medium for communications. The same on-board fiber network that transmits communications (and is in place anyway) can double to transmit information on mechanical stress, temperature extremes, etc. In the narrower sense of the word this, too, is communication except it is not humans like captain and flight engineer who are doing the communicating, but rather the engine, the wing, and the computer.

In recent years many countries have massively introduced unmanned aerial vehicles (UAVs) for reconnaissance and also combat. They are known to the general public as drones; the US military alone has several thousand of them. Figure 12.9 shows an early version on which fiber-optic sensing was tested (image taken 2007). While in manned vehicles, pilots directly sense and report events like bird strike, lightning strike, impact of runway debris, or just unusual vibrations, in UAVs sensors must be used. Fiber-optics has become the standard for sensing

Fig. 12.9 NASA's Ikhana, a modified Predator B unmanned aircraft adapted for civilian research, is being used to test advanced, fiber optic-based sensing technology to monitor structural integrity. Six fibers on the wing's top surface provide more than 2000 strain measurements, thus providing full information of the wing shape in real time. They add merely 1 kg of weight and do not appreciably affect aerodynamics. The data gathered improve safety, but the ultimate goal is to develop active control of wing shape so that the aerodynamics can be adapted to take-off, cruising, and landing. Such capability could dramatically improve efficiency and performance. From [5]

and real-time structural health monitoring due to its obvious advantages: At small size and minimal weight this technology provides high bandwidth, is immune to electromagnetic interference, and is, after some trials, easily embedded. The same advantages also apply to other aircraft of all types, and also to all other types of vehicles. With respect to military applications it is a safe bet to assume that all recent constructions of aircraft, ships, and submarines are equipped with a lot of fiber even when this is not disclosed due to classification.

Now consider the numerous electrical cables in automobiles. Beginning in the luxury car segment and gradually filtering down to medium-class vehicles, copper cables are replaced with fibers because of space and weight savings and immunity to electromagnetic interference, straightforward insertion of some fiber-optic sensors, and because the tremendous bandwidth allows the introduction of advanced on-board sound and infotainment systems. Initially plastic optical fibers (see Sect. 5.4.4) were deemed appropriate as distances are short and data rates were moderate. This now begins to change, and silica fibers are being discussed as increasing data volumes are used.

Fault tolerance refers to the ability of equipment to keep working even when part of it is damaged. This is an important requirement everywhere, and in combat aircraft it is absolutely vital. It can be met by linking all on-board components through a web-like structure, rather than linear point-to-point connections, because then in the event of any local damage, data flow is not interrupted but can be rerouted. This is a strategy well known to power utility companies and telephone service providers alike, and it applies to on-board fiber-optic networks, too.

The paradigm of a web structure is the internet which, as is well known, was designed with the idea in mind that it should be nearly indestructible. What cannot be killed even by a nuclear blast is obviously quite robust. Some crooks exploit this robustness by sending us zillions of spam messages or trading in unsavory material, all the while relying on the notion that they are almost unstoppable. This activity is detestable, but it does illustrate the point.

12.4 The Status Today

These days structures such as dams, bridges, tunnels, mines, storage tanks, and towers are more and more often equipped with sensors, and fiber-optic types are used increasingly and command an increasing share of the total sensor market. The most frequently used types are based on fiber-Bragg gratings, Raman and Brillouin scattering, and mechanical or thermal length variation.

Fiber-optic sensors are always in competition with existing technology, and must assure a definite advantage before they are adopted. There is the difficulty that on one hand scientists working in research labs are fully prepared to respectfully treat novel technology with care, but that on the other hand in the environment of a major construction site the hard hat-wearing crowd has little patience for the fragility of delicate fibers. If one embeds a fiber in concrete, one should take great care not to break the fiber end at the place where it sticks out of the concrete: If it is broken, it may be useless (and is indeed a "lost fiber"!).

Nonetheless, fiber-optic sensors are big business now: Various market researchers place the current (2016) global consumption across all types near 3 billion US $/year. If current growth continues, that number will exceed 4 billion US $/year by 2018 [6]. Continuous distributed sensors are the fastest growing segment, and already make up ⅔ of the value. By value, applications are in the military/aerospace/security sector, followed by petrochemical/energy, then civil engineering/construction and structural health monitoring. By-value shares are skewed towards those sectors requiring long fibers; biomedical and other scientific applications may be relevant by number, but require short fiber lengths and are thus minor by value.

References

1. Telegärtner Gerätebau GmbH, 01774 Höckendorf, Germany. See www.telegaertner.com
2. C. I. Merzbacher, A. D. Kersey, E. J. Friebele, *Fiber Optic Sensors in Concrete Structures: A Review*, in: *Optical Fiber Sensor Technology* Vol III: Applications and Systems, K. T. V. Grattan and B. T. Meggitt (Eds.), Kluwer Academic Publishers, Dordrecht (1999)
3. AOS Advanced Optical Solutions GmbH, Ammonstraße 35, 01067 Dresden, Germany. See www.aos-fiber.com
4. Smart Fibres Limited, Bracknell, UK. See www.smartfibres.com
5. NASA Dryden Flight Research Center Photo Collection: Photo ED07-0186-13. Photographer: Jim Ross. See www.dfrc.nasa.gov/Gallery/Photo/Ikhana/HTML/ED07-0186-13.html

6. ElectroniCast Consultants (Aptos, CA, USA): *Fiber Optic Sensors Global Market Forecast & Analysis*, February 2014. See www.electronicastconsultants.com/files/Announcement_-_Fiber_Optic_Sensor_Market_Forecast_-_February_2014__ElectroniCast.pdf

7. J. A. Bucaro, T. R. Hickman, *Measurement of Sensitivity of Optical Fibers for Acoustic Detection*, Applied Optics **18**, 938 (1979)

8. P. L. Fuhr, D. R. Huston, *Multiplexed Fiber Optic Pressure and Vibration Sensors for Hydroelectric Dam Monitoring*, Smart Materials and Structures **2**, 260 (1993)

9. H. Henschel, M. Körfer, K. Wittenburg, F. Wulf, *Fiber Optic Radiation Sensing Systems*, TESLA Report No. 2000–26 September (2000)

10. H. Holbach, *Faseroptische Temperaturmessung*, brochure by Polytec GmbH, Polytec-Platz 1–7, 76337 Waldbronn, Deutschland. Figs. 12.4 and 12.5 first appeared in *Application Note AN86-MF02* and *Advances in Fluoroptic™ Thermometry: New Applications in Temperature Measurement (presented at Digitech '85)* by Luxtron Corporation, 3033 Scott Boulevard, Santa Clara, CA 95054-3316 (USA), which is represented in Germany by Polytec GmbH.

11. N. Lagakos, J. A. Bucaro, J. Jatzynski, *Temperature-Induced Optical Phase Shifts in Fibers*, Applied Optics **20**, 2305 (1981)

12. C. I. Merzbacher, A. D. Kersey, E. J. Friebele, *Fiber Optic Sensors in Concrete Structures: A Review*, Smart Materials and Structures **5**, 196 (1996)

13. S. J. Wilson, *Temperature Sensitivity of Optical Fiber Path Length*, Optics Communications **71**, 345–350 (1989)

Part VI
Appendices

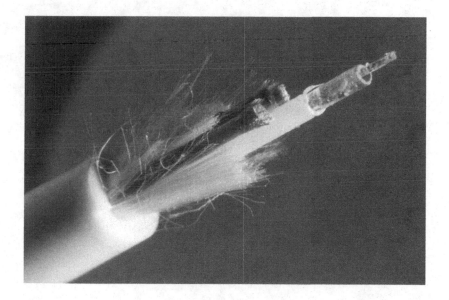

Beyond the fiber itself, a fiber-optic cable contains a complex structure of mechanical elements for protection against abrasion and stress. The picture shows the tube that contains the fiber; not visible is the gel filling. Further outside there are strands of Kevlar fiber acting as stress members. Kevlar is a resilient fibrous material from which, among other things, bulletproof vests are made.

Chapter 13
Decibel Units

The measurement units of "decibels" are in widespread use in electrical engineering; both physicists and engineers are expected to use them proficiently. They constitute a logarithmic measure useful for gain factors, attenuation factors, etc. The advantage of using a logarithmic measure is that in a transmission chain, there are many elements concatenated, and each has its own gain or attenuation. To obtain the total, addition of decibel values is much more convenient than multiplication of the individual factors.

13.1 Definition

One decibel is the tenth part of 1 B. The name refers to Alexander Graham Bell; for historic reasons his name is truncated to "Bel." One Bel designates a ratio of 10:1 between two quantities which have the dimension of a power. Let us call them P_1 and P_0:

$$\delta[\text{Bel}] = \log_{10} \frac{P_1}{P_0}.$$

It is also common to use this definition for quantities that are proportional to a power, such as energy, work, energy density, or intensity (power per area). Of course, both quantities involved must be of the same type so that the argument of the logarithm is dimensionless.

In contrast to standard practice in the SI system of units, neither the unit Bel itself nor its combination with prefixes such as milli- and micro- is used. The decibel is the only form in use, and it is abbreviated as dB. In fact, one would rather speak of one hundredth of a decibel than a millibel. A decibel is defined as the tenth part of a Bel, so that

$$\delta[\text{dB}] = 10 \log_{10} \frac{P_1}{P_0}.$$

© Springer-Verlag Berlin Heidelberg 2016
F. Mitschke, *Fiber Optics*, DOI 10.1007/978-3-662-52764-1_13

For example, inserting the output power of some amplifier as P_1 and its input power as P_0, $\delta[dB]$ is the gain factor. Negative gain factors imply attenuation.

It is prudent to remember a few selected numbers: 10 dB imply a factor of 10, 20 dB a factor of 100. 3 dB pretty closely corresponds to a factor of 2, and correspondingly, 6 dB to a factor of 4.

13.2   Absolute Values

Up to now we described the use of the dB as a relative unit. It can also be used for absolute values when a standardized reference value for P_0 is agreed upon. Most frequently, this is the value of 1 mW. Whenever power is measured with reference to 1 mW, the letter "m" is appended to the dB to produce dBm:

$$\delta[dBm] = 10 \log_{10} \frac{P}{1\,mW}.$$

For example, the maximum output power of amplifiers is frequently quoted in dBm. Thus, 40 dBm are a fancy way to say 10 W.

Another variant of using dB units for absolute values is in widespread use: $dB\mu V$ implies dB referred to a reference voltage of one microvolt.

Carried away by the convenience of the notation, some authors use dB for just about all kinds of quantities. There have been sightings of, e.g., the ratio of two resistances in dB. This author strongly recommends against such practice.

13.3   Possible Irritations

When using decibel units, novices frequently get confused in either one of two circumstances.

13.3.1   Amplitude Ratios

The first source of confusion is that decibel is not always used for powers. In electrical engineering in particular, measurement typically yields not power directly, but rather a voltage U or a current I. As a consequence, when providing a gain factor of some device, one needs to specify whether it is a voltage gain, a current gain, or a power gain factor. This distinction disappears when decibel units are used.

According to Ohm's law and assuming a standard load resistance R,

$$P = UI = \frac{U^2}{R}.$$

Similarly, in optics, the power of a light signal is proportional to the square of the electrical field amplitude. Again, one needs to specify whether the amplification/attenuation is referred to the field amplitude or to the power.

Since

$$\log \frac{U_1^2}{U_0^2} = 2 \log \frac{U_1}{U_0},$$

decibels can be used without conflict with the above definition when

$$\delta[\text{dB}] = 20 \log_{10} \frac{A_1}{A_0}$$

is observed. Here A denotes an "amplitude type" quantity such as voltage and field strength, which is proportional to the square root of a "power type" quantity as described above.

13.3.2   Example

Consider an amplifier that boosts some input signal from 1 to 20 mV; source and load impedance are equal. When specifying the gain factor, one needs to make the distinction between voltage gain and power gain: $G_{\text{amplitude}} = 20$, $G_{\text{power}} = 400$. Using decibels the gain factor is simply

$$\delta = 20 \log_{10} \frac{20\,\text{mV}}{1\,\text{mV}} = 26\,\text{dB}$$

or

$$\delta = 10 \log_{10} \frac{400\,\text{mV}^2/R}{1\,\text{mV}^2/R} = 26\,\text{dB}.$$

Two different numerical values are replaced with a single value in dB. On first encounter, students tend to find this irritating, but in practical usage it is a real simplification. Of course we had to assume equal impedances at input and output, but in radio engineering that is very frequently true.

13.3.3   Electrical and Optical dB Units

The second source of confusion occurs when light is converted to an electric signal. As described in Sect. 8.10, common photodetectors such as, e.g., photo diodes, convert the impinging light power to a proportional electric current. The electric power delivered by the detector is thus proportional to the *square* of the received optical power. One therefore needs to specify whether one speaks of optical or electric dB. Consider a statement about the dynamic range of some detector, i.e., the ratio of maximum received power before severe distortion sets on and the smallest detectable power that is not masked by noise. One "dB_{opt}" corresponds to two "dB_{elect}".

13.4   Beer's Attenuation and dB Units

When light is impinging on a more or less transparent material, the reduction of power $P(L)$ with increasing penetration depth L is described by the well-known Beer's law:

$$P(L) = P_0 \exp\left(-\alpha L\right).$$

Here, α is called Beer's absorption coefficient; its reciprocal value is that particular penetration depth where the initial power P_0 has decayed to the fraction $1/e \approx 37\,\%$.

By taking the logarithm, one immediately sees that

$$\alpha L = -\frac{1}{\log_{10} e}\ \log_{10}\frac{P}{P_0}.$$

On the other hand, using the definition of the dB we can write

$$\alpha_{dB} L = 10\ \log_{10}\frac{P}{P_0}.$$

Comparing terms yields the conversion formula

$$\alpha_{dB} = -\alpha 10 \log_{10} e \approx -4.34\alpha.$$

Chapter 14
Skin Effect

When alternating current flows through a conductor, the current density is not necessarily constant across its entire cross-section. When $J(0)$ denotes the current density at the surface, the current density at some depth x below the surface is given by

$$J(x) = J(0)e^{-x/\delta}e^{-ix/\delta}.$$

Here, δ is a characteristic penetration depth, i.e., that depth where the current density is reduced by a factor of $1/e \approx 37\%$ in comparison to the surface. At the same time, at this depth there is a phase shift of 1 rad. This depth is given by

$$\delta = \sqrt{\frac{2\rho}{\omega\mu_0\mu_r}}. \tag{14.1}$$

Here, in turn, $\mu_0 = 4\pi\,10^{-7}\,\text{Vs/Am}$ is the vacuum permeability, μ_r the relative permeability of the material, ρ the specific resistance of the material, and ω the current's angular frequency.

It is straightforward to realize that at a depth of $\pi\delta$, there is a phase shift of 180°. This implies that at this depth the current flows in opposite phase to that at the surface. This is, of course, a hindrance for the current flow through the conductor and is felt as an effective increase of resistance. Somewhat paradoxically, a *massive* conductor, such as a solid wire, conducts current less well than a *hollow* conductor (a tube) of the same outside diameter!

The effective resistance is given by

$$R = l\frac{\rho}{\delta s} \tag{14.2}$$

with l the length and s the circumference of the conductor. This should be compared with the usual expression for direct current resistance,

$$R = l\frac{\rho}{A},$$

© Springer-Verlag Berlin Heidelberg 2016
F. Mitschke, *Fiber Optics*, DOI 10.1007/978-3-662-52764-1_14

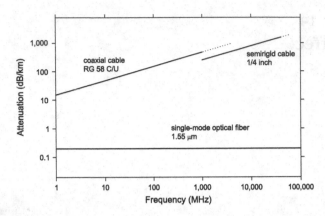

Fig. 14.1 Skin effect causes the effective resistance of a cable to rise with increasing frequency. Optical fiber is superior, in particular at high frequencies

where A is the conductor's cross-sectional area. In effect, only a surface layer of thickness ρ contributes to the current flow.

Inserting Eq. (14.1) into Eq. (14.2) yields

$$R(\omega) = \frac{l}{s} \sqrt{\frac{\omega \rho \mu_0 \mu_r}{2}}.$$

For our consideration the relevant feature is the relation

$$R(\omega) \propto \sqrt{\omega}.$$

The effective resistance grows with increasing frequency (Fig. 14.1). This limits the usefulness of electrical conductors at very high frequencies. Optical fibers do not suffer from a comparable limitation.

Chapter 15
Bessel Functions

Bessel's differential equation reads

$$x^2 \frac{d^2y}{dx^2} + x\frac{dy}{dx} + (x^2 - m^2)y = 0.$$

We are only interested in solutions for integer m, denoted as J_m, N_m, and $H_m^{1,2}$.
The modified Bessel's differential equation reads

$$x^2 \frac{d^2y}{dx^2} + x\frac{dy}{dx} - (x^2 + m^2)y = 0.$$

Solutions for integer m are denoted by I_m and K_m.

15.1 Terminology for the Various Functions

	J_m	Bessel function 1st kind	Cylinder function 1st kind
Bessel's equation	N_m, Y_m	Bessel function 2nd kind	Cylinder function 2nd kind Weber's function Neumann function
	$H_m^{1,2}$	Bessel function 3rd kind	Cylinder function 3rd kind Hankel function
Modified Bessel's equation	I_m	Modified Bessel function 1st kind	
	K_m	Modified Bessel function 2nd kind	Modified Hankel function McDonald's function

© Springer-Verlag Berlin Heidelberg 2016
F. Mitschke, *Fiber Optics*, DOI 10.1007/978-3-662-52764-1_15

15.2   Relations Between These Functions

$$N_m(x) = \lim_{k \to m} \frac{J_k(x) \cos(k\pi) - J_{-k}(x)}{\sin(k\pi)},$$

$$H_m^{1,2}(x) = J_m(x) \pm iN_m(x),$$

$$I_m(x) = i^{-m} J_m(ix),$$

$$K_m(x) = \lim_{k \to m} \frac{\pi}{2} \frac{I_{-k}(x) - I_k(x)}{\sin(k\pi)}.$$

15.3   Recursion Formulae

With Z_m denoting some cylinder function of type J_m, N_m, H_m, I_m, or K_m and with $Z'_m(x)$ the derivative of $Z_m(x)$ with respect to the argument x,

$$xZ'_m(x) = mZ_m(x) - xZ_{m+1}(x),$$

$$xZ'_m(x) = -mZ_m(x) + xZ_{m-1}(x).$$

15.4   Properties of J_m and K_m

The functions J_m describe standing waves. J_0 "looks like" cosine, J_1 like sine. Asymptotically (for very large x) the following approximations hold:

$$J_0(x) = \sqrt{\frac{2}{\pi x}} \cos(x - \pi/4),$$

$$J_1(x) = \sqrt{\frac{2}{\pi x}} \sin(x - \pi/4),$$

$$J_m(x) = \sqrt{\frac{2}{\pi x}} \cos\left(x - \pi/4 - n\pi/2 + \mathcal{O}\left(\frac{1}{x}\right)\right).$$

Functions with negative index are defined as

$$J_{-n}(x) = (-1)^n J_n(x).$$

There is a relation with angular functions of the form

$$\cos(\alpha + \beta \sin \gamma) = \sum_{n=-\infty}^{+\infty} J_n(\beta) \cos(\alpha + n\gamma),$$

which is relevant in context with frequency modulation (see Sect. 11.1.2.2).

Functions K_m describe decaying cylinder waves and "look like" decaying exponential functions. Asymptotically (for very large x), for all m the following approximations hold:

$$K_m(x) = \sqrt{\frac{\pi}{2x}}\, e^{-x} \left(1 + \mathcal{O}\left(\frac{1}{x}\right)\right).$$

As long as the argument is larger than 2, the error is less than 5 %.

15.5   Zeroes of J_0, J_1, and J_2

J_0	J_1	J_2
2.4048	0.0000	0.0000
5.5201	3.8317	5.1356
8.6537	7.0156	8.4172
11.7915	10.1735	11.6198
14.9309	13.3237	14.7960
...	...	...

15.6   Graphs of the Most Frequently Used Functions

See Fig. 15.1.

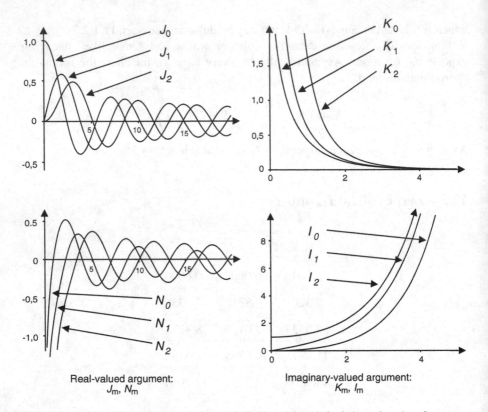

Fig. 15.1 Graph of Bessel functions of types J, K, N, and I, each for index values $1\ldots3$

Chapter 16
Optics with Gaussian Beams

The general public holds the notion of a laser beam as a cylindrical bundle of rays that travels arbitrary distances without any change of its diameter so that it delivers the same power density wherever it hits. Take note, James Bond: There is no such thing as a cylindrical beam. The laws of diffraction make sure that any beam with a finite diameter widens as it propagates; the wider the beam starts out, the more gradual is its spreading, but it is always there.

16.1 Why Gaussian Beams?

For the sake of a quantitative description let us consider a Gaussian beam, i.e., a light beam with a transverse distribution of power following a Gaussian bell-shaped curve. Such beams are ubiquitous in laser physics; they originate from laser resonators for which the Gaussian profile is the lowest transverse mode. As light is coupled out, one automatically gets a Gaussian beam. The first concise treatment of the situation, still worthwhile reading today, is given in [1]; for a particularly transparent description see [4] or [3].

One might think of a Gaussian beam as created in the following way: Start with a plane wave and send it through an amplitude mask (like a photographic slide). Let the mask have a circularly symmetric gray scale so that a bell-shaped beam is carved out of the plane wave.

Once a beam has a Gaussian profile, it will stay that way. Propagation across free space will change its size, but not the functional form. In other words, the Gaussian is invariant (except for scale factors) under diffraction. The reader is reminded that in the far-field limit of diffraction, the field distribution acquires a shape given by the Fourier transform of the initial shape. It is well known that the Fourier transform of a Gaussian is again a Gaussian. Therefore, the near field Gaussian profile is preserved in the far field—and not just in the limiting cases, but everywhere in between, too.

© Springer-Verlag Berlin Heidelberg 2016
F. Mitschke, *Fiber Optics*, DOI 10.1007/978-3-662-52764-1_16

A Gaussian beam is a diffraction-limited beam. It is even better: Among all diffraction-limited beams, it is the one with the least change of shape.

In analogy to the quantum mechanic uncertainty relation and thinking of light as a stream of photons, one can multiply the transverse localization error (the beam radius) with the transverse photon momentum (a measure of the beam's divergence) and obtain a product that cannot get arbitrarily small. The minimum product is fulfilled by the Gaussian beam.

If a Gaussian beam is transmitted through common optical elements—lenses, curved mirrors—it will get deformed. Nonetheless it will maintain its Gaussian shape. (This is true if we idealize that the lenses are well centered on the beam.) It takes stronger actions to destroy the Gaussian profile: Nonaxially symmetric elements such as cylindrical lenses render it into an ellipsoid version, which can still be considered a generalization; absorbing elements such as apertures can destroy the Gaussian profile altogether. A knife blade inserted halfway into the beam *will* alter the beam shape.

16.2 Formulae for Gaussian Beams

It is a convention to take the beam radius w (as in *width*) as that distance from the axis where the field amplitude is reduced to $1/e$ of the on-axis value; this is also the radius where the intensity has rolled down to $(1/e)^2$ of its on-axis value.[1]

The beam is never cylindrical: Its cross-section varies. At some location the beam has a minimum transverse extent. This is called the *beam waist* and serves as an important point of reference. Let us identify the propagation direction with the z direction; we conveniently place the zero point at the waist. Then the beam radius is described by

$$w(z) = w(0) \sqrt{1 + \left(\frac{z}{z_0}\right)^2}.$$

Here z_0 is a characteristic length called *Rayleigh range*. It indicates the distance after which the beam radius has increased by $\sqrt{2}$ and is given by

$$z_0 = \frac{\pi w_0^2}{\lambda}.$$

As one might have expected, z_0 is referred to as the only length scale of relevance for wave phenomena, i.e., the wavelength λ.

---

[1]Confusingly, in some old texts one can find other conventions, like intensity drop to $1/e$.

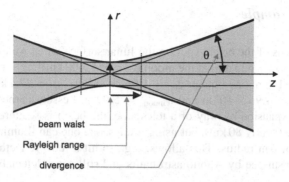

Fig. 16.1 The contour of a Gaussian beam at $r = w(z)$ takes its narrowest width w_0 at the waist ($z = 0$). In the far field, the beam radius $w(z)$ increases at a fixed angle θ, called the beam divergence

The Rayleigh range marks the transition from near field to far field: For distances $z \ll z_0$, the beam propagates approximately without change of radius, whereas for $z \gg z_0$ the radius grows in proportion to distance or at a fixed angle. This divergence angle θ is found as

$$\theta = \text{arctg}\frac{w(z)}{z} \approx \frac{1}{z}w_0\frac{z}{z_0} = \frac{w_0}{z_0} = \frac{\lambda}{\pi w_0}$$

with $w_0 = w(0)$. The reader should note that for a fixed wavelength $\theta \propto 1/w_0$: A beam with wide waist will widen only gradually. Figure 16.1 illustrates the relations between w, z, and θ.

As a Gaussian beam propagates and widens, the wavefront does not remain plane. Its curvature can be described by the associated radius, $R(z)$.

$$R(z) = z\left[1 + \left(\frac{z_0}{z}\right)^2\right].$$

Obviously,

$$\lim_{z \to 0} R(z) = \infty \quad \text{and}$$

$$\lim_{z \to \infty} R(z) = z.$$

This implies that the wave fronts are indeed plane at the waist; at very large distance they form segments of spheres centered around the waist. At some intermediate distance, the curvature has a minimum: This is the case at $z = z_0$ where $R(z) = 2z$.

16.2.1 Example

Find the radius of the bright spot on the lunar surface when we aim the beam of a He–Ne laser ($\lambda = 633$ nm) at the moon ($z = 384,000$ km)!

At $w_0 = 1$ mm, one obtains $z_0 = 4.96$ m and $w_{moon} = 77.4$ km; at $w_0 = 1$ m, one gets $z_0 = 4.96 \times 10^6$ m and $w_{moon} = 5.99$ m. Lesson learned: In the absence of a beam expansion by way of a telescope, the beam is scattered about as to be undetectable (nearly 80 km), but using a telescope one can illuminate a reasonably small spot of 6 m radius. This allows, e.g., to hit a retro reflector such as placed on the lunar surface by Apollo astronauts and still get a detectable back-reflected signal.

16.3 Gaussian Beams and Optical Fibers

Fibers are waveguides. Indeed, the waves are *weakly guided* because the index contrast between core and cladding is very small. In this situation, the fundamental mode profile is *nearly* Gaussian. Consider as a thought experiment that the index difference shrinks to zero: Then one would expect the same shape as in free space. It is of course simpler to deal with a Gaussian, rather than the complicated composition of Bessel functions described in Chap. 3. Therefore, this approximation is popular and sometimes good enough.

When a beam is coupled from free space into a fiber, one is usually faced with the matching problem between a true Gaussian beam in free space and a less-than-exact almost-Gaussian profile of the fiber's fundamental mode. (However, it is usually safe to assume that in a perpendicularly cleaved fiber surface the wave fronts are plane, so that for incoupling as well as outcoupling of light the fiber face can be identified with the position of a beam waist.) The mismatch causes a reduction in coupling efficiency. Even in the presence of ideal lenses without aberrations, this limit cannot be surpassed [2]. Part of the light winds up in the cladding and is eventually scattered out of the fiber, rather than guided in it.

If light propagating inside the fiber is coupled out at the other end and hits a screen at some distance, the pattern on the screen again is similar to a Gaussian. This is because the far-field pattern is the Fourier transform of the near-field pattern as discussed in Sect. 7.3.2 where it was also shown how deviations from the Gaussian pattern can be exploited to gauge the mode profile.

References

1. H. Kogelnik, T. Li, *Laser Beams and Resonators*, Proceedings IEEE **54**, 1312 (1966)
2. E.-G. Neumann, *Single Mode Fibers*, Springer Series in Optical Sciences Vol. **57**, Springer-Verlag, Berlin (1988)
3. F. L. Pedrotti, L. M. Pedrotti, L. S. Pedrotti, *Introduction to Optics*, 3rd ed., Benjamin-Cummings, Upper Saddle River, New Jersey (2006)
4. B. E. A. Saleh, M. C. Teich, *Fundamentals of Photonics*, 2nd ed., John Wiley & Sons, New York (2007)

Chapter 17
Relations for Secans Hyperbolicus

The function *Secans Hyperbolicus* (hyperbolic secant) is defined as

$$\text{sech}(x) = \frac{2}{e^x + e^{-x}}.$$

For convenience we introduce a numerical factor

$$\mathcal{Z} = \cosh^{-1}(\sqrt{2}) \approx 0.881373587,$$
$$2\mathcal{Z} = \cosh^{-1}(3)$$
$$= \ln(3 + \sqrt{8}) \approx 1.762747174.$$

$\cosh^{-1}$ refers to the inverse function arcosh, not the reciprocal of the function. One finds the following special values (Fig. 17.1):

$$\text{sech}(0) = 1,$$
$$\text{sech}(1) \approx 0.6480542737,$$
$$\text{sech}(2) \approx 0.2658022288,$$
$$\text{sech}(\mathcal{Z}) = \frac{1}{2}\sqrt{2}.$$

A light pulse with envelope $U(t) = \hat{U}\,\text{sech}(t/T_0)$ has the power profile $P(t) = \hat{P}\,\text{sech}^2(t/T_0)$. The following special values hold (Fig. 17.2):

$$\text{sech}^2(1) \approx 0.4199743416$$
$$\text{sech}^2(\mathcal{Z}) = \frac{1}{2}.$$

© Springer-Verlag Berlin Heidelberg 2016
F. Mitschke, *Fiber Optics*, DOI 10.1007/978-3-662-52764-1_17

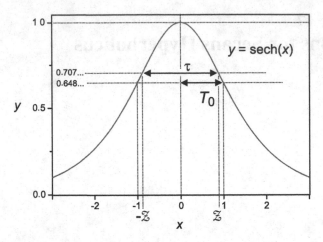

Fig. 17.1 $y = \mathrm{sech}(x)$

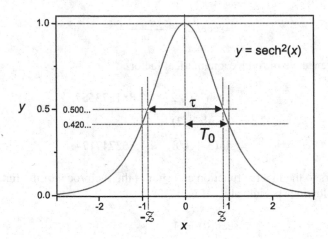

Fig. 17.2 $y = \mathrm{sech}^2(x)$

The pulse duration, taken as the full width at half-maximum (FWHM), is

$$\tau = 2\mathcal{Z}T_0.$$

The pulse energy is

$$E = \int\limits_{-\infty}^{\infty} P(t)\, dt = \hat{P} \int\limits_{-\infty}^{\infty} \mathrm{sech}^2\left(\frac{t}{T_0}\right) dt$$

$$= 2\hat{P}T_0$$

$$= \frac{1}{\mathcal{Z}}\hat{P}\tau.$$

Chapter 18
Autocorrelation Measurement

18.1 Measurement of Ultrashort Processes

It would be a straightforward task to measure duration and shape of picosecond or femtosecond pulses if detectors (and oscilloscopes) with temporal resolution better than the pulse duration would exist.

However, the fastest photodiodes are restricted to temporal resolutions of several picoseconds. A dramatic advance of technology is not anticipated, because the finite mobility of charge carriers themselves inside the solid-state detectors defines the limitation.

For this reason, an entirely different method is used for temporal measurements on ultrashort time scales. The central idea is charming: The pulse to be measured is referenced to *itself*. The technique is known as autocorrelation measurement. It does indeed work—the price to pay is that one does not obtain the full unambiguous information about the pulse shape. To understand the principle, let us first briefly discuss the mathematical concept of the autocorrelation function, without getting too formal.

18.1.1 Correlation

The word "correlation" describes a similarity. If A does something and B does the same at the same time, then the actions of A and B are correlated. If B consistently does the opposite of what A does, then they are anticorrelated. If B acts independently of A, both are uncorrelated.

Speaking more specifically, we will compare two real functions of time, $f(t)$ and $g(t)$. How can we establish a similarity?

First, we take the product of both functions, $f(t) \times g(t)$. It should be obvious that this product is non-negative when both functions have the same algebraic sign (no matter which one) at all times—whenever one changes sign, so does the other.

© Springer-Verlag Berlin Heidelberg 2016
F. Mitschke, *Fiber Optics*, DOI 10.1007/978-3-662-52764-1_18

On the other hand, the product is negative whenever both functions have opposite sign (again, no matter which function has which sign).

In the event that $f(t)$ and $g(t)$ are, say, independent random functions, either will change sign at random times that usually do not coincide with the moments at which the other changes sign. The product, then, will be positive at some times, negative at others—and in the long run, either possibility will occur about half of the time. On average the product is zero.

This brings us to the correlation which is the long-term average of the product:

$$\text{corr} = \int_{-\infty}^{+\infty} f(t)g(t)\,dt.$$

For uncorrelated functions this quantity will tend to zero.

What happens if the functions are related? The strongest possible correlation of $f(t)$ and $g(t)$ occurs when both are identical (up to a scale factor), or $f(t) = ag(t)$ with a some positive real constant. In that case, corr will take its (positive) maximum value. If, on the other hand, $f(t) = -ag(t)$, corr takes the same value but with a negative sign. This constitutes the strongest possible anticorrelation.

18.1.2   Autocorrelation

For those trained in Latin and Greek, the word "autocorrelation" is immediately clear: it describes a correlation of something with itself. The autocorrelation function is the temporal average of the product of two functions, which are identical except for a temporal shift τ:

$$\text{autocorr}(\tau) = \int_{-\infty}^{+\infty} f(t)f(t+\tau)\,dt.$$

It is convenient, and customary, to normalize this expression to its maximum possible value, which occurs, as argued above, for $\tau = 0$:

$$\text{ACF}(\tau) = \frac{\int_{-\infty}^{+\infty} f(t)f(t+\tau)\,dt}{\int_{-\infty}^{+\infty} f(t)^2\,dt}.$$

$\text{ACF}(\tau)$ has the following properties:

- $\text{ACF}(0) = 1$ for any function $f(t)$ (which is not zero everywhere); this is due to the normalization.
- $-1 \leq \text{ACF}(t) \leq +1$ for all t and any function; the case $ACF(t) > 1$ is impossible for all t.
- $\text{ACF}(T) = \text{ACF}(-T)$: ACF is symmetric.

- If $f(t)$ = const., then ACF(t) = 1 for all t, independent of the value of the constant (disregarding the case of zero).
- If $f(t)$ = $f(t + T)$, then ACF(T) = 1: Periodic functions have a periodic autocorrelation function with the same period. Phase is irrelevant, though.

To develop a feeling for this, consider a few selected functions.

Sine function. If $f(t)$ = $A\sin(\omega t + \varphi)$, then ACF($\tau$) = $\cos(\omega t)$. This is independent of φ. Thus, it is also true for $\varphi = \pi/2$, that is, for cosine instead of sine.

Noise. ACF(0) = 1 is true for any function; on the other hand, we noted above that for a random signal, ACF(τ) = 0. There is no contradiction. At τ = 0, ACF of noise is equal to 1 and then drops very rapidly to zero as τ grows. The range of τ over which ACF decays and is still different from zero indicates the correlation time of the signal.

The inverse of the correlation time is the bandwidth of the noise; there is no such thing in physics as noise with infinite bandwidth or zero correlation time. Nonetheless that limiting case is important conceptually and is studied by theorists due to its nice mathematical properties. It comes by the name of δ-correlated white noise. Any physical noise has a correlation time that is different from zero because nothing in nature ever acts infinitely fast. White noise with infinite bandwidth violates the law of energy conservation.

Gaussian. A signal with a temporal variation according to a Gaussian (a Gaussian pulse) has an autocorrelation function that is Gaussian again, but is wider by a factor of $\sqrt{2}$. We encountered the spatial case in Sect. 7.3.1: Measurement of a mode profile in the near field by the transverse offset method yields the autocorrelation function of the mode profile. As long as one can justifiably approximate it as a Gaussian, one can take the measured radius and simply divide by $\sqrt{2}$ to obtain the correct modal radius.

18.1.3 Autocorrelation Measurements

Let us return to our task to measure the duration of ultrashort optical pulses. The procedure involving the autocorrelation function is now easy to understand. In the setup of an autocorrelator, the light beam carrying the pulses is first split at a partially reflecting mirror with reflectivity R = 50 %. Thus, there are two paths; on each there is one replica of the pulses to be measured. Both replicas are recombined in a nonlinear crystal. The crystal is chosen such that it can generate the second harmonic of the light [a $\chi^{(2)}$ effect; compare Eq. (3.18)]. In this case there always is a term containing the product of two electrical fields. Each partial beam by itself creates such a product of its field with itself, i.e., $|E_1|^2$ and $|E_2|^2$; of more interest for us is the combination term of $E_1 E_2$, which is also generated. (In a popular variant called background-free autocorrelation, these three contributions can be geometrically separated and only the combination term is used.)

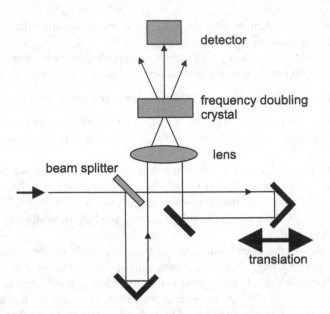

Fig. 18.1 Sketch of an optical autocorrelator setup. Two replicas of the pulse stream to be measured arrive at a frequency-doubling crystal through different paths. Only if both arrive at the same time can there be appreciable power in the combination product which is monitored by the detector. By variation of the path-length difference, one can map out the temporal pulse shape

Finally, a temporal integration is performed. In practice, one does not need to push the integration limits to ±∞; it fully suffices when the integration interval is much longer than the pulse duration. Ironically, a slow photodetector is not only good enough here: it is actually required to be slow!

Both replicas of the incoming beam travel similar, but not necessarily equal, path lengths before they are recombined at the focal point of a lens, which is inside the nonlinear crystal (Fig. 18.1). By fine-tuning the path-length difference, one can arrange that both replicas arrive simultaneously so that a maximum combination product term is generated and registered on the detector.

The measurement is performed in the following way: The path difference is scanned while the detector signal is monitored. Typically one provides the path difference information to the horizontal input of an oscilloscope, and the detector signal to the vertical input. As the path difference is varied, the pulse profile (or more precisely, its autocorrelation function) is mapped out and appears as a trace on the oscilloscope screen. If this is done at a repetition rate of at least 30 Hz, the eye perceives a flicker-free representation of the pulse shape.

18.1.4 A Catalogue of Autocorrelation Shapes

Since it is not the pulse shape directly which is measured, there is always the question of finding the pulse shape that corresponds to the measured autocorrelation function. This is not a unique relation, but in many cases one has some extra information to reduce the ambiguity and gets away with it. We list the relation in Table 18.1, and in Fig. 18.2 by way of symbolic schematic representations.

Table 18.1 Table of some selected pulse shapes and the corresponding autocorrelation functions

Pulse shape	Corresponding autocorrelation
Rectangular, width ± 1	Triangular, width at pedestal ± 2
Gaussian, width T	Gaussian, width $\sqrt{2}\,T$
$\text{sech}^2(t)$	$3\,\dfrac{t\cosh(t) - \sinh(t)}{\sinh^3(t)}$
Two equal pulses separated by T	Three-pronged fork. Prongs separated by T. Center component is twice as high as off-center components. Width of components: ACF of original pulses

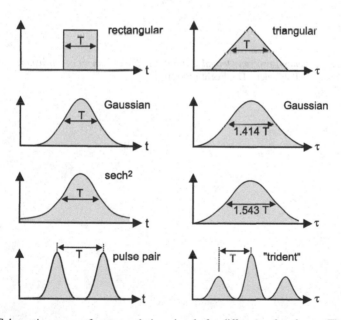

Fig. 18.2 Schematic survey of autocorrelation signals for different pulse shapes. The ACF of a sech^2 pulse is given in Table 18.1. A pulse pair is represented by a "three-pronged fork"; the center peak is twice as high as the off-center peaks

It is important to note the following facts: Autocorrelation measurements allow to assess pulse durations down into the few-femtosecond regime; it is conceivable that this can be pushed even further. This exceeds the temporal resolution of direct detection by several orders of magnitude.

On the other hand, autocorrelation measurement do not yield the pulse shape directly. The pulse shape cannot be uniquely reconstructed from the autocorrelation function. (The other way around it would work; alas, that is of little help.) In particular, phase information about the pulse (such as, the existence of a chirp) is lost.

As a consequence, an ambiguous determination of the pulse shape implies an ambiguity in the pulse duration. Absent a better solution, one typically makes educated guesses about the pulse shape based on independent information, then calculates the pulse width based on that. This is certainly less than exact science. There are more involved procedures giving more detailed information (see, e.g., [1]), but they are not always available. However, as long as the procedures used are stated clearly (like, the width is quoted as "FWHM assuming a Gaussian shape" or so), this is acceptable, and indeed widespread practice. On the other hand, one should not attribute a precision to the widths thus determined, which is not warranted.

Reference

1. R. Trebino, *Frequency-Resolved Optical Gating: The Measurement of Ultrashort Laser Pulses*, Kluwer Academic Publishers, Dordrecht (2000)

Legal Note

Whenever names of manufacturers or vendors are given, this serves information purposes only and can in no way be construed to imply an endorsement or recommendation.

Similarly, web addresses are given for information purposes only. Neither the author nor the publisher mean to imply an endorsement of products or views described on those websites. Future existence of these websites is beyond the control of author and publisher.

Author and publisher have undertaken any reasonable effort to obtain copyrights for all material used in this book. However, in some cases the rights owner could not be identified, or reached. These persons are invited to contact the publisher.

© Springer-Verlag Berlin Heidelberg 2016
F. Mitschke, *Fiber Optics*, DOI 10.1007/978-3-662-52764-1

Glossary

Akhmediev breather (p. 188)   Special solution of the → Nonlinear Schrödinger
equation involving a continuous wave background from which a perturbation
arises, culminates in a series of power peaks, then decays again. This is the
mechanism behind → modulational instability.

Amplifier (p. 153)   In optics, a device which increases the power of a light wave
passing through it. Amplifiers are a central element of any → laser. In optical
telecommunications, mostly semiconductor optical amplifiers and doped-fiber
amplifiers are used.

Autocorrelator (p. 331)   Device to measure the duration of ultrashort pulses
down to the few-femtosecond regime. The light beam is split into two; both
parts are brought together again with variable delay in a nonlinear medium. The
mixing signal is detected; the detector does not have to be very fast. The resulting
signal, mathematically the autocorrelation function of the pulse shape, allows
conclusions about pulse duration and shape.

Avalanche diode (p. 169)   Special type of → photodiode, in which a high bias
voltage is applied to accelerate charge carriers to the point that they in turn
generate new carriers. In an avalanche process, an amplification of the primary
photocurrent is obtained.

Bandwidth (p. 250)   Frequency interval over which a certain signal contains
energy. Usually stated as the difference between highest and lowest signal
frequency.

Bending loss (p. 89)   When an optical fiber is tightly bent, additional loss occurs.
A fiber that carries visible light can be observed to shine brightly at tight bends;
here, some of the guided light is lost.

Birefringence (p. 75)   Phenomenon in anisotropic materials. Light of different
linear → polarization is subject to different → refractive index.

Bragg grating (p. 142)   A periodic array of scatterers (a grating, in the widest
sense) can reflect a wave when a certain relation between grating constant
(grating period) and wavelength is fulfilled; named after William Henry Bragg
and William Lawrence Bragg (father and son), who shared a Nobel prize in 1915.

© Springer-Verlag Berlin Heidelberg 2016   339
F. Mitschke, *Fiber Optics*, DOI 10.1007/978-3-662-52764-1

Channel (frequency channel) (p. 256)   A frequency band reserved for a specific signal is also called a channel. Using several channels, different signals can be transmitted simultaneously; this is well known for radio and TV. In fiber optics one such channel may also be called a → WDM channel to clarify this use of the term.

Channel (transmission channel) (p. 261)   General term for an arbitrary transmission medium such as a cable and radio link. which provides a certain → bandwidth. This results in a certain → channel capacity.

Channel capacity (p. 262)   According to a theorem by C. Shannon, there is a maximum rate with which information can be successfully transmitted over a given → channel; this rate is known as channel capacity.

Chirp (p. 184)   Term denoting a slide of carrier frequency within a short pulse of light. The product of spectral and temporal width can be equal to or larger than a certain constant; in the presence of chirp it is larger.

Circulator (p. 148)   A device to steer light signals between several ports. It lets light beams pass in one direction. Light beams traveling in the opposite direction are redirected to a third direction.

Cladding (p. 19)   The zone in an optical fiber which surrounds the → core. In most commercially available fibers the outside diameter of the cladding is $125\,\mu m$.

Core (p. 8)   The innermost zone in the structure of an optical fiber. In the case of → single-mode fibers the radius is several micrometers. Most of the light is guided in the core.

Coupler (p. 149)   Device for coupling of two fibers, so that signals traveling in them can be split or combined.

Cutoff wavelength (p. 47)   The shortest wavelength at which a fiber supports only a single mode. Occasionally also used for the limit of existence range of higher-order → modes.

Dispersion (p. 10)   Wavelength dependence of some optical characteristic of a signal. This may be the → refractive index of a glass or the deflection angle of a prism ("angular dispersion"). In fiber optics the term usually refers to the group velocity dispersion.

Fabry–Perot interferometer (p. 138)   Arrangement in which light passes back and forth between two mirrors. When the round trip distance equals an integer multiple of the wavelength, a resonance occurs. Fabry–Perot interferometers are often used to select specific wavelengths, e.g., in laser resonators. The name derives from Charles Fabry and Alfred Pérot (Marseille, ca. 1890).

Fiber (p. 6)   Spelled *fibre* in Great Britain. Here the term refers to optical fibers, thin flexible strands of glass which can conduct light.

Fiber laser (p. 157)   A type of → laser, in which the → amplifier (gain medium) is formed by a fiber which is doped with active substances. In optical telecommunications, it is particularly the Erbium-doped fiber which finds widespread use.

Fused silica (p. 6)   Chemically, silicon dioxide, but in glassy rather than crystalline form. The corresponding crystal is called quartz.

Gaussian beam (p. 174) Light beam which contains a single spatial → mode. It is characterized by a transverse power profile which takes the form of a Gaussian. Gaussian beams are diffraction-limited, i.e., their spread is minimal. They are typically generated in lasers, and can propagate in free space. In fibers, the fundamental → mode is only approximately Gaussian.

Gradient index profile (p. 68) In some fibers the → refractive index in the → core is not constant but varies continuously in the radial direction, typically in a parabolic way. In → multimode fibers, such a profile reduces → modal dispersion.

Higher order soliton (p. 200) Solution of the → nonlinear Schrödinger equation similar to a fundamental → soliton but with an amplitude that is larger by a factor $N \geq 1.5$. Can be understood to be a compound of several fundamental solitons; they beat with each other during propagation; hence an oscillating pulse shape. They come apart in the process of → soliton fission.

Holey fiber (p. 79) The → cladding of this type of fiber contains voids, i.e., cylindrical hollows which run the entire length of the fiber. This lowers the effective → refractive index of the cladding and enables the guiding of light.

Isolator (p. 146) In optics, an arrangement which allows light to pass in one direction, but blocks it in the opposite direction.

Kerr effect (p. 175) Also known as "quadratic electro-optic effect," named after John Kerr (1875). By the Kerr effect the → refractive index of a material is modified in proportion to the square of the amplitude of an applied electric field. In fibers the "optical Kerr effect" occurs in which the light field takes the role of the applied field. Then the refractive index is modified in proportion to the intensity of the light.

Laser (p. 6) The acronym stands for "light amplification through stimulated emission of radiation." A light source capable of producing coherent light. The laser principle relies on stimulated emission in a material which is used as an optical → amplifier. Energy must be supplied for the amplification; in the example of → diode lasers, this is done by running a current through the device.

Laser diode (p. 160) Type of laser, in which the → amplifier (gain medium) is formed by a semiconductor device of diode structure. Energy is supplied by an operating current.

Latency (p. 289) The total propagation time from transmitter to receiver, represents a delay by which the signal is received. It is fundamentally limited by the geographical distance times the fiber's group index, divided by the speed of light. That amounts to ca. 100 ms from some point on earth to its antipodal point, measured along the great circle. On top of that there can be additional delay due to processing for error correction, etc. Low latency is relevant for timing-sensitive information, in particular in financial markets.

LED (p. 159) Acronym for *light-emitting diode*, also known as luminescent diode. A semiconductor device producing light when an operating current passes through. Simpler in structure than a → diode laser; also, the light is not coherent. Often used for indicator or pilot lights in electronic equipment of all kinds.

Increasingly used for general illumination as LED technology proceeds because LEDs are much more power-efficient than light bulbs.

Material dispersion (p. 55)   Phenomenon based on the frequency dependence of the → refractive index. It lets short pulses of light widen as they propagate through a fiber. It also causes chromatic abberations in lens-based imaging and enables prisms to spread white light into colors.

Modal dispersion (p. 23)   In → multimode fibers different → modes propagate at different speed. This causes a scatter in the arrival time at the receiving end. This spreading of a signal pulse is called modal dispersion and is typically measured in ps/km.

Mode (p. 26)   Throughout physics there is an important concept of elementary oscillations known as modes. Resonators of a given geometry support specific modes which can be obtained from the geometric constraints. For example, a violin string has a fundamental oscillation and harmonics, each with its own characteristic frequency and oscillation pattern. In optical fibers, the constraints select certain field distributions and propagation constants known as the modes of the fiber. Fibers can be designed to be → single-mode or → multi-mode fibers.

Mode coupling (p. 26)   Energy can be exchanged between the → modes of a fiber at perturbations of the geometry, like in tight bends.

Mode division multiplex (p. 290)   Technique to use several modes of a few-mode fiber to carry independent data streams. Related to → space division multiplex.

Mode locking (p. 160)   The phases of longitudinal modes of a laser can be locked together to generate very short pulses of light.

Modulation (p. 120)   In optics, the controlled modification of amplitude, phase, frequency, or polarization of a light wave in order to impress information on it which is then carried along.

Modulational instability (p. 187)   Phenomenon in some materials exhibiting → nonlinearity, in which a continuous wave becomes unstable and forms a more or less periodic modulation. In fibers this can happen by the interplay of → Kerr effect and anomalous → dispersion. See also → Akhmediev Breather.

Multimode fiber (p. 8)   Type of fiber which supports several → modes. Due to → modal dispersion this is useful only for moderate data rates and short distances. Plastic optical fibers are almost always multimode fibers. The total power of the light signal is distributed over all participating → modes. This distribution may fluctuate; then *mode partition noise* is generated which can be a nuisance in many contexts including fiber-optic → sensor applications. The safest fix is the use of → single-mode fiber.

Nonlinearity (p. 11)   The phenomenon that a property of a device or material which has an influence on the signal may not be constant but affected by the signal. In fiber optics the most relevant nonlinearity is that the → refractive index of the fiber depends on the light intensity by way of the → Kerr effect.

Nonlinear Schrödinger equation (p. 181)   Wave equation describing propagation of light in optical fibers in the simultaneous presence of both → dispersion and → nonlinearity, each represented by the leading term (second order group velocity dispersion; optical → Kerr effect).

Normalized index step (p. 22)   A metric for the difference of → refractive index between → core and → cladding of a fiber. In most fibers this difference is in the range from 0.001 to 0.01. Bend loss tends to be lower for fibers with large values.

NRZ (p. 255)   Acronym for *no return to zero*: A binary coding format in which the light power stays constant throughout the entire clock period. In a succession of several logical "1"s, the light power stays on for several clock cycles, without returning to zero in between. Compare → RZ.

Numerical aperture (p. 21)   A metric for the acceptance angle of a fiber, i.e., the angle of the cone within which light can be coupled into a fiber. The same cone also appears for light leaving the fiber.

OTDR (p. 131)   Acronym for *optical time domain reflectometry*: Procedure to measure the time after which a light pulse returns from the fiber and to evaluate for the position of loss from bends, splices, damage, etc.

Peregrine soliton (p. 207)   Special solution of the → Nonlinear Schrödinger equation involving a continuous wave background from which a perturbation arises. It grows until it forms a singular peak which then disappears again. Discussed in the context of → Rogue Waves.

Photo diode (p. 165)   Semiconductor device for the detection of light. The photoeffect creates free charge carriers inside the photodiode; these give rise to a current which can be measured.

Photonic crystal fiber (p. 79)   Similar to → holey fiber, voids run the entire length of the fiber in the → cladding zone. Here the holes are located precisely in a periodic pattern so that by a → Bragg effect it acts as a reflector. This generates a strong guiding of light so that the → core can even have a *lower* → refractive index than the → cladding, without compromising the guiding.

PMD (p. 56)   Acronym for *polarization mode dispersion*. In → birefringent fibers, parts of the signal with different → polarization propagate at different speed; this causes a distortion of the signal.

Polarization (of matter) (p. 30)   Under the action of an external electrical field as provided by a light wave, electrons in a material experience Coulomb interaction forces. This distorts the atomic orbitals. Do not confuse with → polarization of light.

Polarization (of light) (p. 74)   Orientation of the oscillation in a wave. The oscillation can take place longitudinally (e.g., in sound waves in air) or transversally (in light waves). If it is transversal, there are several choices for the direction: The oscillation can be linear (two orthogonal directions, and their linear combinations) or circular (two directions of rotation, and their linear combinations). Ordinary lamp light or sunlight is often called "nonpolarized"; here the state of polarization changes extremely rapidly so that over time all possibilities are represented with equal probability. Do not confuse with → polarization of matter.

Polarization-maintaining fiber (p. 76)   A type of fiber in which by design the → birefringence has been made large. To maintain polarization requires that the light be linearly polarized along one of the birefringent axes.

Polarizer (p. 145)   Device which selects the component of a desired polarization from a light beam with arbitrary → polarization.

Preform (p. 107)   Intermediate state in the production of optical fibers.

Refractive index (p. 6)   Also index of refraction: An important quantity in optics to characterize a material. The refractive index is a complex function of wavelength. The real part indicates how much the speed of light is reduced in comparison to vacuum. It also governs the angle of refraction when light passes through an interface between different media and is therefore responsible for the function of prisms and lenses, among other things. Its frequency dependence gives rise to → material dispersion. The imaginary part describes the attenuation of the light wave. Since attenuation can often be neglected in typical materials encountered in optics (air, glass, etc.), the term "refractive index" is often used for the real part alone.

Rogue wave (p. 207)   Concept from water waves on the ocean surface, referring to rare events of huge waves occurring suddenly and disappearing immediately again. Analogous situations have been discussed in which isolated power peaks are formed in optical fibers due to interaction with → dispersion and → nonlinearity. Typically occurs in the context of → supercontinuum generation. See also → Peregrine soliton.

RZ (p. 255)   Acronym for *return to zero*: A binary coding format in which a light pulse signals a logical "1," and its absence a logical "0." The pulse duration is shorter than the clock period so that at the beginning and end of each clock period the intensity is zero in any event. Compare → NRZ.

Self phase modulation (p. 184)   Process in optical fibers in which → nonlinearity (→ Kerr effect in particular) generates a → chirp in light pulses.

Sensor (p. 13)   Device which assesses some physical (or chemical, etc.) quantity and transfers the value to some easily evaluated format, such as an electrical voltage. Fiber-optic sensors gain acceptance and sometimes can do things which other sensors cannot.

Single-mode fiber (p. 8)   Fiber which supports only a single → mode. Speaking strictly, this mode is doubly degenerate (and may therefore be counted as two) due to polarization effects. Single-mode fibers are indispensable for the transmission of very high data rates over long distances.

Soliton (p. 190)   A light pulse which maintains its shape during propagation in the presence of both → dispersion and → Kerr effect. Sometimes referred to as fundamental soliton, to distinguish from → higher-order soliton.

Soliton fission (p. 239)   The process of splitting a → higher-oder soliton into its constituent fundamental → solitons. One of the mechanisms to start the generation of → supercontinuum.

Space division multiplex (p. 290)   In order to increase the data-carrying capacity of fibers, the concept of multiple core fiber places different data streams across the fiber's cross section. The concept is related to the use of fibers that admit several modes; these modes are used to carry independent data streams (mode division multiplexing).

Splice (p. 139)   Low-loss joint between two fibers. Most often, fusion splices are used: The cleaved surfaces of two fibers are put together, heated, and melted together.

Step index fiber (p. 19)   Optical fiber consisting of → core and → cladding; either zone has a fixed → refractive index. This results in a radial step in the index profile.

Supercontinuum (p. 206)   In optics, supercontinuum describes light with a very wide spectrum, roughly one octave or more. It can be generated from irradiated laser light in the presence of nonlinear interactions in suitable media. This works particularly well in → holey fiber. The spectral power density of supercontinuum can easily exceed that of thermal light sources, as it is not limited by the Planck distribution.

TDM (p. 253)   Acronym for *time division multiplex*. Format for the simultaneous transmission of several signals which are interleaved into each other so that one falls into the pauses of the other.

Total internal reflection (p. 17)   Phenomenon at the interface between two materials with different → refractive index. The medium with the higher index is often called "optically more dense." If a light ray inside the more dense medium ($n = n_a$) hits the interface with the "thinner" medium ($n = n_b < n_a$) under a sufficiently flat angle, it gets totally reflected. The limiting angle is given by $\alpha_{crit} = \arcsin(n_b/n_a)$.

Waveguide dispersion (p. 62)   Contribution to a fiber's total → dispersion which is specific to the geometry of a waveguide.

WDM (p. 152)   Acronym for *wavelength division multiplex*. Format for the simultaneous transmission of several signals by spreading them out over the available → bandwidth of the → transmission channel.

Index

© Springer-Verlag Berlin Heidelberg 2016
F. Mitschke, *Fiber Optics*, DOI 10.1007/978-3-662-52764-1

Printed in the United States
By Bookmasters